The Sociol

The rapid e   s
for man   r
example   ns.
Howeve   o-
moting   ts,
especially

This pi   ensi   ese complex
tourism is: ogi al p   ective.   theoretical and
empirical appr ches are introduced and the following issues are then
discussed:

- Identifiable and stable forms of touristic behaviour and roles,
- Social divisions within tourism,
- Whether tourism reinforces underdevelopment,
- The interdependence of tourism and social institutions,
- The effects of transnational tourism on the ecosystem and commodification,
- The sociology of tourism as a sub-discipline.

The Sociology of Tourism has sixteen contributions from nine different countries, including the UK, USA, Canada and Australia. It brings together the most noted theoretical and empirical studies written in this area and enriches them with diverse experiences and perspectives.

Dr Yorghos Apostolopoulos is Research Associate Professor in the Department of Sociology at Arizona State University, USA. **Dr Stella Leivadi** is Assistant Professor of Statistics and Research Methodology at the University of Northumbria at Newcastle, UK. **Dr Andrew Yiannakis** is Professor and Director of the Laboratory for Leisure, Tourism and Sport at the University of Connecticut, USA.

THE SOCIOLOGY OF TOURISM

Theoretical and Empirical Investigations

Edited by Yorghos Apostolopoulos, Stella Leivadi and Andrew Yiannakis

London and New York

First published 1996
by Routledge
11 New Fetter Lane, London EC4P 4EE

Simultaneously published in the USA and Canada
by Routledge
29 West 35th Street, New York, NY 10001

Reprinted 1999, 2000, 2001

First published in paperback 2002

Routledge is an imprint of the Taylor & Francis Group

© 1996, 2002 Selection and editorial material Yorghos Apostolopoulos,
Stella Leivadi and Andrew Yiannakis; individual chapters the contributors.

Printed and bound in Great Britain by
St Edmundsbury Press, Bury St Edmunds, Suffolk

British Library Cataloguing in Publication Data
A catalogue record for this book is available from the British Library

Library of Congress Cataloging in Publication Data
A catalog record for this book has been requested

ISBN 0-415-13508-7 (Hbk)
ISBN 0-415-27165-7 (Pbk)

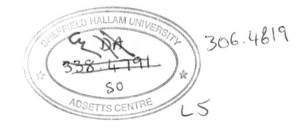

CONTENTS

CONTENTS

FIGURES

TABLES

CONTRIBUTORS

Dr Yiorgos Apostolopoulos, University of Miami, Coral Gables, Florida, USA, and Tourism Research & Consulting Group, Athens, Greece

Dr Stephen Britton, University of Auckland, Auckland, New Zealand

Dr Erik Cohen, The Hebrew University of Jerusalem, Jerusalem, Israel

Dr Malcom Crick, Deakin University, Geelong, Victoria, Australia

Dr Graham H.S. Dann, The University of the West Indies, Bridgetown, Barbados

Dr C. Michael Hall, University of Canberra, Canberra, Australia

Dr Michael Ireland, College of St Mark and St John, Plymouth, Devon, UK

Dr Cecilia A. Karch, The University of the West Indies, Bridgetown, Barbados

Dr Joseph P. Kopachevsky, University of Prince Edward Island, Charlottetown, PEI, Canada

Dr Maria Kousis, University of Crete, Rethymno, Greece

Dr Giuli Liebman Parrinello, University of Rome, Rome, Italy

Dr Linda K. Richter, Kansas State University, Manhattan, Kansas, USA

Dr Maurice Roche, Sheffield University, Sheffield, UK

Dr John Urry, Lancaster University, Lancaster, UK

Dr G. Llewellyn Watson, University of Prince Edward Island, Charlottetown, PEI, Canada

FOREWORD

Tourism, in all respects, is an inspiring and fascinating venture. The tourism and travel industry, together with the oil and automobile industries, comprise the three largest international economic activities of our time. In 1994, 6.5 per cent of the world's workforce was employed in tourism activities, while internationally, tourism generated 528 million arrivals and $321 billion in receipts. This further secures its trend towards becoming the primary international economic activity by the beginning of the next century.

It is, therefore, natural that the addition of the present book to the international academic literature on tourism constitutes a substantial contribution to the study of such an important subject. This publication acquires an even greater importance as it approaches tourism from the sociological angle, something ignored so far to a large extent, with the dominant perspective being economic.

The present publication will help us see tourism as a social phenomenon consisting of many facets which can, and should, become subjects for research. The mutual understanding between peoples from various countries; their familiarization with other civilizations, cultures, art, history, traditions, and usages; the influence tourism has brought to bear on customs, habits, values, and the way of life of local populations; as well as its negative social repercussions for the inhabitants of receiving countries, and many other related topics should be studied from a sociological point of view. The present book constitutes an incentive and a challenge for such an orientation of scientific research.

We also should not forget that tourism is an extremely sensitive and vulnerable product, subject to human motivation and behaviour. Prospective travellers react immediately to world news. A piece of information or news flashed worldwide by the media, even if it is exaggerated – and we know how often the media exaggerate – has the power to generate a flow of arrivals to a certain country or create mass cancellations to another. News about local wars, terrorist activities, epidemics, natural disasters, and so on is sufficient to alter completely the tourism prospects of a country and cause catastrophic results for travel, hotel accommodation, investments, and promotion budgets. It is imperative, therefore, for national tourist administrations, as well as private enterprises, to take into consideration the 'unforeseen' and make provisions for mechanisms to meet and remedy such contingencies.

I would also like to single out another phenomenon which becomes increasingly important in international developments in the field of tourism. This is the concentration of extensive economic power in the hands of only a few multinational 'mega-enterprises'. This leads to a powerful vertical concentration of economic activity where a holding concern can simultaneously control tour operating companies, airlines, travel agencies, hotel chains, and the like. These powerful interests may direct or uphold tourist arrivals to one or another country, decisively influencing international tourism development. There exists, of course, international competition, but margins for such competition become narrower and narrower as the international market becomes more and more oligopolistic.

Another important related phenomenon is the invasion of the computer sciences into the field of tourism, extending from the simple computerization of hotels and travel agencies to the booking of holidays from one's armchair. This final development has important implications. The ability to select, reserve, and confirm travel and holiday plans from home, with the use of computers and modems, brings with it enormous changes for the role and function of travel agencies, as well as for the tourism industry as a whole. Could this development be a response to the needs of the client who is unprotected from the brainwashing of advertising and, moreover, from the immense power of the tourist 'megaenterprises' mentioned above?

Concluding this brief wandering into the future developments of tourism, I wish to single out another positive, yet ignored, aspect – the contribution of tourism to architecture and the arts. In ancient times, art was associated with the ruling authorities, both secular and religious. The most significant works of architecture were palaces and castles, temples, cathedrals, and monasteries. The aesthetic expression of a certain period was evident in the rhythms of its palaces and cathedrals, and their sculptures, mosaics, frescoes, icons, and paintings. Today, with the power of the earthly ruler and the bishop restricted, and with palaces and cathedrals built to a lesser and lesser extent, the role of creating impressive works of architecture, of modern 'monuments', is handed over to a great extent to large hotels and congress halls, where, instead of individual needs, the functional needs of groups of people are met. It is in these modern 'monuments' that architects, scientists, and artists have the opportunity to express themselves with imagination and inspiration. The contribution of tourism to the arts is a parameter that should not be ignored. Many years in the future, people who study the aesthetic expression of our era will also turn to the art and architecture of our hotels and congress halls. Some will do so with admiration and others with scepticism.

Tourism, in all respects, is an inspiring and fascinating venture.

Professor Marios Raphael
President, Executive Council of the World
Tourism Organization
President, University of Piraeus Centre of
Studies & Education for Tourism, Greece

December 1994

PREFACE

Tourism is a multibillion-dollar international industry that rivals in scope a number of world economies. It is an international enterprise that employs millions of people and contributes in significant ways to the economies of many nations. It is hardly surprising, therefore, that a complex infrastructure has evolved around the production, promotion, and delivery of the tourism product to the public. One aspect of this infrastructure is concerned with education and training.

In academe, the education and training of tourism personnel reflect two very different perspectives. The professional perspective emphasizes the study and application of tourism as a business enterprise. Graduates from such programmes acquire the skills to manage, promote, and service the needs of the tourist market. The emphasis in many such programmes is to produce well-trained professionals who will then take their place in the tourism industry. To support the training of such professionals a vast amount of literature has been produced in the form of textbooks, journals, videos, and the like. While some of these programmes also include, as part of their professional focus, an academic component, this component is relatively underdeveloped and often lacks intellectual rigour.

A second approach to the study of tourism is an academic one. This approach emphasizes the examination of the phenomenon of tourism as a social phenomenon worthy of study, *sui generis*. As such, this approach stresses the need to describe, explain, and understand the phenomenon and, ultimately, to generate theoretical insights that transcend the immediate concerns of the industry practitioner or the phenomenon under investigation. While insights generated from the academic study of tourism often translate into application with clear benefits for industry professionals, the primary goal of such an orientation is explanation, analysis, and understanding from a liberal arts perspective.

Despite the ubiquitousness and the vast economic and sociocultural impacts of tourism, the study of the phenomenon as an academic discipline has been slow in coming. In fact, it is only in the last twenty or thirty years that we have witnessed a concerted effort by social scientists in sociology, psychology, economics, anthropology, geography, and politics, among others, to develop the field as a legitimate area of scholarly investigation.

While many of the earlier sociological works reflect varying degrees of theoretical rigour, the majority of the literature is relatively detached from a substantive theoretical base. A lot of the early works are of a mostly distinctive nature and those that embrace higher levels of analysis reflect the relatively underdeveloped nature of the field both theoretically and substantively. Thus, owing to the general lack of the use of theory and more sophisticated methodological approaches, the sociology of tourism as a formal sociological speciality (such as the sociology of religion or the family) is yet to be recognized within the discipline of sociology. As a consequence, no single sociological perspective reasonably claims a monopoly in the study and analysis of tourism. In fact, it has been argued that at present, the sociology of tourism is in a state of flux, in search of an 'all-embracing theory of tourism' and an identity. It is hardly surprising, therefore, that no textbook of an academic nature exists which explores, in a comprehensive manner, the sociological dimensions of tourism.

It is our intent, ambitious though it may appear, to fill this gap and further encourage the development and formal recognition of the sociology of tourism. For, regardless of whether tourism is viewed as a form of 'commercialized hospitality', 'modern leisure activity', 'democratized travel', a 'modern pilgrimage', or a 'form of neocolonialism', it remains a complex sociocultural, political, and economic phenomenon that is worthy of systematic sociological investigation.

This book is intended to serve the needs of those with a newfound interest in the field, and those who already possess a more in-depth understanding of the phenomenon as a sociocultural process. However, the book's primary purpose is to assist the reader in developing a more thorough understanding and appreciation of the study of tourism, from a sociological perspective – a perspective which, up till now, has received relatively little recognition within academe, the tourism industry, and society.

ACKNOWLEDGEMENTS

This book is the product of the frustration experienced by its senior editor in his search for an appropriate textbook with which to teach a course on Tourism and Development. The lack of a comprehensive text on both theory and research in the area and a shared interest in the sociological approach to tourism led the co-editors of this book on their mission to fill this gap with the critical selection and compilation of noted works in the area.

Books are never completed by individuals, however. The work and thinking of a number of people have gone into producing the chapters that follow. The scholarly contributions of fifteen individuals, providing their valuable insights and expertise on the sociological perspective on tourism, are the heart of this text. The calibre of their work made our selection process a much more enjoyable task. We would also like to extend our gratitude to the publishers of the included material for their timely cooperation in providing copyright permissions. Thanks must also go to Lorrine Basinger, a doctoral candidate of sociology at the University of Miami, who as the research assistant of the senior editor contributed significantly with her challenging and valuable input throughout all phases of the book's production. Dr Lyn Mac-Corkle, of the University of Miami's Richter Library, must also be thanked for her diligent efforts and expertise as an information specialist. We are also deeply grateful to all the professors, mentors, and family members who have supported us and contributed to our intellectual development at various stages of our lives.

Finally, we wish to thank Sally Close, Alan Jarvis, Laura Large, Caroline Law, Katy Lyall and Francesca Weaver of the Business and Management Department at Routledge Publishers. Their encouragement and tireless support made the entire publication process a pleasure.

1

INTRODUCTION

Reinventing the sociology of tourism

Yorghos Apostolopoulos

THE MAGNITUDE OF TOURISM

Tourism, the 'largest peacetime movement of people' (Greenwood 1972), presents itself as a challenging sociocultural phenomenon. Since the Second World War, the growth of the tourism industry and its promotion by international financial organizations as an agent of quick economic development and change have been unprecedented. Is tourism 'blessing or blight', 'trick or treat', 'boom or doom', 'panacea or a new slave trade', 'mirage or strategy for the future' (Lanfant and Graburn 1992)? Is tourism a passport to the socioeconomic development of developing nations (de Kadt 1979)? Is tourism the means for resolving all the developmental problems of an ill-planned economy or is it a well-planned and organized economy that makes tourism a profitable enterprise (Apostolopoulos 1996)? Do the development and implementation of, and investment in, tourism have trickle-down effects for the lower strata and the disadvantaged parts of host societies? Is tourism another form of imperialism or neocolonialism, perpetuating inequalities in the capitalist world system, accelerating the ecological degradation of the planet, and destroying the most fragile and marginal cultures (van den Berghe 1992)? Does dependence on tourism lead to a social, economic, political, and cultural dependency (Apostolopoulos 1995b; Britton 1982)? Are tourists 'barbarians' and 'suntanned destroyers of culture' who seek only 'sun, sea, sand, and sex' (Crick 1989; Turner and Ash 1976)? Who is the tourist; are there different types of tourists who adopt different roles seeking different optimal experiences (Cohen 1979b; Leivadi and Apostolopoulos 1996; Yiannakis *et al.* 1991)? Can tourism be understood without viewing it as a 'megasystem' that generates and receives simultaneously in the context of interdependent structures and forces (Jafari 1989)? Or, can contemporary tourism be comprehensively studied without stressing the transformation of the industry to a transnational industry having established ascendancy in the developing world through the internationalization of capital, international trade, and international politico-military systems (Apostolopoulos 1995b; Dunning 1989; Enloe 1990; Vandermerwe and Chadwick 1989)? Is the emergence and implementation of 'alternative' or 'appropriate' forms of tourism (Cohen 1989; Smith and Eadington 1992) the only saviour of host societies from the adverse consequences of mass tourism influx?

The international tourism industry – the largest industry in the world – holds a very influential position in the world economy and belongs to those economic sectors offering realistic potential for long-term growth, especially with the immense worldwide expansion of the service sector (Faketekuti 1987; Giarini 1987; Richter 1987). In 1994, international tourism produced approximately $3.5 trillion in gross output (6.1 per cent of global GDP) and employed close to 130 million people, 6.8 per cent of workers worldwide (Waters 1995; World Travel & Tourism Council 1995). For the same year, the tourism industry accounted for 12.3 per cent of consumer expense, absorbed 75 per cent of the total capital investment, and payed almost 6 per cent of total tax payments (World Travel & Tourism Council 1995). The multiplier effects of the industry are even more impressive: the industry fuels an economic chain reaction that begins with purchases from other industries and extends to the spending of profits, dividends, and employee compensations. Are all these benefits evenly distributed among the populations of host countries or the different parts of the world? Does this tourist influx contribute to an actual development involving not only changes of aggregate economic indicators but also changes in the 'reorganization and reorientation of entire economic and social systems' (Todaro 1989)?

As with several aspects of modern life, the development and expansion of the tourism industry has brought both 'blessings' and 'curses' to the socioeconomic and sociocultural spheres. In the socioeconomic sphere, tourism has dramatically affected 'foreign exchange, income, employment, prices, the distribution of benefits, ownership and control, development, and government revenue' (Apostolopoulos 1993; Cohen 1984), while in the sociocultural sphere, tourism has affected 'community involvement in wider frameworks, the nature of interpersonal relations, the bases of social organization, the rhythm of social life, migration, the division of labor, stratification, the distribution of power, deviance, and customs and the arts' (Apostolopoulos 1993; Cohen 1984). In fact, no other contemporary industry has had such a crucial and far-reaching impact on so many facets of society.

Tourism,[1] 'the study of man away from his usual habitat, of the touristic apparatus and networks, and of the ordinary (home) and (the) nonordinary (touristic) worlds and their dialectic relationship' (Jafari 1987), can only be understood if studied holistically and if a 'knowledge-based platform' (Jafari 1989) is adopted. The 'advocacy-based', 'cautionary-based', and 'adaptancy-based' platforms currently in use are one-dimensional approaches to tourism. They focus primarily on either the positive or negative aspects of the industry, and are not mutually exclusive or able to replace one another. The 'knowledge-based' platform, on the other hand, takes into account the multidimensionality of tourism, presenting a balanced perspective which bridges various existing views and is intended to contribute to the formulation of a scientific body of tourism. Such a perspective presents a fairer treatment of the touristic phenomenon and its impacts on tourists, hosts, tourist corporations, governments, and other involved parties.

OBJECTIVES

In the past thirty years, a number of sociologists (among others, see Apostolopoulos 1993, 1995a, b; Boorstin 1992; Britton 1982; Cohen 1972, 1974, 1979a, b, 1984, 1988a, b; Dann 1977, 1981; Dann and Cohen 1991; de Kadt 1979; Forster 1964; Graburn 1989; Karch and Dann 1981; Lanfant 1980; MacCannell 1976, 1992; Machlis and Burch 1983; Pearce 1982; Turner and Ash 1976; van den Berghe 1992; Young 1973) have been intrigued by the growth of tourism as an economic and cultural phenomenon and by its dramatic impact on both developed and developing nations. Although their research has utilized various sociological theories and perspectives,[2] the majority of sociological work on tourism has been relatively detached from a substantive theoretical examination. Most studies have focused only on a descriptive level, and those which have included theoretical ideas have most often done so in an implicit or underdeveloped manner. Owing to the general lack of grounded theory, sophisticated methodological/statistical treatment of tourism data, and adequate contextualization in broader applied domains (Dann and Cohen 1991), the sociology of tourism as a formal sociological speciality (similar to the sociology of education, sociology of the family, and so forth) has not yet been established within the discipline of sociology. As a consequence, no single sociological perspective reasonably claims a monopoly in the understanding of tourism. At present, the sociology of tourism is in a state of flux, in search of an 'all embracing theory of tourism' and in search of identity (Dann and Cohen 1991).

It is no surprise, then, that there has been no single work in book format (textbook in particular) in the international (English language) literature which attempts comprehensively to explore the sociology of tourism in terms of all its basic dimensions. However, books have been published on the anthropology, history, marketing, management, economics, and geography of tourism. While these disciplines as a whole provide a broad coverage of tourism, the topic remains in need of the unifying perspective a sociological approach can provide. It is the present authors' intent both to present this perspective and provide a push towards establishing the formally recognized sociological examination of tourism. Furthermore, the transition of the social structure from an industrial to a 'post-industrial' or 'modern' type, along with the fact that leisure is displacing work as the centre of modern social arrangements,[3] currently offer tourism the potential to surpass any previous records in volume of travellers and economic output. Such a dynamic process can always be examined in retrospect; now is the time to explore it sociologically as it unfolds.

The sociologists whose seminal works have provided the foundation for a sociological approach to tourism are primarily Cohen, MacCannell, Urry, and Lanfant. Their individual contributions are both comprehensive and profound, and efforts to expand their work with a strong empirical approach would boost even further the emerging sociology of tourism. This anthology will begin to fill a gap in the literature through its comprehensive presentation of the sociological picture of tourism. A truly holistic, pluralistic sociological approach to tourism will attempt to address all that affects and is affected by tourism. Tourism service-sector development is as important a focus as tourist/resident interaction. Regardless of whether tourism is considered or treated as a 'commercialized hospitality', 'democratized travel', 'modern leisure activity', 'modern variety of the traditional pilgrimage', 'expression of basic cultural themes', 'acculturative process', a 'type of ethnic relations', or a 'form of neocolonialism' (Cohen 1984), it remains a complex sociocultural, political, and economic phenomenon that demands systematic sociological investigation.

The Sociology of Tourism: Theoretical and Empirical Investigations is being presented, quite simply, for those interested in the study of tourism's role in society. This book should serve the interests of both those with a newfound interest in the field and those who already have some understanding of tourism in the sociocultural and socioeconomic process. Its most fundamental goal is to assist the reader in developing a sociological perspective related to the emerging and dynamic institution of tourism. Whether a member of the social science community or of the tourism industry itself, from service to management, the reader will benefit from a sociological approach to tourism with a broader, more comprehensive understanding of the industry and his or her role within it.

ORGANIZATION AND THEMES

This anthology is a pioneering work aimed at introducing and legitimizing the sociological study of tourism in international academe. It is intended to 'reinvent' an area first presented three decades ago. The anthology is based upon the major theme of tourism as a far-reaching transnational industry that continues to affect crucially the 'post-industrial' world. The book is divided into seven parts, organized around various major sociological themes. The themes have been selected with attention to continuity and to the provision of a unifying structure for the collection as a whole. It presents a variety of theoretical and methodological approaches in order to provide a more pluralistic sociological picture.

Most studies on tourism can be classified as 'impact' research, incorporated in the literature of the sociology of leisure, migration, or travel. The present anthology treats tourism as an autonomous, dynamic entity which has marked and continually defines an era. Tourism is addressed in a conventional sociological-text manner with all the major societal areas covered as they relate to the touristic phenomenon. This was selected as the best approach, taking into account the complexity of the touristic phenomenon and the lack of systematic sociological work on the subject. In this way, the seven parts comprise an anthology that can serve as a skeleton text as well. The following sections are summaries of the material and perspectives presented in each chapter.

Part I Towards a sociological understanding of contemporary tourism

How does society really work? Who really has power? Who benefits under the existing social arrangements and who does not? Sociology concerns itself with answering such all-important questions. This introductory part provides a collection of articles which exemplify a sociological approach to tourism, an approach concerned with 'seeing through' the façades of social structures. The works by Crick and Cohen offer a comprehensive framework for the sociological study of tourism. The structure of the tourism system – the tourist (motivations, attitudes, reactions, and roles), relations and perceptions of hosts and guests, and the study of hosts and hosting – and the socioeconomic and sociocultural impact of the tourism industry provide a unique analytical setting. In addition, it is becoming clear that regardless of the fact that tourism and travel seem to generate contradictory representations in social science research and the fact that the term 'tourist' is increasingly used as a derisive label for someone who seems content with his or her obviously unauthentic experiences, tourism provides an irreplaceable key to understanding the peculiarities of modern culture.

Part II The tourism system and the individual

Through socialization, human beings come to share certain symbols, ideologies, norms, values, and roles in the existing social structure. While there is a tendency for the members of a given society to be alike in these shared characteristics, heterogeneity increases with the size of a society. The sources of the differences include the family, schools, religions, social locations, contradictory influences, and conflicts in role definitions. This part presents an analysis of how the international tourism system influences the prospective tourist and host, and how they interact among themselves despite their inherent cultural differences. The fulfilment of certain needs creates specific motivations which result in related roles, behaviours, and touristic experiences. The works by Parrinello and Cohen present the touristic experience in a comprehensive manner in accordance with the complexity of the phenomenon. In a 'post-industrial' society where the service sector dominates the international economic arena, the motivation for travel and tourism is closely associated with the phase of anticipation. On the other hand, structural theorizing about tourism proposes that social behaviours are governed not only by immediate relations with close kin and friends but also by the collective influence of relational patterns within social networks. This quest for a touristic experience could be distinguished in the recreational, diversionary, experiential, experimental, and existential mode and supports the notion of the existence of identifiable and stable forms of touristic behaviour.

Part III Structures of social inequality in the tourism system

Social inequality refers to the existence of differential degrees of social influence or prestige among individual members of the same society. Social stratification involves the existence within a single society of two or more differentially ranked groups, the members of which control unequal amounts of power, privilege, and prestige. While social inequality has been a universal feature of human societies, social stratification goes beyond this into the realm of differential access to resources, involving groups and not individuals. The central issues in this part focus on the impact mass institutionalized tourism has on tourists and especially on locals and tourism employees along class, gender, race, and ethnicity lines. The works of Urry and Ireland illustrate the immense effect of tourism on the division of labour in various parts of the world. Contemporary tourism's ramifications are placed and examined in a historical context focusing on local, regional, and national tourist developments in terms of macroeconomic policies in a global political economy historically structured by colonialism, post-war European and American hegemony, and the present politico-economic domination of the South by the North. These two sociologists indicate the existence of strong inequalities in the consumption of leisure and tourism where ethnic groups often become 'attractions for outsiders'. Tourism has not only become a gendered institution, but also one where ethnic and racial subordination are present in the context of restructuring the social relations of production and reproduction worldwide.

Part IV Tourism, underdevelopment and dependency

Worldwide differences in levels of economic and social development have long been crucial concerns of social scientists and policy makers. Underdevelopment, the state of those nations having the lowest levels of technological and economic development within the capitalist world economy, is perhaps best addressed in terms of a society's level of GNP. However, numerous other indicators of underdevelopment exist and the most crucial ones include marked levels of economic inequality, high rates of population growth, poor standards of nutrition and health, and a high percentage of the population still employed in agriculture. In addition to an improvement in income and output, an actual development process should involve radical changes in institutional, social, and administrative structures as well as changes in popular attitudes and, in many cases, even in customs and beliefs (So 1990). This part directs attention to a crucial question: does tourism's remarkable growth narrow or widen the gap between the affluent and poor? The papers by Britton and Karch and Dann attempt to shed light on the question of whether or not tourism is advantageous not only in aggregate terms but in terms of to whom exactly advantages accrue. If development includes the goal of reducing inequalities and redistributing social goods according to the priorities of basic needs, then the distributive aspect of the tourism industry is of central importance. The metropolitan control over the tourism industry of developing nations through tourist transnational corporations (tourist plants and their advertising are primarily based on metropolitan capital, transport systems are metropolis oriented, and the structure of the industry tends towards a monopoly), along with relations of dependency in such vulnerable economic activity, result in a structural dependency on developed countries. These effects might include not only economic dependency but also a lack of endogenous development and domination of national political economies. Tourism, along with migration and foreign debt, represent the new modes of dependency between developing and developed nations.

Part V Tourism and social institutions

Every society has created standardized, patterned ways of fulfilling its fundamental needs in the form of institutions. Human beings fashion social institutions in accordance with their society's dominant norms and values. At the same time, well-defined institutional patterns mould and channel individual behaviour and awareness. Social institutions, however, are not monolithic; tensions, defects, and countervailing pressures within them mirror conflicts and controversies within society at large. The present part discusses the role these socially constructed norms and behaviours play in the tourism system and how institutions are influenced by the tourism system. The papers by Urry, Kousis, and Richter discuss the influence of the transnational tourism industry on social institutions and their interdependence with a particular emphasis on the economy, family, and the polity. The changing economics of the international tourism industry, along with its complex and contradictory effects, have changed the industry's identity. Globalization, the package holiday industry, and the unprecedented expansion of tour operations have contributed to the construction of this powerful market-dominating industry. The central question remains: 'What freedom of action do developing countries have?' The 'know-how' and the networks of tour operators considerably limit the freedom of action and bargaining power of developing countries. On the other hand, tourism's international influence has been and can still be exploited by both democratic and non-democratic regimes to their own political advantages. Tourism has been used as a means of 'selling martial law' to attract investment capital or internationally prestigious conferences, to promote peace and brotherhood, and to legitimize the objectives of a government. The tourism industry may further become an agent of desirable social change by having a positive impact on the established sex division of labour and on family institutions with a decline in family control over the individual, as has been exercised through marital arrangements and the like.

Part VI Tourism and social change

Social systems, if they are to survive, require two opposite tendencies: stability and change. While stability and predictability are crucial, social change and transformation are more than inevitable. Changes occur from endogenous as well as exogenous forces and may be gradual or abrupt, reforming or revolutionary, deliberate or accidental. The present part illustrates some of the adverse effects of tourism-industry development and expansion in the sociocultural change scene, especially when inadequate planning has taken place. The works of Hall and Watson and Kopachevsky reaffirm that tourism is much more than a traveller's game and is not the panacea that hasty planners proclaimed it to be forty years ago. Beyond the easily quantifiable effects of tourism development, tourism's mark is left on the physical environment, human values and relationships, and culture. Specifically, tourism has adversely affected developing nations' people through commodification of cultural manifestations and the dramatic emergence of crime, prostitution, and lack of

7

authenticity. These disruptive forces on local communities, depending on the socio-economic and geographic organization of transnational tourism, lead to profound consequences for the bases of social organization, the rhythm of social life, and the nature of interpersonal relations. The present part highlights the fact that modern transnational tourism's ramifications will be better understood in the contexts of the commodification process and contemporary consumer culture.

Part VII Towards a 'new' sociology of tourism

A multitude of issues have been subjected to sociological examination, as evidenced in the various established subdisciplines such as the sociology of education, political sociology, medical sociology, and even the sociology of sociology. All these perspectives are taught in various collegiate departments of sociology and a great deal of scholarly research and publication has taken place in these areas. With such a wide range of formal sociological foci, the incorporation of the new and important *sociology of tourism* into mainstream academic and research programmes – in the context of socioeconomic development – seems natural and logical. A phenomenon which so crucially affects our culture, the welfare of millions of people around the world – especially those of developing nations – our environment, and our values should not remain without a recognized sociological identity. The two works by Dann and Cohen and Roche contribute to the establishment of this identity. A formal sociological approach to tourism will serve as the long-overdue vehicle for moving beyond a limited examination of the touristic phenomenon to a comprehensive understanding of all its socioeconomic, political, and cultural effects through a careful integration of grounded theory and empirical investigation.

THE SOCIOLOGY OF TOURISM: A NEED FOR A PROGRAMMATIC CALL?

The International Sociological Association, the most prestigious sociological organization in the world today, included, for the first time ever, an autonomous section on the sociology of tourism in the XIIth World Congress of Sociology (Madrid, Spain, 1990). This debut was accompanied by the formation of a new Thematic Group involving over forty tourism researchers who dealt with 'Tourism in the World: Unity and Diversity'. Stimulated by the success of its deliberations, the Thematic Group decided to apply for Working Group status within the association with the ultimate goal of becoming a permanent Research Committee in the International Sociological Association. For the XIIIth World Congress of Sociology (Bielefeld, Germany, 1994), the sociology of tourism was upgraded to Working Group status, involving eighty social scientists in thirteen thematic sessions, presenting approximately sixty papers which addressed how 'International Tourism Displaces Boundaries'.

While the formal and scientific investigation of the sociology of tourism is still in its infancy, it has already generated considerable interest and debate. Despite both this and its acceptance in the International Sociological Association's forums, the sociology of tourism remains unrecognized by other major organizations, such as the American Sociological Association and the Society for Applied Sociology. To date, neither organization has included even a section addressing tourism. The official reports of the American Sociological Association under the areas of 'sociological interest' include such new and non-traditional areas as emotions, visual sociology, and language. The progressive inclusion of such areas of interest broadens and enriches the discipline of sociology. The inclusion of the sociology of tourism would surely do the same, attracting professional interests and scholarly contributions worldwide.

At the other end of the spectrum, international professional organizations and societies on tourism increase in number. The establishment of the International Academy for the Study of Tourism in 1988 under the aegis of the World Tourism Organization has been very active in the promotion of tourism and tourism research from a scholarly viewpoint. The European Association of Development Research and Training Institutes has also established a Working Group on 'Tourism and Development' with a very active research agenda on the sociological aspects of tourism. The Travel and Tourism Research Association also includes a focus on tourism and travel research, though its primary aim is to enhance the effectiveness in tourism marketing, planning, and management, and has continually stressed the significance of the industry and of scholarly research work in the area. In addition, the occurrence of a series of international conferences on tourism sponsored and co-sponsored by Unesco, the European Union, the World Tourism Organization, the Organization for Economic Co-operation and Development, the World Bank, and other national and regional tourism organizations further confirms the global interest in tourism and its multifaceted effects.

In terms of exactly where the sociology of tourism fits, it does not need to fit in, *per se*. The sociology of tourism needs simply to take root and define itself as a unifier of perspectives and as the comprehensive informer of social scientists, tourism practitioners, and citizens on the phenomenon of tourism. The questions have been asked,

> *How is it possible to lay the foundation for a conceptual apparatus capable of seizing a phenomenon (tourism) which touches all levels of society, while propagating itself to all corners of the globe, establishing a new social bond between them? How is it possible to elaborate a theoretical and methodological framework which responds to the imperatives of a global and multidimensional approach? How are we to establish the pluridisciplinary cooperation which these two imperatives call for? What will become of the specificity of sociology in this confrontation?*

(Lanfant 1989: 591, emphasis added)

The answer lies in the addition of the formal sociological study of tourism to the various established perspectives.

NOTES

1 There are many definitions of tourism and tourist (see Cohen 1974, 1979b, 1984; Leiper 1979; Urry 1991) which have a number of limitations. Official statistics normally identify the number of tourists visiting a country simply by counting the number of arrivals at ports and airports. Measured in this way all travellers appear as tourists, but sociologists have approached tourism as associated with pleasure and leisure rather than work.

2 See MacCannell (1976) for a neo-Durkheimian perspective, Turner and Ash (1976) for a conflict perspective, Mayo and Jarvis (1981) for a functionalist perspective, Dann (1977) for a Weberian perspective, Cohen (1979b) for a phenomenological perspective, McHugh *et al* (1974) for an ethnomethodological perspective, Dann (1989) for a symbolic interactionist perspective, and Apostolopoulos (1993) for a world system/dependency perspective.

3 Contemporary estimates see the average person living approximately 640,000 hours, spending 60,000 hours at work and having approximately 280,000 hours of spare time. In the near future, with the decrease in the length of the work day, the average person will work only 40,000 hours and will have 300,000 hours left for leisure and travel (Leisure Studies Association 1993).

REFERENCES

Apostolopoulos, Y. (1993) 'The perceived effects of tourism-industry development: a comparison of two hellenic islands', Ph.D. Dissertation, The University of Connecticut, Storrs, Connecticut.

—— (1995a) 'Transnational tourism and socioeconomic development: issues, implications, and alternatives', unpublished manuscript, University of Miami, Coral Gables, Florida.

—— (1995b) 'The political economy of tourism: a critical assessment of selected developing nations in the mediterranean', *Sociological Quarterly*, forthcoming.

—— (1996) *The Effects of Tourism Expansion in the Greek Islands: Regional Tourism Planning and Policy Lessons for Sustainable Development*, Athens: Papazissis Publishers.

Boorstin, D. (1992) *The Image of Pseudo-Events in America*, New York: Vintage.

Britton, S. (1982) 'The political economy of tourism in the Third World', *Annals of Tourism Research* 9: 331–58.

Cohen, E. (1972) 'Toward a sociology of international tourism', *Social Research* 39: 64–82.

—— (1974) 'Who is a tourist? A conceptual clarification', *Sociological Review* 22: 527–55.

—— (1979a) 'Rethinking the sociology of tourism', *Annals of Tourism Research* 6: 18–35.

—— (1979b) 'A phenomenology of tourist experiences', *Sociology* 13: 179–201.

—— (1984) 'The sociology of tourism: approaches, issues, and findings', *Annual Review of Sociology* 10: 373–92.

—— (1988a) 'Traditions in the qualitative sociology of tourism', *Annals of Tourism Research* 15: 29–46.

—— (1988b) 'Authenticity and commoditization in tourism', *Annals of Tourism Research* 15: 371–86.

—— (1989) '"Alternative tourism" – a critique', in T.V. Singh, H.L. Theuns, and F.M. Go (eds) *Towards Appropriate Tourism: The Case of Developing Countries*, pp. 127–42, Frankfurt am Main: Peter Lang.

Crick, M. (1989) 'Representations of international tourism in the social sciences: sun, sex, sights, savings, and servility', *Annual Review of Anthropology* 18: 307–44.

Dann, G. (1977) 'Anomie, ego-enhancement, and tourism', *Annals of Tourism Research* 4: 184–94.

—— (1981) 'Tourism motivation: an appraisal', *Annals of Tourism Research* 8: 187–219.

—— (1989) 'The tourist as a child: some reflections', *Cahiers du Turisme*, serie c, no. 135, Aix-en-Provence: CHET.

—— and E. Cohen (1991) 'Sociology and tourism', *Annals of Tourism Research* 18: 155–69.

de Kadt, E. (ed.) (1979) *Tourism: Passport to Development? Perspectives on the Social and Cultural Effects of Tourism in Developing Countries*, New York: Oxford University Press.

Dunning, J. (1989) 'Multinational enterprises and the growth of services: some conceptual and theoretical issues', *The Service Industries Journal* 9: 5–39.

Enloe, C. (1990) *Bananas, Beaches, and Bases: Making Feminist Sense of International Politics*, Berkeley, CA: University of California Press.

Faketekuti, G. (1987) 'About trade in tourism services', in O. Giarini (ed.) *The Emerging Service Economy*, pp. 245–51, Oxford: Pergamon Press.

Forster, J. (1964) 'The sociological consequences of tourism', *International Journal of Comparative Sociology* 5: 217–27.

Giarini, O. (ed.) (1987) *The Emerging Service Economy*, Oxford: Pergamon Press.

Graburn, N. (1989) 'Tourism: the sacred journey', in V. Smith (ed.) *Hosts and Guests: The Anthropology of Tourism*, pp. 21–36, Philadelphia: University of Pennsylvania Press.

Greenwood, D.J. (1972) 'Tourism as an agent of change: a Spanish Basque case', *Ethnology* 11: 80–91.

Jafari, J. (1987) 'Tourism models: the sociocultural aspects', *Tourism Management* 8: 151–9.

—— (1989) 'Sociocultural dimensions of tourism: an English language literature review', in J. Bystrzanowski (ed.) *Tourism as a Factor of Change: A Sociocultural Study*, pp. 17–60, Vienna: European Coordination Centre for Research and Documentation in Social Sciences.

Karch, C. and Dann, G. (1981) 'Close encounters in the Third World', *Human Relations* 34: 249–68.

Lanfant, M.F. (1980) 'Tourism in the process of internationalization', *International Social Science Journal* 32: 14–42.

—— (1989) 'World tourism: unity and diversity', *Annals of Tourism Research* 16: 591–2.

Lanfant, M.F. and Graburn, N. (1992) 'International tourism reconsidered: the principle of the alternative', in V. Smith and W. Eadington (eds) *Tourism Alternatives*, pp. 88–112, Philadelphia: University of Pennsylvania Press.

Leiper, N. (1979) 'The framework of tourism: towards a definition of tourism, tourist, and the tourism industry', *Annals of Tourism Research* 6: 390–407.

Leisure Studies Association (1993) *Newsletter* 36: 26–8.

Leivadi, S. and Apostolopoulos, Y. (1996) *The Social Psychology of Tourist Behavior*, in preparation.

MacCannell, D. (1976) *The Tourist: A New Theory of the Leisure Class*, New York: Schocken.

—— (1992) *Empty Meeting Grounds: The Tourist Papers*, London: Routledge.

Machlis, G.E. and Burch, W.R. (1983) 'Relations between strangers: cycles of structure and meaning in tourist systems', *Sociological Review* 1: 666–92.

Mayo, E. and Jarvis, L. (1981) *The Psychology of Leisure Travel*, Boston: CBI.

McHugh, P., Raffel, S., Foss, D., and Blum, A. (1974) *On the Beginning of Social Inquiry*, London: Routledge.

Pearce, D. (1982) *The Social Psychology of Tourist Behavior*, Oxford: Pergamon Press.

Richter, C. (1987) 'Tourism services', in O. Giarini (ed.) *The Emerging Service Economy*, pp. 213–44, Oxford: Pergamon Press.

Smith, V. and Eadington, W. (1992) *Tourism Alternatives*, Philadelphia: University of Pennsylvania Press.

So, A.Y. (1990) *Social Change and Development: Modernization, Dependency, and World-System Theories*, Newbury Park, CA: Sage.

Todaro, M. (1989) *Economic Development in the Third World*, New York: Longman.

Turner, L. and Ash, J. (1976) *The Golden Hordes*, New York: St Martin's Press.

Urry, J. (1991) *The Tourist Gaze: Leisure and Travel in Contemporary Societies*, Newbury Park, CA: Sage.

van den Berghe, P. (1992) 'Tourism and the ethnic division of labor', *Annals of Tourism Research* 19: 234–49.

Vandermerwe, S. and Chadwick, M. (1989) 'The internationalization of services', *The Service Industries Journal* 9: 79–93.

Waters, S. (1995) *Travel Industry World Yearbook*, New York: Child & Waters.

World Travel & Tourism Council (1995) *Travel and Tourism in the World Economy*, Brussels, WTTC.

Yiannakis, A., Leivadi, S., and Apostolopoulos, Y. (1991) 'Some cross-cultural patterns in tourist role preference: a study of Greek and American tourist behavior', *World Leisure and Recreation* 33: 33–7.

Young, G. (1973) *Tourism: Blessing or Blight?*, Harmondsworth: Penguin.

Part I

TOWARDS A SOCIOLOGICAL UNDERSTANDING OF CONTEMPORARY TOURISM

2

REPRESENTATIONS OF INTERNATIONAL TOURISM IN THE SOCIAL SCIENCES*

Sun, sex, sights, savings, and servility

Malcolm Crick

Travel seems to generate consistently ambivalent or contradictory representations. Why is it that Lévi-Strauss opens his travel autobiography *Tristes Tropiques*, which brought him such fame, by declaring that he hates traveling and travelers (111: 15)? Why do so many tourists claim that they are not tourists themselves and that they dislike and avoid other tourists (115: 10): is this some modern cultural form of self-loathing? In *The Innocents Abroad* Mark Twain asserts that "travel is fatal to prejudice, bigotry and narrowmindedness ..." (198, Vol. 2: 407) and yet goes on, page after page, about the daily torture and anxiety involved in foreign travel. Fatigue and the constant annoyance of beggars and guides "fill one with bitter prejudice" (198, Vol. 1: 253), he comments. "Another beggar approaches. I will go out and destroy him and then come back and write another chapter of vituperation" (198, Vol. 1: 269). Unlike Malinowski's mythologizing record of participant observation in his professional works, with embarrassing confessions, ambivalence, and hostility confined to his diary (118), Twain serves up the negative, positive, and contradictory in a single work.

Twain traveled and wrote at a time when the foundations of the modern travel industry were being laid; and if in 19th-century creative literature we have images of "travel," in that of the 20th we find portrayed its contemporary degenerate offspring – mass tourism. "Degeneracy" is an image that keeps surfacing, and so not surprisingly representations of tourism are frequently even more hostile than those of travel. As MacCannell puts it, "The term 'tourist' is increasingly used as a derisive label for someone who seems content with his obviously inauthentic experiences" (115: 94). John Fowles puts it more metaphysically, describing a typical scene with a group of tourists, cameras at the ready, watching a collection of locals and their performing scorpion. "There was some kind of triple blasphemy involved; against nature, against humanity, against themselves – man the ape, all the babooneries, the wrong motives of package travel" (61: 598–9).

* Reproduced with the permission of Annual Reviews Inc. from *Annual Reviews in Anthropology*, 1989, 18: 307–44.

The "blasphemy" Fowles describes is for many other commentators simply appalling bad taste. Surveying literary work on travel in the 1930s, Fussell (63: 37) describes his book as a "threnody" – that is, a mourning for a form of experience now beyond our reach. "Real" travel is now impossible; we can only aspire to "touring." A similar cultivated disdain is expressed by Nancy Mitford. Able to enjoy extensive periods overseas while writing her books, she totally dissociates herself from tourists: "The Barbarian of yesterday is the Tourist of today" (130: 3). An increase in tourists, "far more surely than any war, will be the end of Old Europe" (130: 7).

In his influential book *The Image*, Boorstin also stresses the difference between "traveling" [with its etymological connection to the notion of work (*travail*)] and tourism (the apotheosis of the pseudo, where passivity rather than activity reigns). Tourism is a form of experience packaged to prevent real contact with others (10: 91), a manufactured, trivial, inauthentic way of being, a form of travel emasculated, made safe by commercialism (10: 109). The Age of Discovery, when explorers lived, passed long ago. The aristocratic *Grand Tour* of the 18th century then gave way to the Age of Industrialism, when middle-class travel became possible. Now this world, too, has died, replaced in a post-industrial age by mass tourism. For Fussell, to write about tourism is necessarily to write satire, for the "travel industry" is a contradiction in terms: exploration is discovering the undiscovered; travel is at least intended to reveal what history has discovered; tourism, on the other hand, is merely about a world discovered (or even created) by entrepreneurs, packaged and then marketed (63: 39). According to Fussell, the tourist – a fantasist temporarily equipped with power – is someone whose essential nature has not been grasped by anthropologists (63: 41).

My purpose here is to examine the collective representations of international tourism – often referred to as the "Four S's" – sun, sex, sea, and sand (123: 25) – that exist currently in the social sciences. It may seem derogatory to speak of collective social science representations rather than analyses. I do so to raise the issue of whether we yet have a respectable, scholarly analysis of tourism, or whether the social science literature on the subject substantially blends with the emotionally charged cultural images relating to travel and tourists expressed in the literary views above. Mings (127: 343), for instance, has analyzed the history of academic writing on tourism in terms of an oscillation between two extreme myths – tourism as a godsend and tourism as evil – so the approach via representations has much to recommend it. This view is further supported by the comment of social psychologist P. Pearce, who has written much on tourism, that much of the sociology of tourism simply mirrors popular ideas about the subject (148: 17) and is perhaps only another example of how sophisticated tourists like to laugh at inferior versions of themselves (148: 18). D. Pearce, a geographer examining the literature on tourism, speaks of "weak methodologies and a certain degree of emotionalism" (146: 43). R. A. Britton rightly points out that some work on tourism by geographers, and especially economists, reads like a series of industry press releases (14: 30). Other social scientists can barely disguise their contempt; they disapprove as strongly as the economists once approved (9: 524).

Ambivalence, sweeping generalizations, and stereotypes abound. Lawson (104: 16), commenting on the interminable controversy over the value of tourism to Third World development, writes that the debate has been "intellectually insulting." Titles of well known books on tourism are very revealing. In *The Golden Hordes* (195), Turner and Ash write that tourists are the "barbarians," the suntanned destroyers of culture. Sir George Young, in one of the earliest critical discussions of the subject (213), subtitles his work *"Blessing or Blight?"* Rosenow and Pulsipher, surveying the American scene, write a volume called *Tourism: The Good, the Bad and the Ugly* (167). Britton and Clarke (19), looking at the record of international tourism in small, developing countries, edit a collection called *Ambiguous Alternative*. Tourism is referred to as a highly "deceptive" industry (194: 259) and such phrases as "malevolent potential of unregulated tourism" (44: 122) are not uncommon. Jafari, the founding editor of *Annals of Tourism Research*, wrote: "That 'each man kills the thing he loves,' as Oscar Wilde observed nearly a century ago, should perhaps be engraved over the door of national and regional tourism offices" (93: 210). Special Issue 6(3) of *Cultural Survival Quarterly* (1982) is ambiguously entitled "The Tourist Trap – Who's Getting Caught?" Wrote Valene Smith, a leading anthropologist of tourism: "Just as Rousseau decried the rise of industrialization but was powerless to do other than philosophize, so contemporary scientists must accept tourism as an existent major phenomenon" (184: 16). No wonder a leading sociologist of tourism, Erik Cohen, could say that the social scientific study of tourism was in a state of crisis (35: 5). Ulla Wagner, on the other hand, makes the balanced comment that when looking at an industry involving "individual, local, national, and international levels, as well as economic, social, and cultural aspects, we can hardly expect the impact [on Third World countries] to be uniformly 'good' or 'bad'" (205: 192).

By a number of criteria, tourism has a profound importance in the contemporary world. Before the rise in the price of oil in the early 1970s, tourism was the single largest item in world trade (209: 274), having grown at a rate of approximately 10% per annum since the 1960s. Some believe that by the year 2000 tourism will again be the world's largest industry (98: 16), though others suggest that the ever-expanding leisure of post-industrial society may not be expended in ways defined as "tourism" (51: 13). In the World Tourism Organization (founded in 1975 in succession to the International Union of Official Travel Organizations) was created an international body whose members are sovereign states devoted to the expansion of this single industry. Tourism also represents perhaps the largest movement of human populations outside wartime (74: 81). Since anthropology has been much concerned with culture contact and social change one would have expected tourism – obviously a modern form of acculturation (138: 207–8) – to receive extensive attention. As Jafari says (96: 137), "Today almost every community and nation, large and small, developed or developing, is influenced in varying degrees by tourism." And many a Third World country, of course, has opted for tourism as a central development strategy.

These criteria aside, there is, as MacCannell has argued, much cultural significance in tourism. A trivial activity could not generate such religiously constructed, lopsided, and ambivalent representations as exist about tourism. For MacCannell, tourism is the quest of modern man; the tourist is post-industrial man (114: 599; 115: 1, 4) doing ritual obeisance to an elaborate and experientially fragmenting division of labor (115: 11, 13) that requires the search for authenticity in other cultures (114: 589). Using anecdotal, lopsided imagery MacCannell suggests that the tourist centrally symbolizes the world in which we now live. Carroll (26: 140, 198–9) claims that tourism is a manifestation of the same restless Western spirit with which the founders of the social sciences were concerned. This point is also made graphically by Horne (87: 21), who suggests that "the camera and tourism are two of the uniquely modern ways of defining reality" (87: 121). Others (116: 669) see tourism as an intense case of "that which regularly occurs in the daily life of modern society – the ordering of relations between strangers."

If tourism tells us something vital about the modern world (67: 64), neither this nor its economic, cultural, and demographic magnitude has made it an important focus in social science research or in university social science curricula (67). Jafari might argue that one can scarcely ignore tourism (96: 137), but the fact is that a large number of social science disciplines have not paid it the theoretical or empirical attention it deserves. For instance, the first anthropological study of tourism dates from as recently as 1963 (137). The first conference was held in 1974 (122: 159). Even at the end of the 1970s most tourism research in the social sciences was incidental to other interests (95: 326); there were fewer actual case studies than afterthoughts or spin-offs from other projects (9: 524). Cohen dates the first full length sociological study of tourism in the 1960s (39: 373). Pi-Sunyer (155: 278) comments on how strenuously anthropological monographs seems to avoid mentioning tourists, as if their authors wished to disassociate themselves from other Western intruders. Nuñez (138: 207, 212) comments on the same lack of interest, even though almost everywhere anthropologists go they will find tourists, and even be categorized with them by locals. Finney and Watson-Gegeo report (57: 470) how tourism research has been looked down upon. L. Mitchell comments on how the study of tourism "gets no respect" in the discipline of geography (129: 236). Valene Smith (183: 274) reports how her early research interests were actively discouraged. Leiper (107: 392) suggests that as late as 1979 in academe tourism as a focus of research might well be derided. We do now have a small number of collections of anthropological articles on tourism (71, 185) and some overviews (68, 134, 138), but we still lack full length anthropological monographs on the subject. This is the current situation for a phenomenon "sponsored by governments, regulated by international agencies, and supported by multinational enterprises" (181: 3–4).

For L. Turner, international tourism is "simultaneously the most promising, complex, and understudied industry impinging on the Third World" (194: 253). How are we to explain such a widespread lack of attention to such an evidently significant phenomenon? If it is not a matter of the complexity, then such neglect is itself a fruitful area for the sociology of knowledge. In the case of anthropology, is it that anthropologists, because we study "them" and not "us," have regarded tourists as someone else's concern? Is it that academic personalities find it difficult to take as a serious area of research a phenomenon so bound up with leisure and hedonism? A social psychologist has suggested that the relative neglect of tourism in the behavioral sciences relates to deeply embedded values in Western society concerning work and play (148: 1–2). But is there an even more basic emotional avoidance at work for anthropological researchers – namely, that tourists appear, in some respects, to be our own distant relatives? Sensing overlaps here, do we find it too unsetting to pursue the matter further (42)?

Quite apart from the *absence* of attention, we must consider the nature of the collective images of tourism that do exist. Boissevain (9: 525) has in this regard usefully set out four distinct types of bias in the academic literature on tourism. One is the grossly inadequate framework of economic analysis. Another is the lack of the local voice. (An anthropologist is bound when reading the literature produced by economists, geographers, and so on, to notice an almost complete silence about what tourism means to the people involved in it – i.e. both those on the receiving end and the tourists themselves.) A third bias is the failure to distinguish the social consequences of tourism from other processes of change going on in a society independently. The fourth bias is the noble savage syndrome. Anthropologists with a possessiveness about "their" people and an oversimplified idea of traditional culture look askance at social change and the hordes of Western intruders "queering their pitch." The anthropologist bemoaning the rampages of tourism, expressing sympathy with the host population and hostility towards tourists (138: 212), is, in this view, a Rousseauesque voice bemoaning the romp of technological civilization over traditional ways of life.

In commenting that the social sciences have failed to accord tourism the importance it deserves I am not arguing for the development of a field of tourological studies or the establishment of a tourism science (cf 208: 153). Nor do I believe (cf 22: 110) that all the literature on tourism should be synthesized into a coherent framework. Over a decade ago Sir George Young commented on the absence of a unified bibliography on tourism, and on the "scrappy" state of the literature scattered through a range of disciplines (213: 3). Since tourism is a highly complex system, countless disciplines, from anthropology and sociology, through recreation geography, social psychology, marketing economics, hotel management studies, and so on, all have a view of what goes on. Although social science bibliographies now exist (94), and academic journals such as the *Annals of Tourism Research* concentrate on tourism as a specific subject, I do not envisage a change in the fragmented, multidisciplinary nature of the field. Perhaps tourism is an outcome of the intersection of a number of wider phenomena, themselves the strategic units for analysis – leisure, play, conspicuous consumption, etc. If so, then studies of tourism should certainly not evolve into a science of tourism; instead, touristic phenomena should be absorbed into "leisure studies" or some similarly construed field. I personally doubt that even this will occur, because touristic systems have so many diverse features. At the moment, the approaches of, for instance, economic geography, social psychology, sociology, and econometrics are radically diverse and incommensurable. What is of interest to the specialist in one discipline may be of no interest to investigators in another. Synthesis in such a situation is out of the question.

19

A final fundamental uncertainty remains – namely, about what a tourist is. There exists an array of definitions and taxonomies (see 14: 80–4; 107; 122: 10). The new animal – a tourist – was first named early in the 19th century, and for most statistical purposes definitions derived from the 1963 United Nations statement are used. Even at that statistical level, though, variations exist, and so quantitative tourism data are not always comparable. Edwards notes acerbically that "there are far better data on . . . canned fish . . . than on tourism" (51: 15). From the viewpoint of the social sciences, however, "passport"-type definitions are of little use (39: 374). The notion that a tourist is someone away from his/her residence for over 24 hours, who is traveling for either business or pleasure, conflicts substantially with the normal understanding that pleasure and leisure are tourism's basic motivations (122: 12). We have another obvious difficulty: does it make sense to call everyone who engages in leisure travel a tourist (39: 378)? The hippy, the FIT (Free Independent Traveller), and the working-class family on a cheap package tour for the annual fortnightly holiday, for example, exhibit a vast range of motivations – fun, relaxation, adventure, learning, escape, etc; and each kind of traveler generates a different set of socioeconomic consequences. Typologies abound based on different motivations, levels of affluence, lengths of stay, methods of organization, and so on.[1] Typically, though, these taxonomies are incommensurable, leave out obvious distinctions, and separate phenomena that are clearly fuzzy or overlapping. In addition, as Hermans notes, no one's taxonomy has yet compelled use by others (82: 10). Turner and Ash add (195: 14) that any unitary phenomenon that may underlie tourism is not in any case constituted by the tourists and their motives but rather by a highly complex set of interlocking structures. Thus to concentrate on the tourist is to miss a great deal about the international tourism system. There is even argument about whether or not tourism is essentially a modern phenomenon, and here perhaps social historians have a vital role in research. Is tourism new? Is it a distinctively 20th-century phenomenon, completely different from the travel, pilgrimage, migration, exploration, and so on, of earlier ages? One encyclopedic survey of travel in the ancient world (27: 262, 274, 279, 321–2) suggests that "touristic" motives and behavior have not really changed significantly in 2000 years. If this is so, we must still ask detailed questions about, for instance, the democratization of travel. Are there no significant differences between pre-industrial, industrial, and post-industrial work, play and leisure? Or, is tourism just a contemporary form of an activity that occurs in all societies at all times? One's approach to tourism research depends to some extent on how one answers such questions.

It is therefore something of an understatement to say that the anthropology of tourism is still in its infancy (134: 461). Some authors, like Nash (134: 467), call for a theoretical framework and want generalizations to start emerging. Others, calling for more detailed comparative empirical studies, protest premature general analytical frameworks. Buck argues that in a new field of study scholars tend to go their own way, so that an integrated body of knowledge is unlikely to arise (23: 326). De Kadt argues in regard to the highly evaluative literature on tourism that both pro- and anti-attitudes are worthless without detailed evidence (48: xiv), and Cohen claims that all the generalizations about the sociocultural repercussions of tourism are premature since there is no such creature as *the* tourist (36: 31). We have to know the particular tourist types involved, the numbers present, and the specific ecocultural niches involved before we can come to any sensible conclusions (35: 7, 9).

Given this array of views, I urge theoretical caution. Detailed studies are needed to break the hold of powerful yet insufficiently examined images, and to create bodies of social scientific data distinct from other kinds of cultural representations. Without that effort our academic writings might amount to little more than detailed outpourings simply slipped into preformed and highly evaluative moulds. We should also be wary of any disciplinary imperialism. International tourism is a highly complex system. Indeed, Jafari has commented that there is such a diversity of goods and services involved that tourism is not an industry in the normal sense (91: 84). This complexity must be respected, and one way of doing this is to acknowledge that a large range of academic disciplines have an interest in it. The complexity will be fruitfully registered if these disciplines pursue unabashedly their own interests and utilize their own distinctive methodologies.

In the remainder of this article I examine several sets of images and approaches to tourism that have been prominent over the last two decades in the social science literature. Given my anthropological focus, I do not attempt an even, discipline-by-discipline approach (see 14: 17–54). Instead, I divide the discussion into three areas: a broadly political–economic view of international tourism, including the issue of development; tourism in relation to meanings, motives, and roles; and lastly, images of tourism as a force in sociocultural change. I emphasize international tourism, particularly in the Third World, even though in monetary terms tourism within the affluent industrial countries is far more significant.

TOURISM, ECONOMIC DEVELOPMENT, AND POLITICAL ECONOMY

An approach to tourism that is recognizably both political and economic has surfaced in recent years. This combination is particularly striking because the first academic representations of tourism were almost exclusively economic – pure and (with hindsight) very simple. International tourism during the 1960s was seen largely in terms of economic development and thus almost entirely in a positive light. This was a time of considerable difficulty for many Third World countries, whose primary produce exports were experiencing a long-term decline in value. International tourism was portrayed as a panacea for the less developed countries, as "manna from heaven" (52, 209). The Organization for Economic Cooperation and Development (OECD) spoke of the almost limitless growth potential in tourism (141: 11–15), and both the World Bank and the United Nations promoted tourist industries in developing countries. The United Nations declared 1967 to be International Tourism Year. The leisure and travel habits of those in the wealthy countries were to open the doors to the economic advancement of those in the poor nations; foreign exchange for the developing countries could be directly tied, in other words, to the increasing affluence of the developed world (102). Tourism was represented as an easy option

21

for development because it relied largely on natural resources already in place – e.g. sand, sun, friendly people – and therefore required no vast capital outlays for infrastructure (92: 227). Some tourism advocates even argued that certain Third World countries might, via tourism, advance from a primary sector-based situation to one based on an expanding service sector, omitting the normal industrial phase of economic growth. Responding to such glaringly positive images, a number of developing countries embarked upon tourism development without adequate feasibility studies, without any sense of opportunity costs (i.e. with no sense of what development might be achieved by the employment of resources in alternative ways), and with little planning to integrate tourism into national development more generally. As Mathieson and Wall comment (122: 178), a great deal of subsequent tourism policy has been aimed at shutting the stable door after the horse has bolted.

In the last 20 years we have become increasingly aware of the political dimension of such seemingly simple encouragement. Much attention has been devoted to tourism's frequently adverse sociocultural consequences. In addition, scholars have begun to realize that even the economic arguments for tourism are not as sound as was first claimed (14: 251). We therefore now read in introductory tourism texts that, in its present form, tourism development is not desirable in many areas (122: 177). Tourism is not a secure growth industry (108: 753). Not only are there the obvious seasonal fluctuations in arrivals, but the developed economies themselves also go through economic cycles; and during recessions, demand for overseas travel declines (14: 150). Vacationing is price elastic, and costs are unstable, given the politics of oil marketing. Pricing in general is beyond the control of the destination countries. Tourists are also faddish in their tastes, so the general growth of international tourism does not mean that any particular Third World destination has a secure future (143). Most Third World tourism destinations are mutually substitutable; travel organizers can easily reroute their clients, leaving many people out of work and much accommodation under-occupied.

Even in good times the record of international tourism is far less spectacular than the original representations predicted (21, 168). To begin with foreign exchange, there is normally a large discrepancy between gross and net receipts as a result of numerous leakages – e.g. repatriation of profits on foreign capital invested. For some economies, like that of Mauritius where the tourism industry is largely dependent on overseas capital, there have been leaks approaching 90% of foreign exchange (30: 31–2, 49). Some countries lose substantial quantities of foreign exchange though black market operations. The high level of vertical integration in the tourist industry (14: 168), where foreign airlines own hotel chains and local rental car firms, and so on, means the economic gains to many developing destination countries are much reduced. In 1978, for instance, a mere 16 hotel chains owned over one third of the hotels in developing countries (28: 11). Indeed, given the nature of inclusive package holidays where payment for airfares, accommodation, food, and services is made in advance, much foreign exchange does not even reach the destination country (195: 116). In many countries, luxury tourism facilities still tend to attract expatriate management, and a high proportion of foreign exchange is expended to import the foodstuffs and facilities the clientele of such establishments expect. Luxury tourism especially may require substantial investment in infrastructure (buildings, transport, etc) that will be little utilized by local people, meaning that locals, in fact, subsidize the holidays of affluent foreigners (213: 2, 152). Also, of course, national tourism authorities must spend foreign exchange overseas in order to advertise themselves. Given that many developing countries do not need tourism as a central growth strategy and can thus allocate scarce resources to other projects, observers (72: 122–3) have pointed out what a reprehensible waste international tourism developments can often represent. We must also remember that attracting foreign capital and business normally requires Third World governments to offer very generous financial incentives, such as tax-free profits (209: 280–2) for a number of years. Such situations are often believed to lead to widespread political corruption and fraud (18: 71–2). Given such factors, plus the fact that the receipt of foreign exchange does not necessarily lead to economic growth, one can see why the foreign exchange argument is not so confidently used nowadays.

Similar difficulties have arisen with regard to other supposed economic benefits. Contrary to predictions, tourism has often proved to be a capital-intensive industry, at least in its development phase (49: 549; 194: 257). If countries borrow overseas capital to build infrastructure, the continuing interest payments promote dependency rather than the reverse (45: 307). Nor has employment been stimulated to the degree expected. Most of the jobs generated are unskilled; tourism can thus breed what one critic (24: 2) calls "flunkey training." Another area where tourism's record has not lived up to the original promise is that of intersectoral linkage and regional diversification. Tourism, like other economic activities, has ramifications in an underdeveloped economy different from those in a developed one. Where the intersectoral linkages are weak, the multiplier effects expected in a developed system will not occur (113: 16). Thus, in Third World conditions, tourism does not always stimulate local agricultural production; indeed, a rise in the value of land and in the price of food, together with the higher wages often to be found in the tourist sector even for menial work, may lead to the sale of agriculturally productive land, an exodus of labor from the fields, food shortages, and malnutrition (200). Regionally, too, tourism may establish small localized enclaves of activity without affecting activities nearby, thus reproducing the dualistic structure, the plantation system, of the colonial economy (45: 314; 48: 139). Employment prospects in the immediate locality may not improve if labor is brought in from elsewhere (e.g. to staff high-quality hotels). The lesson has not yet been adequately learned by national tourism authorities that in tourism development Schumacher's dictum "small is beautiful" applies. Grass roots developments are far more likely to lead to local employment, the stimulation of other local activities, and the avoidance of capital indebtedness to overseas concerns that the standard hotel-based industry involves (14: viii).

Tourism may also differentially affect different classes, increasing inequalities of wealth and social stratification (58: 225) and thus retarding broadly based national development. Unlike some development strategies, tourism is normally a conservative choice (52: 72). Benefits from tourism "unlike water, tend to flow uphill" (165: 7).

In other words, not only does tourism in the Third World "inject the behavior of a wasteful society into the midst of a society of want" (11: 27), but the profits go to the elites – those already wealthy, and those with political influence. Such facts led to analyses of the situation in Fiji, for instance, which suggested that because tourism reinforces existing economic patterns it can only have a negative effect on national welfare (203: 96–100). International tourism is a kind of potlatch in someone else's country (Sessa, quoted in 199: 85); it is conspicuous consumption in front of the deprived. One American travel writer brings himself to admit that "everywhere in the world, of course, you'll come across occasional gaps in living standards and amenities" (quoted in 14: 177); for others, "No printed page, broadcast speech, or propaganda volley can emphasise the inequity in the global distribution of wealth as effectively as tourism can" (14: 258). If local people feel resentment at the display of wealth by foreigners, tourism also fuels class resentment (9: 523). The poor find themselves unable to tap the flow of resources while the wealthy need only use their existing assets (e.g. ownership of well-positioned real estate, political influence) to gain more. Often small operators in the tourism industries of the Third World face increasingly daunting competition with the enormously powerful local elites and multinational conglomerates (79: 110–11); and national tourism authorities, in a desire to maintain a favorable reputation for their countries, frequently sweep human "litter" off the streets; touts, beggars, and street hawkers are treated as so much refuse spoiling things for the visitors (41).

In the 1960s, international agencies spoke of tourism as a force for economic growth and international understanding. By the mid-1970s Levitt and Gulati (112: 326–7) were alleging that a powerful metropolitan tourist lobby operates at the national and international level, through the agency of professional consulting firms, essentially to con international organizations like the World Bank into hoodwinking everyone else about the supposed benefits of international tourism. The World Tourism Organization is an "informed cheerleader for the industry and governments interested in tourism." As Richter notes, the WTO is unlikely to be a critic of the industry (163: 18–19). Much technical analysis, too, is highly suspect. For instance, multiplier analysis is used in the literature to generate highly misleading claims about the beneficial effects of tourism on employment and economic growth (30: 5; 21: 74). Economic representations of this type are attacked now not only for what they omit, but because of the political naïveté inherent in their one-dimensional, asocial, conceptual world. Underneath elaborate cost-benefit analyses there often lurk thinly disguised political values (186).

Commentators now increasingly stress the fact that developing countries must decide on their own tourism objectives (30: 103–4, 168) and must integrate them within an overall development plan, for otherwise the industry will get out of control and redress will then be impossible. Some developing areas, particularly small islands, may have no alternatives to tourism (29: 294); but where other uses of resources are possible, governments are now being urged to consider the opportunity cost of tourism (82: 5). Reports are increasingly hostile to the notion that tourism is beneficial. Harrell-Bond, for instance, claims (80) that the Gambia benefits little from the industry that the World Bank had been energetic enough to set up for it, that such industry rests upon derogatory racist stereotyping of the locals, and that it is a "charlatan development programme" (see also 53, 205). It has, of course, been hard for governments in underdeveloped countries to resist the temptations of tourism. After all, there must be something good in it if the UN, UNESCO, and the World Bank spend vast sums on it and encourage its adoption (53: 9). But Third World governments are now somewhat wary. It is indicative of this new skeptical orientation that a modern collection on tourism (48) is quizzically entitled "Passport to Development?" The recommendations made by the participants at the seminar from which the book derives are, for the most, expressions of concern (48: 339–47).

Since the 1960s we have realized that concepts (or, perhaps images) like "development," "modernization," and "growth" are no longer as clear as was once thought (48: xi–xii). What is development? Development for whom? Are development and growth the same thing? Is the concept of growth of any value without consideration of the distribution of wealth and a whole host of other political and social issues? R. A. Britton, who has written about international tourism from an explicitly politico-economic viewpoint, notes that to define development simply in terms of indexes such as rising GNP or extra foreign exchange is problematic where other indexes, such as levels of education, health care, sanitation, etc, remain static (14: vii). Also, whether a development is beneficial depends on whether one sees it at an international, national, or local level (122: 6). Moreover, a country's attempt to tie itself to the affluence of Europe and North America now appears particularly naïve if it is precisely the forces producing their affluence that maintain the underdevelopment of the Third World. The structural dependencies are visible in the case of tourism. One of the rationales for tourism development in the 1960s was export diversification away from reliance on primary products. That reliance itself was largely the consequence of the deformation created in the colonial economy by the many countries now sending tourists. Tourists do not go to Third World countries because the people are friendly, they go because a holiday there is cheap; and that cheapness is, in part, a matter of the poverty of the people, which derives in some theoretical formulations directly from the affluence of those in the formerly metropolitan centers of the colonial system. That affluence now produces conditions of work and life such that leisure activity is prized. And with high levels of disposable income, that leisure can be spent in the impoverished Third World, the source of many of the surpluses that established the affluence. No wonder one representational framework for the analysis of international tourism is that of neo-colonial political economy. Advocates of various forms of "alternative," "fair," or "just" tourism (64, 78, 86, 140, 169, 191), from which local people will benefit while a situation of intercultural understanding develops, suggest that international tourism in the Third World currently enables neither of these things. The 1977 Pacific Area Travel Association slogan "The consumer – the only person who matters" (quoted in 210: 567) states exactly what Third World, academic, and religious critics of international tourism increasingly deplore. Whether "alternative" tourism can do much to change the overall nature of international tourism is debatable (40).

I discuss academic and political criticism further below. Here comments from religious organizations deserve noting, for they are by no means simply expressions of moral outrage. Indeed, as the Ecumenical Coalition on Thrid World Tourism claims (50: 3), the prostitution and drugs involved in this industry are not really matters of personal morality at all, being instead bound up with racial exploitation. The Coalition, which monitors developments and produces the journal *Contours*, grew out of meetings of churches concerned with international tourism, the first having occurred at Penang in 1975 (86: 6). A second church group, the Christian Conference of Asia (140), met in Manila in 1980, at the same time and place where the Sri Lankan Minister for Tourism told the World Tourism Conference that tourism was a force for peace and understanding. The church conference claimed that, on the contrary, tourism had "wreaked more havoc than brought benefits to recipient Third World countries. . . . In its present form, linked as it is with transnational corporations, ruling elites and political hegemonies, and totally unmindful of the real spiritual, economic and political and socio-cultural needs of recipient countries, the Workshop seriously questioned whether tourism as it is could be salvaged" (140: 3). International tourism had to be rethought outside the normal materialistic framework so that benefits would be shared more equitably. The conference also described the idea that international tourism builds up peace and friendship among people as a "contemporary myth" (140: 18). Biddlecomb, another religious writer, suggests that the myth of international amity might have some basis in fact for elites (8: 16). The Manila Declaration on tourism produced by the World Tourism Organization spoke of tourism as a part of everyone's heritage, indeed as a "fundamental human right" (211, para. 15). It encouraged all, and especially the young and less affluent, to partake of international tourism, suggesting that this would contribute to a "new international economic order" (212: para. 14). The WTO stressed that international tourism would help to eliminate the widening economic gap between developed and developing countries and would contribute substantially to social development and general progress in the developing countries. The Chiang Mai workshop of the Ecumenical Coalition in 1984, by contrast, was claiming that tourism as it presently exists "is a violation of human rights and the dignity of people" (86: 14).

The religious organizations that regularly comment are not simply condemnatory. They are also optimistic that negative factors can be overcome, and that travel can be made an uplifting experience. But a genuinely beneficial outcome, it is argued, involves equality among participants, and this, according to the Chiang Mai workshop, is bound to involve some conflict between governments and the travel industry (86: 18). In one of the earliest general works on tourism, Sir George Young made clear his view that there were a number of ways the tourist industry would have to change substantially – and uncomfortably – if it were to have a future (213: 81). One such change is precisely the shift to alternative forms of tourism, involving less foreign capital and thus more local people, food, and architecture. It is clear from Sri Lanka (78) and elsewhere that these smaller-scale ventures are not without their practical difficulties; as Hiller comments for the Caribbean, it is difficult to persuade tourism policymakers that a more authentic form of tourism is viable (83: 57).

International tourism is political, since the state must be involved in foreign relations, the expenditure of large quantities of capital, and large-scale planning (30: 51). But the emerging political-economy framework makes a different point. As S. G. Britton (16, 17) correctly remarks, what is wrong with many portrayals of tourism and development by economists is that they provide no sociohistorical context to explain the economic inequality between the tourist-generating and tourist-destination countries. In other words, tourism has not commonly been analyzed within the framework of underdevelopment. But as Hiller notes, tourism does tend to "represent the way the powerful nations perceive and relate to the rest of the world" (83: 51). Many of the specific relationships between "hosts" and "guests" in tourism are only comprehensible in the context of these wider international relations between the developing world and the affluent West (100: 250). Indeed, for some critics of standard international tourism, like those confronting tourism with indigenous value systems such as *ujamaa* in Tanzania, the piecemeal analysis of tourism without the political–economic overview is typical of bourgeois social science and is a strategy often used to avoid real social issues (176: 15).

Remedying the shortcomings of such academic representations produces the sort of framework which declares that to "invest in international tourism is to invest in dependence" (85: 81) or that "decolonisation and tourist development have antithetical implications" (120: 197) – in short, that to opt for tourism as a growth strategy is to ask for continued control by overseas forces (133: 35, 45). R. A. Britton sums up this type of representation well when he states that the tourist industry is the "opposite of self-reliant development" (14: 207). In certain countries tourist development wholly contradicts proclaimed anticolonial ideologies (210: 578, n. 49), and as the African socialism debate in Tanzania in the early 1970s makes clear, not all the critics in this vein are Western academics. To cite Shivji:

> Tanzania has proclaimed *ujamaa* as its goal. Tourism, therefore, cannot be evaluated in isolation from this goal and from an overall general development strategy to achieve such a goal. Again, cost-benefit analysis does not help much. For economic development in a colonially structured economy calls for radical structural change. Many of these structural changes are inevitably political decisions. . . . To separate politics and economics is therefore a grave error. The justification for tourism in terms of it being "economically good" though it may have adverse social, cultural and political effects, completely fails to appreciate the integrated nature of the system of underdevelopment (176: x).

As the Tanzanian Youth League went on to comment, "It is not a matter of coincidence but a matter of class interest that both the national and international bourgeoisie should show interest in the same field of economic activity. . . . What is more, nothing could be more dangerous than an alliance and consolidation of the class interests mentioned . . . [since] such alliance can completely change the course of our history, and surely not in the direction of socialism." Maluga, as part of the

27

same debate, comments how international tourism contravenes the objectives of self-reliance set out in the Arusha declaration.

> Investment in tourism is a lopsided one; it serves a sector that is hardly related to the economic structure of the country. . . . It is above all a risky and temporary industry whose viability and continuity depend on the good will of the tourist generating countries. . . . In order to have a continuous influx of tourists our policies ought to be widely acceptable by the public in the tourist generating countries. If we commit ourselves to make tourism one of the leading industries in the country as it is envisaged in the proposed "ten year plan for tourism," our independence will be at stake. . . . Since the success of tourism depends primarily on our being accepted in the metropolitan countries, it is one of those appendage industries which give rise to a neo-colonialist relationship and cause underdevelopment (119).

Kanywanyi adds interrogatively, ". . . are we not building socialism but helping Western monopoly capitalism to more effectively keep our stagnant economy at a standstill?" (99: 65).

I have quoted at length above because the issues have rarely been better stated. If they strike the reader as unduly alarmist, note that the vice-president of Edgar Rice Burrows Inc. once expressed his desire to purchase the whole nation of The Gambia so as to build a series of Tarzan vacation villages. "This is not pie in the sky. We've been talking to the Rothschild Bank in Paris about this. There are a number of very very small African countries that have absolutely nothing. No economy, nothing. All they have is their independence and their UN ambassadors, and the thought is to merchandise the entire country . . . take it over, change the name . . ." (quoted in 14: 58–9).

Ruth Young has argued that the structural form of any tourist development necessarily parallels the preexisting socioeconomic structure in a country (214: 157). Inevitably, then, the very way a tourism industry is planned and shaped will recreate the fabric of the colonial situation. It is no wonder that, to some, international tourism is pictured as the recreation of a foreign-dominated enclave structure reacting to metropolitan interests and entirely unrelated to the local economy (16: 10; 133: 5–8; 150: 476). In the West Indies, for instance, local writers describe tourism as a reexperience of the race and labor relations of the past, as a meeting of Fanon's "wretched of the earth" and the wealthy (190: 217). Whereas tourism makes the Third World a "Garden of Eden" for some (as a local newspaper remarked as far back as 1938), for locals, it is a "Perverted Hell" (190: 221, n. 55). Locals, according to the Mayor of Honolulu (quoted in 115: 165), are "peasants in Paradise," and many features of the sociology of colonial situations are resurrected. Locals are denied access to their own beaches, people are given jobs according to racial stereotypes (170: 205), and humble service roles predominate. In the passage from the cane fields to hotel lobbies the pattern remains basically the same (120: 197). We have, in short, "leisure imperialism" (18: 38, 84; 45: 305; 133: 37–8; 159), the hedonistic face of neocolonialism. International tourism recapitulates a historical process (150: 480); areas of one's country are given over to the pleasure of foreigners, and the rhetoric of development serves as a defence (150: 474). To the extent that tourism can undermine the national identity of newly independent countries, it is a case of the cultural imperialism condemned decades ago by such writers as Albert Memmi and Frantz Fanon. As Shivji has remarked, tourism is part of a continuation of the cluster of attitudes that make up colonial social psychology – submissiveness, arrogance, and so on. He goes on to point out that many hotels are in fact Fanon's "settler towns" (176: ix). Indeed, Fanon even spotted the sexual exploitation that would emerge as part and parcel of much tourism development in the Third World.

> The settler's town is a strongly built town, all made of stone and steel. It is a brightly-lit town: the streets are covered with asphalt, and the garbage-cans swallow all the leavings, unseen, unknown, and hardly thought about. The settler's feet are never visible, except perhaps in the sea; but there you're never close enough to see them. His feet are protected by strong shoes although the streets of his town are clean and even, with no holes or stones. The settler's town is a well-fed town, an easy-going town: its belly is always full of good things. The settler's town is a town of white people, of foreigners. . . . The national bourgeoisie will be greatly helped on its way toward decadence by the Western bourgeoisies, who come to it as tourists avid for the exotic, for big-game hunting and casinos. The national bourgeoisie organises centres for rest and relaxation, and pleasure resorts to meet the wishes of the Western bourgeoisie. Such activity is given the name of tourism, and for the occasion will be built up as a national industry. If proof is needed of the eventual transformation of certain elements of the ex-native bourgeoisie into the organisers of parties for their Western opposite numbers, it is worthwhile having a look at what has happened in Latin America. The casinos of Havana and of Mexico, the beaches of Rio, the little Brazilian and Mexican girls, the half-bred 13-year-olds, the ports of Acapulco and Copacabana – all these are the stigma of this depravation of the national middle class. Because it is bereft of ideas, because it lives to itself and cuts itself off from the people, undermined by its hereditary incapacity to think in terms of all the problems of the nation as seen from the point of view of the whole of that nation, the national middle class will have nothing better to do than to take on the role of manager for Western enterprise, and it will in practice set up its country as the brothel of Europe (55: 30, 123).

Fanon's analysis was written many years ago, but accurately pinpoints many features of the system of contemporary international tourism. That system does not, for a start, normally express local needs or aspirations, although there are "indigenous collaborative elites" (166) who in tourism, as in other imperialist situations, are linchpins by means of which foreign interests maintain their hold in poor countries. Those with political sway, able to hand out contracts and the like, are the beneficiaries. Local elites may well identify with the consumerist life-style of international tourists rather than with the aspirations of their own people, indeed they may themselves be part of the international jet set. When members of Third World elites encourage tourism as economically beneficial, they may not be suffering from a delusion: they may themselves benefit substantially. However, their own gain and the interests of the nation as a whole are two different things (151: 142). As in the previous era, therefore, the periphery is structurally tied to the needs of the metropole; the local and foreign elites gain while deprivation continues to be the lot of the masses (177, 178). Third World politicians who collaborate with metropolitan interests necessarily underwrite underdevelopment (150: 478). For S. G. Britton, describing the situation in Fiji, tourism has simply been "grafted onto a once colonial economy in a way that has perpetuated deep-seated structural anomalies and inequalities." The initiative comes from foreigners and "local political and commercial elites in close liaison with foreign capital. . . . The articulation of international tourism with Fiji was based upon the interaction of foreign and local elites in pursuit of their own interests and mutual benefit" (18: v, 2, 31). It was inevitable that overall long-term planning for economic growth would take place in a system where these alliances were already firmly established (18: 194–5). This is the sociopolitical reality lying behind the series of *ad hoc* measures taken to shape the international tourism industry but that seem neither fully to support nor effectively to guide the industry (18: 119–20).

An essentially imperialist imagery is evident in much of the recent language social scientists use about international tourism. Bugnicourt, for instance, writes that tourism represents the demands of consumers who, having ruined their own environment, desperately need to take over another (24). MacCannell writes of the affluent middle class that systematically "scavenges" the earth in search of new experiences (115: 13). Some see tourism as an expression of that same expansive thrust in Europe that lay behind geographical discoveries and colonialism (68: 18). For Cohen "the easy-going tourist of our era might well complete the work of his predecessors, also travellers from the west – the conqueror and colonialist" (31: 82). Biddlecomb asks: "If from colonialism and tourism the same implicit models and thought patterns are perpetuated, then what if anything has actually changed in the West's way of relating to the third world?" (8: 37). If John Bright was right (quoted in 142: 26) in 1859 to claim that imperialism was a "gigantic system of outdoor relief for the aristocracy," one can see why many Third World critics see tourism as just more of the same thing – except that they now have to put up with the wealthy slumming around and the working class lording it over them. And we should not think that this imperialist imagery concerns only relations between the West and the Third World. Many critics of tourism in Thailand and the Philippines see the now rampant Japanese "sex tourism" as a repetition of Japanese military aggression, showing absolutely no respect for the local people. As a Filipino protester put it, "We would like to forget Japanese military imperialism. But now instead of military uniforms, the men come in business suits, dominating Asia through a pernicious form of socio-economic imperialism, which tramples on the Asian peoples' right to human dignity" (quoted in 139: 29).

Not only, then, may few people in a destination country benefit from international tourism, they may be forced to "grin and bear it." In the West Indies, as in other areas, national tourism authorities launched "courtesy campaigns," in which citizens were instructed how to be civil to tourists, and beggars were swept out of sight (164: 250). International tourism requires, above all, peace and stability. Governments may therefore crack down on the local people in order not to upset a growing tourism industry, suppressing signs of civil disorder and of animosity towards tourists themselves (14: 199; 45: 316; 163). The argument is sometimes put that the tourism industry tends to support right-wing regimes (14: 190–1). Conrad Hilton is famous for his remark that "each of our hotels is a little America." He added: "We are doing our bit to spread world peace, and to fight socialism" (quoted in 139: 50). For the Philippines, L. Richter has shown how the rapid development of tourism facilities after the imposition of martial law by President Marcos in 1972 acted as a message to the international community that life was normal in that country (161; 162: 122, 127–8). The World Bank Conference of 1978 was held in Manila. In 1976 12% of the funds of the Philippines Development Bank was devoted to financing hotel room construction, resulting in windfall commissions for those with political influence (14: 197). This level of expenditure on hotels in 1976 was 40 times that on public housing. At that time crimes against tourists carried more severe punishments than those against locals, and journalists criticizing tourism risked dismissal (162: 244, n. 36, 245, n. 54). Fortunes were made by Marcos supporters in charge of the implementation of tourism policy. Governments sometimes use rough tactics against their own people to safeguard tourism. In Indonesia, for instance, when the government decided to create tourist facilities around the Borobodur Temple, there was much local protest at the sacrilege this involved. Local people were simply moved away and rehoused; then land values skyrocketed (101). In countries like Australia and the United States, tourism has become a force for internal colonialism as peoples of the so-called Fourth World (e.g. Australian Aboriginal groups and American Indian communities) are represented as tourist attractions (62: 24; 156). In 1962, for instance, an Australian tourist advisor suggested the removal of a group of Aborigines to a reserve half a day's drive from Adelaide to function as a tourist spectacle (156: 88).

Modern political–economic analyses of international tourism suggest that the developing countries have little or no choice. Even where small-scale tourism (in which the less wealthy are actively involved and from which they benefit) makes sense, its success is made unlikely by the industry's high level of vertical integration. Third World economies lack control over the world prices of international tourism (14: 124). Although many Third World countries have a "tourism on *our* terms" policy, demand is largely engendered by tourist agencies and a whole industrial network of image makers overseas. When Tunisia in the early 1970s attempted to better its cut in international tourism *vis-à-vis* the overseas operators, in the following year the major European travel agencies simply diverted their customers to other but comparable destinations. Even if developing countries attempted regional cooperation (128: 12; 158) in shaping the nature and size of their tourist sectors, international tourism exists primarily to meet the needs of those in the affluent countries (97: 232), where most of the control remains (28). For those who hope that the Third World might acquire more control of tourism, or that alternative tourism might begin to encourage cultural understanding, the words of Bugnicourt are worth savoring: "There is no doubt whatever that a change in the overall economic and social relations between industrialized and Third World countries and a consequent evolution of behaviour will be needed before there can be any real prospect of a tourism which no longer leaves itself open to the charge of colonialism, but brings people closer and offers the enriching discovery of new environments and different civilisations" (quoted in 14: 350).

TOURISM, MEANINGS, MOTIVATIONS, AND ROLES

International mass tourism today is made possible by some basic material facts about modern industrial societies – among them levels of affluence that free resources for leisure pursuits, compulsory paid annual holidays, and entrepreneurship that invades leisure as well as the work sphere (123). But there is also here an important area for social science research into meanings and motives. What do tourists say about their leisure experiences? What do they learn from other cultures? Why do they go on overseas holidays? In what ways are their ideas and attitudes changed by these experiences? Much of the extant social science literature on tourism does not ask such questions because in most disciplines contributing to the field tourists themselves are not the object of study. Sociology has done some work on meanings but anthropology has tended to concentrate on cultural repercussions in the destination country, and disciplines like geography and economics seldom mention human beings at all. A large area in tourism research has been neglected, although increasing attention is being paid to it by social psychology (56: 124; 148).

Clearly the simplistic push and pull factors set out in introductory tourism textbooks (e.g. 88: 35) inadequately represent the complexities of tourist behavior, let alone more subjective matters of attitude, learning, and meaning. We have not only the motivational differences between distinct subtypes of tourist (e.g. the wanderer, the person on the package tour, and so on) but also such elements of tourist behavior as play, regression, ritual, and so on, which are ripe for detailed empirical investigation. A recent summary of the contributions of social psychology to the study of tourism shows that the psychological way of approaching meaning, motivation, and human behavior, even after the "humanizing" changes within the discipline over the last decade or so, differs from the ways anthropologists would approach the same subject. Anthropologists, however, have not yet studied tourists or the countries that generate them (134: 465; 138: 209). The omission has long been recognized by perceptive travelers, in fact. Aldous Huxley, for instance, subjected to "infantile" behavior during a tourist cruise in the Gulf of Mexico, wrote "My objection to anthropologists is the same as my objection to missionaries. Why do these two classes of people waste their time converting heathens and studying the habits of blackamoors, when they can find, in their own streets, men and women whose beliefs and behaviour are at least as strange as those of the M'pongos and, so far as we are concerned, painfully and dangerously more significant? Anthropology, like charity, should begin at home" (89: 11).

Human migration is often associated with stress, whether resulting from social pressure or natural disaster; and tourism, though voluntary and reversible, is nevertheless a form of migration. Not surprisingly, the images of escape from pressure, alternation, and regression are common in the literature. To be a tourist is to opt out of ordinary social reality, to withdraw from everyday adult social obligations (43: 417; 204). Instead of duty and structure one has freedom and carefree fun. In one obvious sense the spectrum of organizational possibilities for tourism – from individual wandering, to an all-inclusive package tour – might represent the ways different personalities endeavor to cope with this alternation, although economics is also relevant here, too. On the one hand, there is the hedonistic regression to drugs and nude sunbathing by tourists in the midst of people who disapprove of such behavior. For others, winding down involves such anxieties that a highly structured package is required, relieving them of decisions and at the same time providing them with an "environmental bubble" (31: 166) to prevent confrontation with anything alien at all. With an air-conditioned coach, an expatriate guide, a group of travelers from one's own country, and a stay in a star-classified hotel, the tourist need not feel threatened (136: 135; 171: 446). Foreign travel can provoke anxiety in many ways: consider the strain of uncertainty, of getting ill, of finding accommodation, and so on (26, 73, 188: 222). Moreover, the tourist is "one of the world's natural victims" (195: 238). It is no wonder that some forms of tourism cushion the traveler to such an extent that commentators sometimes ask, "Why go at all? Where is the novelty?" This is where the widespread collective representations of triviality, lack of authenticity, and so on, arise.

Such motivations as relaxation, conspicuous spending, having fun, and so on, obviously pose problems for those who represent a search for cultural authenticity as the single meaning of tourism, let alone for those who espouse such grander themes as world peace and understanding. This theme of travel as ennobling and mind broadening was enunciated by the United Nations Conference on International Travel and Tourism in 1963 (188), and it has been uttered by many other dignitaries before and since (American presidents and popes included) (14: 154). The International Union of Official Travel Organizations spoke of travel as "a most desirable human activity deserving the praise and encouragement of all peoples and all governments" (90: 105). But the idealistic and the more mundane are intricately tied up in the imagery of tourism. The slogan "World peace through world travel" adopted by the Hilton International company is in fact only a borrowing of a slogan previously used by the IBM (193: 188).

Little detailed empirical work has been done on the effects of travel on attitudinal change (147: 163), but a study by two educational anthropologists (12) concludes with serious reservations about the educational benefits of tourism. Tourists, for a start, are poor "culture-carriers" (108: 756), being stripped of most customary roles through which their culture could be understood by others. In any case, for most people tourism involves more hedonism and conspicuous consumption than learning or understanding (195: 89–90). T. S. Eliot's dictum that human beings cannot bear too much reality certainly applies in the tourism context (206: 430). Wagner raises the issue of how tourism can be about understanding culture when the behavior of so many tourists is so deeply offensive to the people among whom they stay (205). Tourism is very much about *our* culture, not about *their* culture or our desire to learn about it (192: 187). This explains the presence in guidebooks of sites and signs that have little genuine historic or living connection to a culture but that exist simply as markers in the touristic universe. As Barthes remarks perceptively, travel guidebooks are actually instruments of blindness (4: 76). They do not, in other words, tell one about another culture at all.

Some regard the "peace and understanding" line simply as high-sounding rhetoric camouflaging economic self-interest (153). During a conference in Manila in 1980, for instance, as delegates enunciated noble themes and spoke of the need to preserve Philippine culture, the city in which the conference was held was estimated to contain 10,000 prostitutes at the disposal of international tourists and members of the local elite (177; 178: 3). Delegates to that conference were shielded from the poverty of the local population by huge, white-painted boards that obscured the vast slums that line the roadway from the airport into the city. R. A. Britton points out a further twist to the logic linking reality and image in this industry in his comment that the internationalism bannered by the industry "is consistent with their self-interest since a world without borders is far more conducive to the unfettered movement of capital, manpower, and technology . . ." (14: 155).

One might argue that tourism is actually an activity by means of which stereotypes are perpetuated and even reinforced, rather than broken down (180: 68). As a commentator on the Fijian situation stated, "today travel, far from broadening the mind, is actually contrived to shrink it" (R. J. Scott, quoted in 170: 212): Travelers blindly indifferent to the social reality of their hosts promote mutual contempt, not understanding. No matter how often international tourism is represented as a force for understanding, the empirical evidence suggests that with increasing numbers individual perceptions are replaced by stereotypes. Pi-Sunyer (154: 154–5) explains how such national stereotypes of mass tourists in the Catalan area deprive them of an essential human status: they become less than ordinary folk, and this diminution in turn legitimizes hostility towards them, cheating, dual price levels, and so on: "Contacts between villagers and outsiders have never been greater, but the barriers to understanding have probably never been higher . . . If tourism commoditises cultures, natives categorise strangers as a resource or a nuisance rather than as people" (154: 155). To be sure, the totality of international tourism is not grasped by pointing out the prostitution, servility, exploitation, and so on that are certainly an important part of it. But neither can the "peace and understanding" rhetoric be swallowed wholesale by social scientists studying what goes on, as if this magical phrase captured the essence of the phenomenon. Such rhetoric, disseminated by tourism promoters and some national tourism authorities, should be seen for what it is – a mystifying image that is part of the industry itself, and not an empirically well-founded comment upon its nature. It would be a disaster, in the early stages of social scientific study, if such images as "peace and understanding" obstructed a realistic and empirical analysis of this industry and its consequences.

The imagery of international tourism is not, for the most part, about socio-economic reality at all. It is about myths and fantasies, and in this sense it can harm a country's development efforts precisely because its own image-making creates a false picture of the Third World (15). As Whealen stated (quoted in 14: 202), tourism "is a way of providing a simulacrum of [the] world." The places in the glossy brochures of the travel industry do not exist; the destinations are not real places, and the people pictured are false. The Bahamas become the "playground of the Western world"; South America becomes "an enchanted forest where Walt Disney's Bambi lived" (in 14: 177); and by a deft piece of geopolitical legerdemain, a Greek fishing village grows up in the Caribbean advertised as "the Best of the Mediterranean on Mexico's Pacific" (14: 176). One cannot sell poverty, but one can sell paradise. Those on the receiving end have not always been impressed with how their country's image has been manipulated by overseas commercial interests. As the Premier of St. Vincent once said: "To Hell with Paradise" (quoted in 15: 324). In studying tourism one can investigate in concrete detail the links between power and knowledge, the generation of images of the Other, the creation of "natives" and "authenticity," the consumption of images, and so on. These are basic to the tourism industry as, indeed, they are to the anthropological researcher's ethnographic industry. Foucault (60) and Baudrillard (7) have written at length about these processes, and international tourism would appear to be a rich area in which to extend our insights.

In many areas of the Third World – the West Indies is a leading example – tourism is associated strongly with servility; it reawakens memories of the colonial past (103: 139) and so perpetuates resentments and antagonisms (13: 271–2). Clearly this background imparts a distinctive characteristic to relationships between tourists and locals. But even where this history of race relations is not so evident in the representations of social interaction in the tourism arena, one still finds characteristic behavior of a very specific kind. While it is important to examine relations here at a concrete level it is also important to see that these specific interactions are particular manifestations of larger state, class, and international politico-economic structures. For a start, the organization of the tourist industry (certainly when one is dealing with packaged tours) generally prevents the normal array of social relationships. This is also true of the informal sphere, though to a lesser extent (81: 25–6). Van den Berghe has referred to the links between tourists and locals as a "parody" of a human relationship (202: 378). Depending on one's values, many social ties could be called parodies of human relationship, but the point here is that use of such an extremely evaluative term is commonplace in tourist studies.

The question of what sort of social relationships grow up in tourism encounters can only be answered by detailed and descriptive studies. Attention to a culture's meaning structures is certainly required, for we need to know how people in other cultures perceive and understand tourists as a species of foreigner, what motivations they attribute to their behavior, and how they distinguish among types of tourist (187: 359). When one knows how tourists are classified, one can investigate the rules for relating to such people and compare them to those that structure other social interactions. In the Trobriand Islands the only category the local people had for tourists was *sodiya* (soldier) (106: 357). In the Seychelles the word "tourist" was heard as *tous riches* (all wealthy). Other areas of the world present equally interesting examples of classification, and an important area for research is the overlap between tourist identities and the identities established in previous historical periods. In some cultures there may be an explicit parallel drawn to the colonial era, which may significantly affect the way tourists are treated.

The task of documenting the semantics of tourist–local interaction has only begun. We have, for the most part, taxonomies of tourist types and vague generalizations. For instance, the first anthropological collection on tourism was entitled *Hosts and Guests*, but as has been pointed out (155, 181) both terms in the title may be of dubious value. If tourism is a new activity, one may not find in it anything like customary hospitality or any of the moral norms that apply between hosts and guests. Because of the fleeting nature of tourist relations, a tourist does not become part of any long-term reciprocity structure (47: 62). While there are rules for behavior towards strangers in a culture (157), tourists are not of the culture at all and usually know few local rules. Tourists are, as Cohen insists, not guests at all, but outsiders not part of the visited culture's moral fabric (37: 220). The concept of stranger has been discussed fruitfully (if erratically) on many occasions in the social science literature (59, 173, 175, 179), and it may be valuable to apply some of the general formulations developed there to the specific situation of tourism to ask what types of stranger tourists may be viewed as (110: 31; 133: 40) and what rules apply.

No matter how often tourism industry brochures speak of the natural friendliness of people, generosity usually has little to do with the provision of tourism services. As Boudhiba states (in 47: 63), hospitality "is just another technique of selling." This assertion may, of course, be far too general and may only be the popular cynical representation that exists alongside the glowing image. Social scientists need to know that rules may be different in different cultures. Different types of tourist may be treated differently, for after all, different types of tourist do affect a culture differently; almost certainly, different classes of people in a culture treat tourists according to different standards. And then there is the "development cycle" aspect of tourist systems. Over the years as the nature of tourism, or the type of tourists, or the quantity of tourists in any area changes, the rules for tourist–local interaction may undergo profound transformation. Hills and Lundgren (84) use the idea of an "irritation index" to monitor the levels of adverse reaction of local people to the influx of tourists over time. Like other analysts, they point to the often cyclical nature of tourism. The way tourists are treated today may differ from the way they were perceived and treated 15 years ago. Apart from the possibility that a different type of tourist may be present, one must remember that tourism, along with a host of other forces, may change the culture itself, create new roles and norms, and so on. As Cohen points out (38: 242; cf 154: 149), it is not long in most destinations before talk of guests and hospitality becomes inappropriate. Van den Berghe has christened each individual in the touristic arena a "touree." The touree identity is brought into being specifically by the presence of tourists (202: 378–9), where norms may differ greatly from those operating in other domains in a culture.

The tourist–local relationship is odd in many ways. One member is at play, one is at work; one has economic assets and little cultural knowledge, the other has cultural capital but little money. Not surprisingly, the general image of "cultural brokerage" has established itself firmly in the literature for the activities of a host of middlemen, entrepreneurs, and cultural transformers who try to structure to their own advantage transactions between the two systems brought together by international tourism (38: 246; 54: 192–3; 138: 209–11). In some instances such middlemen may themselves be ethnically marginal (182: 69). In the Peruvian case described by van den Berghe, for instance, tour guides are frequently *mestizo*, so tourism is simply another area where they bring cultures together and exploit the indigenous people (202: 385–6). Clearly, such entrepreneurship and brokerage in tourism (125) deserves detailed ethnographic investigation.

Anthropology has often been defined as the study of human beings in culture and society. Tourism is thus an odd anthropological object, because international tourists are people *out* of culture in at least two senses. First, they do not belong to the culture of the destination country, and second, they have stepped beyond the bounds of ordinary social reality, into what has sometimes been referred to as a "ludic" or "liminoid" realm (109). Wagner expresses the same notion with her phrases "out of place" and "out of time" (204). Tourism, as a UNESCO report once stated, is "life in parenthesis" (199: 85). The semantics and politics of the industry image-makers alter tourists' experience of space and time (2: 67, 102). In more theoretical language, tourism consists of meta-social processes, among which ritual and play are common (131: 207–8). Wealthy Americans on holiday may play at being "Peasant for a Day," while poorer tourists might like to be "King for a Day" (65). The tourist's world is constructed of many inversions – from work to play, normal morality to promiscuity, conspicuous spending rather than saving, freedom rather than structure, and indulgence rather than responsibility. For some (124: 143–4, 152), travel is an escape from real social ties and being communal; it is to be without commitment, to be anywhere rather than somewhere.

Given these inversions (68: 21), it is not surprising that a currently prominent representation in the anthropology of tourism should be tourism as a sacred quest – i.e. as a journey similar to and as significant as the pilgrimages of old (66, 68). In a world where play and freedom replace work and structure, such cosmological interpretations have an understandable appeal. Graburn, whose name is closely associated with this approach, sees tourism not as a frivolous pursuit but as "re-creation." He even suggests a neurological foundation: Like play and ritual, it might be a "right-side" brain phenomenon (71: 11).

Over the years, others have likened tourism to pilgrimage (see 87: 10; 105: 359; 197: 20).[2] Although some who comment on international tourism from a religious perspective find this parallel unpalatable (139: 59), the ethnographic exploration of overlaps between work and play, or among pilgrimage, ritual, play, and tourism [see Graburn's work on Japan (69)], may be of great value. We should particularly note here the importance of the play concept in post-structuralist thinking. It is interesting that our most stimulating and original general discussion of tourism today – MacCannell's *The Tourist. A New Theory of the Leisure Class* (115) – marshals a vast array of approaches: Marxism, semiotics, dramaturgy, and so on, in what is a very modern-looking analysis. Yet, as Thurot and Thurot claim (192: 174), not only is MacCannell behind the times empirically in terms of the contemporary nature of much international tourism, he is also theoretically out of date (201: 4). Marxian theory may have been original, but Marx himself retained a somewhat 18th-century concept of human nature in terms of labor and production. We now live in a post-Marxian world of the "political economy of the sign"; the emphasis has shifted away from production itself to image, advertising, and consumption. We are now interested in what Baudrillard has termed the "mirror of production" (6), and tourism, being so much a matter of leisure, consumption, and image, is an essentially (post-)modern activity.

There is a problem, however, in elevating notions of play or sacred quest into a general explanatory framework. Indeed, the complexity of tourism is such that, as Nash felt obliged to state forcefully (135: 504–5), scholars will do a "disservice" to the study of tourism if they opt for a single conceptual scheme that may obscure other vital perspectives. As P. Pearce puts it (148: 22), we need detailed empirical work on tourist behavior and motivation, not ideological and committed debates about the meaning of tourism. Play itself is a difficult theoretical notion, involving, as Bateson long ago recognized (5: 102, 155; see 196), essentially human creative and reflexive powers. The anthropology of play is still an undeveloped field. Harm might easily be done to the study of play and tourism if they are too closely aligned in the early stages as theoretical leitmotifs. At this stage a very different set of issues must be addressed: for *whom* does tourism mean a sacred quest? Might the appearance of this image represent the anthropologist's craving for meaning rather than a well-thought-out empirical investigation of a highly complex phenomenon? Doesn't such a schema postulate an entity – *the* tourist – instead of looking realistically at the varied clientele of the international travel industry? Besides, at the concrete level, it is well known that in tourism one does not find neat reversals from ordinary time to structureless communitas. The world of tourism is rife with the class distinctions of our everyday world. As MacCannell observes, even the Russian national airline, Aeroflot, makes provision for first-and economy-class travelers (115: 177). Some types of tourism, such as the inclusive package tour, involve less freedom and more structure than normal life (171: 446). We have also to remember, as Schwimmer puts it (174: 223), that tourism is the conspicuous consumption of resources accumulated in secular time; its very possibility, in other words, is securely rooted in the real world of gross political and economic inequalities between nations and classes. In fact, according to van den Abbeele (201: 5), international tourism is doubly imperialistic; not only does it make a spectacle of the Other, making cultures into consumer items, tourism is also an opiate for the masses in the affluent countries themselves. The juxtaposition in the essay collection *Hosts and Guests* (185) of Nash's argument that studies of tourism should take place within the conceptual framework of imperialism

(133) and Graburn's that tourism is a sacred journey (66) is a good reminder that we are dealing with a highly complex system. That complexity may be best brought out by deliberately cultivating a diversity of investigative approaches.

TOURISM AND SOCIOCULTURAL CHANGE

During the 1970s the Greek Orthodox church recommended a new prayer:

> Lord Jesus Christ, Son of God, have mercy on the cities, the islands and the villages of this Orthodox Fatherland, as well as the holy monasteries which are scourged by the worldly touristic wave. Grace us with a solution to this dramatic problem and protect our brethren who are sorely tried by the modernistic spirit of these contemporary Western invaders (quoted in 180: 55).

While the creation of a new prayer in response to international tourism may be a rare occurrence, the expression of such hostile sentiments is not. Tourism is unique as an export industry in that the consumers themselves travel to collect the goods (164: 250; 48: x). This presence of the customer creates a set of sociocultural consequences missing from other export activities. Considerable impact (160) may occur even without actual contact between tourists and locals.

We now have several general surveys of the sociocultural repercussions of tourism (14: 252ff; 25; 29: 383–8; 122; 199). Such surveys tend to be critical, contrasting starkly with the earlier optimistic, quantitative accounts by economists from which qualitative cultural data were usually absent (58). It may well be, as Graburn suggests (67), that the other social sciences became interested in tourism precisely because of the inadequacy of the economic approach. No adequate evaluation of tourism can be based simply on economic criteria, let alone on a single indicator such as foreign exchange earnings (3: 66–7). Of course, how to integrate quantified and nonquantifiable material remains a difficulty.

Most academics writing about sociocultural change and tourism from sociological and anthropological viewpoints have adopted a negative stance. In this they contribute to the mounting condemnation of Third World tourism by intellectuals, church leaders, and radicals in the Third World itself, where images of disintegration, pollution, decay, and so on, abound. Is the social science literature distinct from these other representations? If not, what is missing – analytically and descriptively – from the social science work?

For a start, the effects of tourism are rarely convincingly distinguished from those of other contemporary forces for social change (9: 524; 146: 61). Authors write about the repercussions of tourism with little close attention to the historical processes at work and fail to specify precisely the links involved (145: 70). Wood has added that much writing is not just sloppy, it is ethnocentric (210: 564). Given that social change in the Third World is highly complex, the attribution of adverse changes to tourism rather than to to urbanization, population growth, the mass media, etc, often appears arbitrary. As several authors have recently argued (30: 72;

39

56: 134; 117: 365; 181: 13), tourists may have been chosen as conspicuous scape-
goats. Writers claim to observe, for example, a "demonstration effect": locals imitate
the behavior of tourists, to their own detriment. Close analysis, however, reveals
many problems. For instance, some have commented that the frivolity witnessed in
contemporary pilgrimages in Sri Lanka, where youths carry transistor radios and so
on, is evidence of Western contamination of a traditional activity; but as Pfaffenberger makes clear, such behavior is not new: pilgrimages have always been accompanied by ludic activity (152: 61). Likewise, in Bali the presence of tourist money is
sometimes said to be responsible for an upturn over the last decade in ritual performances. Again, close attention suggests other factors may be responsible. For
Acciaioli the ceremonial efflorescence is, to a degree, being encouraged by Jakarta.
The Indonesian state is hostile to regionalist sentiment, and the encouragement of
ceremonials confines expression of such sentiment to the area of aesthetic culture
where it is politically inconsequential (1: 158–62).

Referring to the cultural consequences of the economic changes brought about
by tourism, Turner and Ash (195: 197) claim that tourism is the enemy of authenticity and cultural identity. Others, though less extreme, likewise use emotive labels to
refer to the replacement of traditional life, with its customary exchange and obligation structure, by the cash nexus of industrialized society (58: 222). One proposed
term for this overall cultural process is "commoditization" (76). Forster has referred
to "phoney folk culture" (58: 226) and others to the "staging" of events for tourists.

> Culture is being packaged, priced and sold like building lots, rights of way,
> fast food, and room service, as the tourism industry inexorably extends its
> grasp. For the monied tourist, the tourism industry promises that the world is
> his/hers to use. All the "natural resources," including cultural traditions, have
> their price, and if you have the money in hand, it is your right to see whatever
> you wish.... Treating culture as a natural resource or a commodity over
> which tourists have rights is not simply perverse, it is a violation of the
> peoples' cultural rights (76: 136–7).

For Greenwood, this "commoditization" is simply the logic of tourism as an identifiable example of capitalist development.

Moral and behavioral changes are certainly occurring, but we must be careful not
to indulge in romanticism and ethnocentrism by setting our descriptions against
some Rousseauesque idyll of traditional life. For a start, in most cases we are dealing
with societies with centuries of exposure to a whole range of economic, political,
and cultural influences from the West. Long before tourism, those cultures were
changing, including in directions that reflected their own understandings of the
nature of Western societies (121). Besides, what in a culture is *not* staged? What does
cultural authenticity consist of? As Greenwood states, all cultures "are in the process
of 'making themselves up' all the time. In a general sense all culture is 'staged authenticity'" (77: 27). That being so, if change is a permanent state, why should the
staging bound up in tourism be regarded as so destructive, and why should the
changes be seen in such a negative light? The very concept of authenticity requires

much closer attention in the arena of tourism; it requires close empirical work on tourist behavior, motivations, expectations, and the meanings attributed to experience (149). One might additionally ask what is so abhorrent about inauthentic phenomena? As Simmel, a lucid explorer of modernity, noted, phenomena we are disposed to call inauthentic or superficial very often reveal the nature of social reality (in 171: 465–6). If we turn to another aspect of the negativism surrounding tourism and social change, we must be careful not to contrast an expanding sphere of monetary relations with some ideal image of a nonmercenary, traditional culture. How noneconomic and uncalculating were traditional norms of reciprocity? A similar lack of clarity obtains with expressions such as "demonstration effect." We need detailed work showing how new activities affect cultural behavior and what the particular mechanisms of change are.

When social scientists have turned from general cultural consequences to analyzing specific areas of change, "consistently contradictory" (82: 5, 10) patterns have emerged. For almost any effect of tourism discovered in one case, one can find a counterexample. For instance, tourism ought to have a symbiotic relationship with the environment: an area often becomes a tourist destination precisely because of its scenic beauty, wildlife, and so on (88: 218–19). That attractiveness must survive to lure tourists. Some studies show that tourism indeed preserves wildlife (132: 36), but many others report that tourism has ruined the very environment that created it (207). Likewise tourism is said to weaken tradition; but it may also, by raising historical consciousness, lead to restoration of ancient monuments and the like (33: 218–19). Ethnic art (70) tells a similarly two-faced story, for tourism brings both the degradation of traditional technique for the mass production of airport art, and the reinvigoration of artistic skill (46). McKean, in fact, has protested at the general image of decay and argues, using Balinese material, that tourism can produce a general process of cultural involution whereby tradition, cultural pride, and identity are strengthened and standards of artistic creativity are consciously maintained (126: 103–4). Swain, whose study of the effects of tourism on Cuna women in Panama captures the contradictory aspect of tourism well, remarks that tourism "simultaneously encourages the maintenance of traditions and provides many stimuli for change" (189: 71). In a study of Fijian fire walkers, Brown (20: 224) argues that tourism does not undermine local culture, but rather provides an extra resource with which traditional forms can be continued. (In this sense international tourism subsidizes a form of ethnic conflict between Indians and Fijians.) Other cases indicate further the contradictory potential of international tourism. For instance, some studies argue that tourism distributes money within a community; others emphasize how tourism reinforces existing inequality by channeling money to the elites. Some argue that tourism stimulates domestic agriculture, others that it leads to people leaving the land and, in some areas, to serious malnutrition. Obviously tourism may have different effects in different regions. And also, as was clear in Spain (75), one might get a short-term positive effect on local agriculture, to be followed in the medium term by a set of negative consequences. Other accounts are contradictory not because the facts are so but because authors approach their studies with different values. It is also conceivable that international tourism sets in motion, or at least reinforces, different and even antithetical patterns of change.

Given that tourism has a range of potentials, it can be a source of social divisiveness and conflict. Crystal (44: 119–22) explains how tourism initially strengthened the solidarity of the Tana Toraja in Sulawesi because it emphasized their cultural particularity. Later, however (44: 123), as tourism began to be a force for the commercialization of religious ceremonies, stripping them of local meaning, conflicts between traditionalists and modernists were created. Crystal is ultimately uncertain (44: 125) whether tourism will turn out to be a source of economic development or a prime cause of cultural dissolution. Far more negative is Greenwood's analysis of the *Alarde* ritual in Fuenterrabia. This was a ritual essentially for local participants, not for outsiders. When the Spanish tourist authorities decided that the spectacle could be a tourist attraction, volunteer performers were not forthcoming. If a culture is an integrated system of meanings, Greenwood argued, selling local culture in tourism will be destructive. "Making their culture a public performance took the municipal government a few minutes: with that act, a 350-year-old ritual died" (76: 137). Ironically the *Alarde*, which commemorates local resistance to foreign invasion, crumbled in the modern touristic economy. This case of repeating a ceremony for tourists, like other touristic accommodations – shortening or rescheduling cultural performances so that they are more palatable to tourists in a hurry – is not unique. The authorities could have chosen differently; Crystal notes that the outcome for the Tana Toraja will depend in part on the attitude of the national tourism planners. Tourism clearly opens up the possibility of new conflicts within small communities and between such communities and larger, embracing political structures.

As Cohen insists, only detailed ethnographic study will enable us to compare data on different types of cultures, different types and numbers of tourists, different types of touristic niches, and so on, so that the concrete social processes operating in any particular case can be analyzed. Perhaps tourism is a contradictory phenomenon. Perhaps tourism, like capitalism, has within it the seeds of its own destruction, as two early writers on the subject argued (164: 250; 213: 2). If so, then international tourism is no manna from heaven, no easy passport to development. But we must again be careful to see whose perceptions and evaluations we are dealing with. It is striking that in many social science disciplines (e.g. economics and geography) we rarely hear the local voice on these issues. Nor does that voice often enrich anthropological writings (172: 255). Perhaps this state of affairs would be rectified if tourism became an explicit focus for ethnographic research rather than an incidental afterthought to other projects. Without close attention to the local voice (*voices*, for tourism produces a range of local reactions), our social scientific work risks being descriptively poor and ethnocentric. We need to know the local perceptions and understandings of tourism, we need to know the local perceptions of change and continuity, and we need to recognize that any culture is likely to have contradictory things to say about both. International tourism may be about *our* culture rather than that of the destination country (192: 187); and unless the anthropological approach to international tourism accords a crucial status to the full range of local voices, it risks putting itself in the same position.

ACKNOWLEDGMENT

This paper first appeared in 1988 under a slightly different title in Vol. 1, No. 1, of *Criticism Heresy and Interpretation*, Department of Asian Languages and Anthropology, University of Melbourne.

NOTES

1 For taxonomies of tourist types, see references 31, 32, 34 (organized mass tourist, individual mass tourist, explorer, drifter; or recreational mode, diversionary mode, experiential mode, experimental mode, existential mode), and 181 (ethnic, cultural, historical, environmental, recreational).

2 In my article on the "anthropological self" (42: 82), I constructed a *triangle des déplacements* consisting of tourists, pilgrims, and anthropologists in order to consciously play with the differences and overlaps among these three identities. Leach, writing about anthropologists who attended a Mardi Gras and became participants, speaks of "tourist pilgrims (ourselves included)" (105). The social psychologist P. Pearce (148: 32–4) has explored in detail the differences and similarities in behavior and attitudes of a large field of travelers – tourists, anthropologists, migrants, missionaries, pilgrims, and explorers. Peacock (144: 51–4, 58–65) similarly discusses the overlaps between anthropological field researchers and other travelers – spies, missionaries, explorers, and so on – but does not include tourists on the list.

LITERATURE CITED

1 Acciaioli, G. 1985. Culture as art: from practice to spectacle in Indonesia. *Canberra Anthropol.* 8:148–72

2 Bail, M. 1980. *Homesickness*. Melbourne: Macmillan

3 Baretje, R. 1982. Tourism's external account and the balance of payments. *Ann. Tourism Res.* 9:57–67

4 Barthes, R. 1972. *Mythologies*. London: Jonathan Cape

5 Bateson, G. 1973. *Steps to an Ecology of Mind*. London: Paladin

6 Baudrillard, J. 1975. *The Mirror of Production*. St. Louis: Telos Press

7 Baudrillard, J. 1983. *Simulations*. New York: Semiotext(e) Inc.

8 Biddlecomb, C. 1981. *Pacific Tourism, Contrasts in Values and Expectations*. Suva: Pacific Conf. Churches

9 Boissevain, J. 1977. Tourism and development in Malta. *Dev. Change* 8:523–8

10 Boorstin, D. 1972. *The Image. A Guide to Pseudoevents in America*. New York: Atheneum

11 Boudhiba, A. 1981. Mass tourism and cultural traditions. *People's Bank Econ. Rev.* (Sri Lanka). August: 27–9

12 Brameld, T., Matsuyama, M. 1977. *Tourism as Cultural Learning. Two Controversial Case Studies in Educational Anthropology*. Washington: University Press of America

13 Britton, R. A. 1977. Making tourism more supportive of small state development: the case of St. Vincent. *Ann. Tourism Res.* 4:268–78

14 Britton, R. A. 1978. International tourism and indigenous development objectives: a study with special reference to the West Indies. PhD thesis, Univ. Minnesota

15 Britton, R. A. 1979. The image of the Third World in tourism marketing. *Ann. Tourism Res.* 6:318–28

16 Britton, S. G. 1980. A conceptual model of tourism in a peripheral economy. In *Tourism in the South Pacific*, ed. D. Pearce, pp. 1–17. Paris: UNESCO/Dept. Geography, Univ. Christchurch

17 Britton, S. G. 1982. The political economy of tourism in the third world. *Ann. Tourism Res.* 9:331–58

18 Britton, S. G. 1983. *Tourism and Underdevelopment in Fiji.* Dev. Stud. Cent. Monogr. No. 31. Canberra: ANU

19 Britton, S. G., Clarke, W. C., eds. 1987. *Ambiguous Alternative: Tourism in Small Developing Countries.* Suva: Univ. South Pacific

20 Brown, C. H. 1984. Tourism and ethnic competition in a ritual form. The firewalkers of Fiji. *Oceania* 54:223–44

21 Bryden, J. M. 1973. *Tourism and Development. A Case Study of the Commonwealth Caribbean.* Cambridge: Cambridge Univ. Press

22 Buck, R. C. 1978. Towards a synthesis in tourism theory. *Ann. Tourism Res.* 5:110–11

23 Buck, R. C. 1982. On tourism as an anthropological subject. *Curr. Anthropol.* 23:326–7

24 Bugnicourt, J. 1977. Tourism with no return. *Dev. Forum* 5(2):2

25 Butler, R. W. 1974. The social implications of tourist developments. *Ann. Tourism Res.* 2:100–11

26 Carroll, J. 1980. The tourist. In *Sceptical Sociology*, J. Carroll, pp. 140–9. London: Routledge & Kegan Paul

27 Casson, L. 1979. *Travel in the Ancient World.* London: Allen & Unwin Ltd.

28 Centre on Transnational Corporations 1982. *Transnational Corporations in International Tourism.* New York: United Nations

29 Chib, S. N. 1980. Tourism and the Third World. *Third World Q.* 11:283–94

30 Cleverdon, R. 1979. *The Economic and Social Impact of International Tourism on Developing Countries.* London: Economist Intelligence Unit Ltd.

31 Cohen, E. 1972. Towards a sociology of international tourism. *Soc. Res.* 39:164–82

32 Cohen, E. 1974. Who is a tourist? A conceptual clarification. *Sociol. Rev.* 22:527–55

33 Cohen, E. 1978. The impact of tourism on the physical environment. *Ann. Tourism Res.* 5:215–37

34 Cohen, E. 1979. A phenomenology of tourist experiences. *Sociology* 13:179–201

35 Cohen, E. 1979. The impact of tourism on the Hill Tribes of Northern Thailand. *Int. Asienforum* 10:5–38

36 Cohen, E. 1979. Rethinking the sociology of tourism. *Ann. Tourism Res.* 6:18–35

37 Cohen, E. 1982. Marginal paradises. Bungalow tourism on the islands of Southern Thailand. *Ann. Tourism Res.* 9:189–228

38 Cohen, E. 1982. Jungle guides in northern Thailand – dynamics of a marginal occupational role. *Sociol. Rev.* 30:236–66

39 Cohen, E. 1984. The sociology of tourism: approaches, issues, and findings. *Ann. Rev. Sociol.* 10:373–92

40 Cohen, E. 1987. "Alternative tourism" – a critique. *Tourism Recreation Res.* 12:13–18

41 Collignon, R. 1984. La lutte des pouvoirs publics contre les "encombrements humains" à Dakar. *Can. J. African Stud.* 18:573–82

42 Crick, M. 1985. "Tracing" the anthropological self: quizzical reflections on fieldwork, tourism and the ludic. *Soc. Anal.* 17:71–92

43 Crompton, J. L. 1979. Motivations for pleasure vacations. *Ann. Tourism Res.* 6:408–24

44 Crystal, E. 1978. Tourism in Toraja (Sulawesi, Indonesia). See Ref. 185, pp. 109–25

45 Davis, D. E. 1978. Development and the tourism industry in third world countries. *Soc. Leisure* 1:301–22

46 Deitch, L. I. 1978. The impact of tourism upon the arts and crafts of the Indians of the Southwestern United States. See Ref. 185, pp. 173–92

47 de Kadt, E. 1978. The issues addressed. See Ref. 48, pp. 3–76

48 de Kadt, E., ed. 1978. *Tourism. Passport to Development? Perspectives on the Social and Cultural Effects of Tourism in Developing Countries.* New York: Oxford Univ. Press

49 Diamond, J. 1977. Tourism's role in economic development. The case reexamined. *Econ. Dev. Soc. Change* 25:539–53

50 Ecumenical Coalition on Third World Tourism. 1983. *Tourism Prostitution Development.* Bangkok: ECTWT

51 Edwards, A. 1982. *International Tourism to 1990.* Economist Intelligence Unit, Spec. Ser. 4. Cambridge, MA: Abt Books

52 Erbes, R. 1973. *International Tourism and the Economy of Developing Countries.* Paris: OECD

53 Esh, T., Rosenblum, I. 1975. *Tourism in developing countries – trick or treat? A report from the Gambia.* Res. Rep. No. 31. Uppsala: Scand. Inst. African Stud.

54 Evans, N. H. 1976. Tourism and cross cultural communication. *Ann. Tourism Res.* 3:189–98

55 Fanon, F. 1974. *The Wretched of the Earth.* Harmondsworth: Penguin

56 Farrell, B. H. 1979. Tourism's human conflicts. Cases from the Pacific. *Ann. Tourism Res.* 6:122–36

57 Finney, B. R., Watson-Gegeo, K. A. 1979. A new kind of sugar. Tourism in the Pacific. *Ann. Tourism Res.* 6:469–7

58 Forster, J. 1964. The sociological consequences of tourism. *Int. J. Comp. Sociol.* 5:217–27

59 Fortes, M. 1975. Strangers. In *Studies in African Social Anthropology,* ed. M. Fortes, S. Patterson, pp. 229–53. London: Academic

60 Foucault, M. 1979. *Power Truth Strategy,* ed. M. Morris, P. Patton. Sydney: Feral Publications

61 Fowles, J. 1978. *Daniel Martin,* London: Panther Books

62 French, L. 1979. Tourism and Indian exploitation. A social indictment. *Indian Hist.* 10(4):19–24

63 Fussell, P. 1980. *Abroad, British Literary Travelling Between the Wars.* New York: Oxford Univ. Press

64 Gonsalves, P., Holden, P., eds. 1985. *Alternative Tourism. A Resource Book.* Bangkok: Ecumenical Coalition on Third World Tourism

65 Gottlieb, A. 1982. American's vacations. *Ann. Tourism Res.* 9:65–87

66 Graburn, N. H. 1978. Tourism: the sacred journey. See Ref. 185, pp. 17–31

67 Graburn, N. H. 1980. Teaching the anthropology of tourism. *Int. Soc. Sci. J.* 32:56–68

68 Graburn, N. H. 1983. The anthropology of tourism. *Ann. Tourism Res.* 10:9–33

69 Graburn, N. H. 1983. *To Pray, Pay and Play: The Cultural Structure of Japanese Domestic Tourism.* Ser. B, No. 26. Aix-en-Provence: Cent. Hautes Etud. Touristiques

70 Graburn, N., ed. 1976. *Ethnic and Tourist Arts. Cultural Expressions from the Fourth World.* Berkeley: Univ. Calif. Press

71 Graburn, N. H., ed. 1983. The anthropology of tourism. *Ann. Tourism Res.* 10(1):Spec. Issue

72 Gray, H. P. 1982. The contribution of economics to tourism. *Ann. Tourism Res.* 9:105–25

73 Greenblatt, C. S., Gagnon, J. H. 1983 Temporary strangers. Travel and tourism from a sociological perspective. *Social Perspect.* 26:89–110

74 Greenwood, D. 1972. Tourism as an agent of change: a Spanish Basque case. *Ethnology* 11:80–91

75 Greenwood, D. 1976. The demise of agriculture in Fuenterrabia. In *The Changing Faces of Rural Spain,* ed. J. B. Acheves, W. A. Douglas, pp. 29–44. Cambridge, MA: Schenkman Publ. Co.

76 Greenwood, D. 1978. Culture by the pound. An anthropological perspective on tourism as cultural commoditization. See Ref. 185, pp. 129–38

77 Greenwood, D. 1982. Cultural "authenticity". *Cult. Survival Q.* 6(3):27–8

78 Haas, H. 1984. A decade of alternative tourism in Sri Lanka. See Ref. 86, pp. 1/1–1/17

79 Hannerz, U. 1973. Marginal entrepreneurship and economic change in the Cayman Islands. *Ethnos* 38:101–12

80 Harrell-Bond, B. E., Harrell-Bond, D. L. 1979. Tourism in the Gambia. *Rev. African Polit. Econ.* 14:78–90

81 Hassan, R. 1975. International tourism and intercultural communication. The case of Japanese tourists in Singapore. *Southeast Asian J. Soc. Sci.* 3(2):25–38

82 Hermans, D. 1981. Consistently contradictory. Economic and social impacts of tourism on host societies. Unpublished.

83 Hiller, H. L. 1979. Tourism: development or dependence? In *The Restless Caribbean*, ed. R. Millett, W. M. Will, pp. 51–61. New York: Praeger

84 Hills, T. L., Lundgren, T. 1977. The impact of tourism in the Caribbean. *Ann. Tourism Res.* 4:248–57

85 Hoivik, T., Heiberg, T. 1980. Centre–periphery tourism and self-reliance. *Int. Soc. Sci. J.* 32:69–97

86 Holden, P., ed. 1984. *Alternative Tourism*. Bangkok: Ecumenical Coalition on Third World Tourism

87 Horne, D. 1984. *The Great Museum. The Representation of History*. London: Pluto Press

88 Hudman, L. E. 1980. *Tourism. A Shrinking World*. Columbus, Ohio: Grid Inc.

89 Huxley, A. 1955. *Beyond the Mexique Bay. A Traveller's Journal*. Harmondsworth: Penguin

90 International Union of Official Travel Organizations. 1974. Tourism. Its nature and significance. *Ann. Tourism Res.* 1(4):105–12

91 Jafari, J. 1974. The components and nature of tourism. The tourism market basket of goods and services. *Ann. Tourism Res.* 1(1):73–89

92 Jafari, J. 1974. The socio-economic costs of tourism to developing countries. *Ann. Tourism Res.* 1(7):227–62

93 Jafari, J. 1978. Editor's page. *Ann. Tourism Res.* 5:210–11

94 Jafari, J. 1979. Tourism and the social sciences. A bibliography 1970–78. *Ann. Tourism Res.* 6:149–78

95 Jafari, J. 1981. Editor's page. *Ann. Tourism Res.* 8:323–27

96 Jafari, J. 1982. Comment. *Curr. Anthropol.* 21:137

97 Jenkins, C. L. 1982. The effects of scale in tourism projects in developing countries. *Ann. Tourism Res.* 9:229–49

98 Kahn, H. 1980. Tourism and the next decade. In *Tourism Planning and Development Issues*, ed. D. Hawkins *et al.*, pp. 3–20. Washington, DC: George Washington Univ.

99 Kanywanyi, J. L. 1973. Tourism benefits the capitalists. See Ref. 176, pp. 52–65

100 Karch, C. A., Dann, G. 1981. Close encounters of the Third World. *Hum. Relat.* 34:249–69

101 Kelompok Studi dan Bantuan Hukum/Lembaga Bantuan Hukum. 1982. *Voices from Under Borobodur*. Yogyyakarta: KSBH Info. Courier

102 Krause, W., Jud, G. D., with Joseph, H. 1973. *International Tourism and Latin American Development*. Austin: Bur. Business Res., Univ. Texas

103 La Flamme, A. 1979. The impact of tourism. A case from the Bahama Islands. *Ann. Tourism Res.* 6:137–48

104 Lawson, R. W. 1983. Tourism research – break for thought. *Dev. Forum* 11(5):16

105 Leach, E. R. 1984. Conclusion. Further thoughts on the realm of folly. In *Text, Play and Story. The Construction and Reconstruction of Self and Society*, ed. S. Plattner, E. M. Bruner, pp. 356–64. Washington, DC: Am. Ethnol. Soc.

106 Leach, J. W. 1973. Making the best of tourism: the Trobriand situation. In *Priorities in Melanesian Development*, ed. R. J. May, pp. 357–61. Canberra: Aust. Natl. Univ./Port Moresby: Univ. Papua New Guinea

107 Leiper, N. 1979. The framework of tourism. Towards a definition of tourism, tourist, and the tourism industry. *Ann. Tourism Res.* 6:390–407

108 Lengyel, P. 1975. Tourism in Bali – its economic and social impact. A rejoinder. *Int. Soc. Sci. J.* 27:753–57

109 Lett, J. W. 1983. Ludic and liminoid aspects of charter yacht tourism in the Caribbean. *Ann. Tourism Res.* 10:35–56

110 Levine, D. N. 1979. Simmel at a distance. On the history and significance of the sociology of the stranger. See Ref. 175, pp. 21–36

111 Lévi-Strauss, C. L. 1976. *Tristes Tropiques*. Harmondsworth: Penguin

112 Levitt, K., Gulati, I. 1976. Income effect of tourist spending: mystification multiplied. A critical comment on the Zender report. *Soc. Econ. Stud.* 19:325–43

113 Liew, J. 1980. Tourism and development. A reexamination. In *Tourism in the South Pacific. The Contribution of Research to Development and Planning*, ed. D. Pearce, pp. 13–17. Christchurch: UNESCO/Dept. Geography, Univ Christchurch

114 MacCannell, D. 1973. Staged authenticity: arrangements of social space in tourist settings. *Am. J. Sociol.* 79:589–603

115 MacCannell, D. 1976. *The Tourist. A New Theory of the Leisure Class*. New York: Shocken Books

116 Machlis, G. E., Burch, W. R. 1983. Relations between strangers: cycles of structure and meaning in tourist systems. *Sociol. Rev.* 31:666–92

117 MacNaught, T. J. 1982. Mass tourism and the dilemmas of modernization in Pacific island communities. *Ann. Tourism Res.* 9:359–81

118 Malinowski, B. 1967. *A Diary in the Strict Sense of the Term*. London: Routledge & Kegan Paul

119 Maluga, A. P. 1973. Tourism and the Arusha Declaration. A Contradiction. See Ref. 176, pp. 44–8

120 Manning, F. E. 1978. Carnival in Antigua. An indigenous festival in a tourist economy. *Anthropos* 73:191–204

121 Marcus, G. E. 1980. The ethnographic subject as ethnographer: a neglected dimension of anthropological research. *Rice Univ. Stud.* 66:55–68

122 Mathieson, A., Wall, G. 1982. *Tourism. Economic, Physical and Social Impacts*. London: Longman

123 Matthews, H. G. 1977. *International Tourism. A Political and Social Analysis*. Cambridge, MA: Schenkman Publ. Co.

124 McHugh, P., et al. 1974. *On the Beginnings of Social Inquiry*. London: Routledge & Kegan Paul

125 McKean, P. F. 1976. An anthropological analysis of the culture-brokers of Bali: guides, tourists and Balinese. Paris: UNESCO/IBRD

126 McKean, P. F. 1978. Towards a theoretical analysis of tourism. Economic dualism and cultural involution in Bali. See Ref. 185, pp. 93–107

127 Mings, R. C. 1978. The importance of more research on the impacts of tourism. *Ann. Tourism Res.* 5:340–4

128 Mitchell, F. 1970. The value of tourism in East Africa. *East African Econ. Rev.* 2:1–21

129 Mitchell, L. S. 1979. The geography of tourism. An introduction. *Ann. Tourism Res.* 6:235–44

130 Mitford, N. 1959. The tourist. *Encounter* 13:3–7

131 Moore, A. 1980. Walt Disney World: Bounded ritual space and the playful pilgrimage center. *Anthropol. Q.* 53:207–18

132 Myers, N. 1975. The tourist as an agent for development and wildlife conservation: the case of Kenya. *Int. J. Soc. Econ.* 2:26–42

133 Nash, D. 1978. Tourism as a form of imperialism. See Ref. 185, pp. 33–47

134 Nash, D. 1981. Tourism as an anthropological subject. *Curr. Anthropol.* 22:461–81

135 Nash, D. 1984. The ritualisation of tourism. Comment on Graburn's *The Anthropology of Tourism*. *Ann. Tourism Res.* 11:503–7

136 Nettekoven, L. 1979. Mechanisms of intercultural interaction. See Ref. 48, pp. 135–45

137 Nuñez, T. 1963. Tourism, tradition, and acculturation. *Weekendismo* in a Mexican village. *Southwest. J. Anthropol.* 34:328–36

138 Nuñez, T. 1978. Touristic studies in anthropological perspective. See Ref. 185, pp. 207–16

139 O'Grady, E. 1982. *Tourism in the Third World. Christian Reflections*. New York: Orbis Books

140 O'Grady, R., ed. 1980. *Third World Tourism*. Singapore: Christian Conf. Asia

141 Organization for Economic Cooperation and Development. 1967. *Tourism Development and Economic Growth*. Paris: OECD

142 Palmer, A., Palmer, V. 1976. *Quotations in History*. Harrocks: Harvester Press

143 Papson, S. 1979. Tourism. World's biggest industry in the twenty-first century? *Futurist* 13:249–57

144 Peacock, J. L. 1986. *The Anthropological Lens. Harsh Light, Soft Focus*, London: Cambridge Univ. Press

145 Pearce, D. 1980. Tourism and regional development. A genetic approach. *Ann. Tourism Res.* 7:69–82

146 Pearce, D. 1981. *Tourist Development*. London: Longman

147 Pearce, P. 1982. Perceived changes in holiday destinations. *Ann. Tourism Res.* 9:145–64

148 Pearce, P. 1982. *The Social Psychology of Tourist Behavior*. Oxford: Pergamon

149 Pearce, P., Moscardo, G. M. 1982. The concept of authenticity in tourist experiences. *Aust. N. Z. J. Sociol.* 22:121–32

150 Perez, L. A. 1973. Aspects of underdevelopment. Tourism in the West Indies. *Sci. Soc.* 37:473–80

151 Perez, L. A. 1975. Tourism in the West Indies. *J. Commun.* 25:136–43

152 Pfaffenberger, B. 1983. Serious pilgrims and frivolous tourists. The chimera of tourism in the pilgrimages of Sri Lanka. *Ann. Tourism Res.* 10:57–74

153 Pihlstrom, B. 1980. Greater values at stake. *Ann. Tourism Res.* 7:611–12

154 Pi-Sunyer, O. 1978. Through native eyes. Tourists and tourism in a Catalan maritime community. See Ref. 185, pp. 149–55

155 Pi-Sunyer, O. 1981. Tourism and anthropology. *Ann. Tourism Res.* 8:271–84

156 Pittock, A. B. 1967. Aborigines and the tourist industry. *Aust. O.* 39(3):87–95

157 Pitt-Rivers, J. 1968. The stranger, the guest and the hostile host. Introduction to the study of the laws of hospitality. In *Contributions to Mediterranean Sociology*, ed. J. Persistiany, pp. 13–30. The Hague: Mouton

158 Popovic, V. 1972. *Tourism in East Africa*. Munich: Weltforum

159 Rajotte, F., Crocombe, R., eds. 1980. *Pacific Tourism: As Islanders See It*. Suva: S. Pacific Soc. Sci. Assoc./Univ. S. Pacific

160 Richter, D. 1978. The tourist art market as a factor in social change. *Ann. Tourism Res.* 5:323–38

161 Richter, L. 1980. The political uses of tourism: a Philippine case study. *J. Dev. Areas* 14:237–57

162 Richter, L. 1982. *Land Reform and Tourism Development. Policy-making in the Philippines*. Cambridge, MA: Schenkman Publ. Co.

163 Richter, L. 1984. The political and legal dimensions of tourism. See Ref. 86, pp. 18/1–18/21

164 Rivers, P. 1973. Tourist troubles. *New Soc.* 539(Feb. 1):250

165 Rivers, P. 1974. Misguided tours. *New Int.* Feb. 6–9

166 Robinson, R. 1972. Non-European foundations of European imperialism: sketch for a theory of collaboration. In *Studies in the Theory of Imperialism*, ed. R. Owen, B. Sutcliffe, pp. 117–42. London: Longman

167 Rosenow, J. E., Pulsipher, G. L. 1979. *Tourism. The Good, the Bad and the Ugly*. Lincoln: Century Three Press

168 Sadler, P. G., Archer, B. H. 1975. The economic impact of tourism in developing countries. *Ann. Tourism Res.* 3(1):15–32

169 Saglio, C. 1979. Tourism for discovery. A project in Lower Casamance, Senegal. See Ref. 48, pp. 321–35

170 Samy, L. 1975. Crumbs from the table. The worker's share in tourism. In *The Pacific Way, Social Issues in National Development*, ed. S. Tupouniua, R. Crocombe, C. Slatter, pp. 205–14. Suva: S. Pacific Soc. Sci. Assoc.

171 Schmidt, C. J. 1979. The guided tour. Insulated adventure. *Urb. Life* 7:441–67
172 Schudson, M. S. 1979. Review essay: on tourism and modern culture. *Am. J. Sociol.* 84:249–58
173 Schutz, A. 1976. The stranger. An essay in social psychology. In *Collected Papers*. Vol. 2. *Studies in Social Theory*, ed. A. Brodersen, pp. 91–105. The Hague: Martinus Nijhoff
174 Schwimmer, E. 1979. Feasting and tourism: a comparison. In *Semiotics of Culture*, ed. I. Portis-Winner, J. Umiker-Sebeok, pp. 221–35. The Hague: Mouton
175 Shack, W. A., Skinner, E. P., eds. 1979. *Strangers in African Societies*. Berkeley: Univ. Calif. Press
176 Shivji, I. G., ed. 1973. *Tourism and Socialist Development*. Dar-es-Salaam: Tanzania Publ. House
177 Shoesmith, D. 1980. *Tourism in Asia. Questions of Justice and Human Dignity*. Sydney: Asia Partnership Hum. Dev.
178 Shoesmith, D. 1980. *Tourism in the Philippines: Towards a Case Study*. Sydney: Asia Partnership Hum. Dev.
179 Simmel, G. 1908. The stranger. In *On Individualism and Social Forms*, ed. D. Levine, pp. 143–9. Chicago: Chicago Univ. Press
180 Smith, M. A., Turner, L. 1973. Some aspects of the sociology of tourism. *Soc. Leisure* 5:55–71
181 Smith, V. 1978. Introduction. See Ref. 185, pp. 1–14
182 Smith, V. 1978. Eskimo tourism. Micro models and marginal men. See Ref. 185, pp. 51–70
183 Smith, V. 1978. Hosts and guests. *Ann. Tourism Res.* 5:274–7
184 Smith, V. 1980. Anthropology and tourism. A science–industry evaluation. *Ann. Tourism Res.* 7:3–33
185 Smith, V., ed. 1978. *Hosts and Guests. The Anthropology of Tourism*. Oxford: Blackwell
186 Stewart, F. 1975. A note on social cost-benefit analysis and class conflict in LDCs. *World Dev.* 3:31–9
187 Stringer, P. F. 1981. Hosts and guests. The bed and breakfast phenomenon. *Ann. Tourism Res.* 8:357–76
188 Sutton, W. A. 1967. Travel and understanding. Notes on the social structure of tourism. *Int. J. Comp. Sociol.* 8:218–23
189 Swain, M. B. 1978. Cuna women and ethnic tourism. A way to persist and an avenue to change. See Ref. 185, pp. 71–81
190 Taylor, F. F. 1973. The tourist industry in Jamaica 1919–1939. *Soc. Econ. Stud.* 22:205–28
191 Tevi, L. 1985. Alternative tourism. Some ethical considerations for western tourists visiting Third World countries. *Contours* 2(2):10–17
192 Thurot, J. M., Thurot, G. 1983. The ideology of class and tourism. Confronting the discourse of advertising. *Ann. Tourism Res.* 10:173–89
193 Turner, L. 1974. Tourism and the social sciences. From Blackpool to Benidorm and Bali. *Ann. Tourism Res.* 1:180–205
194 Turner, L. 1976. The international division of leisure: tourism and the Third World. *World Dev.* 4:253–60
195 Turner, L., Ash, J. 1975. *The Golden Hordes. International Tourism and the Pleasure Periphery*. London: Constable
196 Turner, V. 1982. *From Ritual to Theatre. The Human Seriousness of Play*. New York: Performing Arts Journal
197 Turner, V., Turner, E. 1978. *Image and Pilgrimage in Christian Culture. Anthropological Perspectives*. Oxford: Blackwell
198 Twain, Mark, n.d. (orig. publ. 1869) *The Innocents Abroad*. New York: Harper & Row
199 UNESCO, 1977. The effects of tourism on socio-cultural values. *Ann. Tourism Res.* 4:74–105

200 Urbanowicz, C. F. 1978. Tourism in Tonga. Troubled times. See Ref. 185, pp. 83–92

201 van den Abbeele, G. 1980. Sightseers: the tourist as theorist. *Diacritics* 10:3–14

202 van den Berghe, P. L. 1980. Tourism as ethnic relations: a case study of Cuzco, Peru. *Ethnic Racial Stud.* 3:375–92

203 Varley, R. C. G. 1978. *Tourism in Fiji. Some Economic and Social Problems.* Bangor: Univ. Wales Press

204 Wagner, U. 1977. Out of time and place – mass tourism and charter trips. *Ethnos* 42:38–52

205 Wagner, U. 1981. Tourism in the Gambia. Development or dependency? *Ethnos* 46:190–206

206 Weales, G. 1968/9. The stranger came. *Antioch Rev.* (Winter):427–34

207 Wenkham, R. 1975. The Pacific tourist blight. *Ann. Tourism Res.* 3(2):63–77

208 Wild, R. A. 1980. Review of *Hosts and Guests. Oceania* 51:53

209 Wood, R. E. 1979. Tourism and underdevelopment in S. E. Asia. *J. Contemp. Asia* 9:274–87

210 Wood, R. E. 1980. International tourism and culture change in S. E. Asia. *Econ. Dev. Cult. Change* 28:561–81

211 World Tourism Organization. 1980. *Manila Declaration on World Tourism.* Madrid: WTO

212 World Tourism Organization. 1986. Tourism bill of rights and tourist code. *Contours* 2(5):18–22

213 Young, Sir G. 1973. *Tourism. Blessing or Blight?* Harmondsworth: Penguin

214 Young, R. C. 1977. The structural context of the Caribbean tourist industry. A comparative study. *Econ. Dev. Cult. Change* 25:657–71

3

THE SOCIOLOGY OF TOURISM*

Approaches, issues, and findings

Erik Cohen

INTRODUCTION

The sociology of tourism is an emergent specialty concerned with the study of touristic motivations, roles, relationships, and institutions and of their impact on tourists and on the societies who receive them.

The scientific study of tourism originated in continental Europe, which was the first region to experience the impact of mass tourism. The Italian L. Bodio published the first social scientific article on the subject in 1899. The major early contributions, however, were in German (cf Homberg 1978: 36–7).

The first specifically sociological writings on tourism were also in German, beginning with L. von Wiese's (1930) classic article and leading to the first full-length sociological work on the subject by H. J. Knebel (1960). Ogilvie's (1933) book on tourism is the first social scientific treatise on the subject in English; it was followed by Norval's (1936) book on the tourist industry. The subject, however, received little attention until well into the post-World War II period when the rapid expansion of tourism provoked some spirited, critical writings (Mitford 1959; Boorstin 1964: 77–117) and the first empirical studies (Nuñez 1963; Forster 1964).

The study of tourism as a sociological specialty rather than merely as an exotic, marginal topic emerged only in the 1970s with Cohen's (1972) typological essay and MacCannell's (1973) first theoretical synthesis. Since the mid-1970s, the field has grown rapidly, which is attested by the publication of a series of treatises and reviews (Young 1973; L. Turner and Ash 1975; MacCannell 1976; Noronha 1977; de Kadt 1979: 3–76) and general collections of articles (V. L. Smith 1977c, 1978a; Tourismus und Kulturwandel 1978; Cohen 1979d; de Kadt 1979: 77–335; Lengyel 1980; Graburn 1983b).

* Reproduced with the permission of Annual Reviews Inc. from *Annual Reviews in Anthropology*, 1984, 10: 373–92.

CONCEPTUALIZATIONS AND THEORETICAL APPROACHES

The most widely accepted, but technical, definition of the tourist was proposed by the International Union of Official Travel Organizations (IUOTO) in 1963 and approved in 1968 by the World Tourist Organization (Leiper 1979: 393). It states that (international) tourists are "temporary visitors staying at least twenty-four hours in the country visited and the purpose of whose journey can be classified under one of the following headings: (a) leisure (recreation, holiday, health, study, religion and sport); (b) business (family mission, meeting)" (IUOTO 1963: 14).

This definition is useful primarily for "statistical, legislative and industrial purposes" (Burkart and Medlik 1974: 3), but it is unsatisfactory for most sociological work because it is too broad and theoretically barren. There have been numerous efforts to devise a theoretically fruitful, sociological definition of the "tourist", begun by German sociologists (Knebel 1960: 1–6) and leading to Cohen's (1974) and Leiper's (1979) work, and P. L. Pearce's (1982: 28–30) appraisal of touristic taxonomies. None of the general conceptualizations of the tourist has been widely adopted, however. Only P. L. Pearce (1982: 29–37) has studied people's conceptions of "tourist-related roles" empirically, following up on Cohen's approach.

There are considerable differences among students in the field in their general philosophical and ideological perspectives, as well as in their theoretical approaches to tourism; these have produced a variety of conceptual approaches. We have selected eight of the most important ones for consideration.

1 *Tourism as commercialized hospitality:* The focus is on the visitor component (Cohen 1974: 545–6) of the tourist's role. Its proponents conceive of the touristic process as a commercialization of the traditional guest–host relationship through which strangers were given a temporary role and status in the society they visited (von Wiese 1930; cf Knebel 1960: 2). Tourism is thus viewed as a commercialized and eventually industrialized form of hospitality (Taylor 1932; Hiller 1976, 1977; Leiper 1979: 400–3). This approach proved fruitful in studying the evolution and dynamics of relationships between tourists and locals and in analyzing conflicts within roles and institutions dealing with tourists.

2 *Tourism as democratized travel:* The emphasis is on the traveler component of the tourist role; the tourist is viewed as a kind of traveler marked by some distinct analytical traits (Cohen 1974; P. L. Pearce 1982: 28–40). The authors who pioneered this approach saw modern mass tourism as a democratized expansion of the aristocratic travel of an earlier age (Boorstin 1964: 77–117). Though anchoring tourism in an area – namely, travel – that has not been explored by sociologists (Nash 1981: 462), this perspective generated some important work on the historical transformation of touristic roles (e.g. Knebel 1960, L. Turner and Ash 1975).

3 *Tourism as a modern leisure activity:* Tourism is seen as a type of leisure (Dumazdier 1967: 123–38; P. L. Pearce 1982: 20) and the tourist as a "person at leisure who also travels" (Nash 1981: 462). Its protagonists see leisure as an activity free of obligations (Dumazdier 1967: 14), but they usually abstain from investigating the

deeper cultural significance of leisure activities. They take a functionalist view, identifying leisure – and hence tourism – with recreation (e.g. Scheuch 1981: 1099; see also Cohen 1979b: 183–5). This approach informs much of the macrosociological and institutional research on modern tourism (e.g. Dumazdier 1967: 123–38; Scheuch 1981).

4 *Tourism as a modern variety of the traditional pilgrimage:* This perspective focuses on the deeper structural significance of modern tourism and identifies it with pilgrimages in traditional societies; it was proposed by MacCannell (1973: 589). Graburn's (1977) paper, identifying tourism as a form of the "sacred journey," brings the study of tourism even closer to that of the pilgrimage (but see Cohen 1984).

5 *Tourism as an expression of basic cultural themes:* The emphasis here is on the deeper cultural meaning of tourism. Rejecting the general, "etic" approach to tourism (e.g. Nash 1981), its advocates are trying to reach an "emic" understanding of its culture-specific, symbolic meaning that is "based on the views of the vacationers themselves" (Gottlieb 1982: 167; cf Graburn 1983a). The program implicit in such an approach would eventually do away with tourism as an analytic concept and would lead to a comparative study of different, culture-specific varieties of travel.

6 *Tourism as an acculturative process:* Proponents of this viewpoint focus upon the effects that tourists have on their hosts and strive to integrate the study of tourism into the wider framework of the theory of acculturation (Nuñez 1963: 347–78). It has not been very popular, however (but see Nettekoven 1979: 144–5), even though tourists in many remote areas appear to be important agents of an often caricatured form of Westernization.

7 *Tourism as a type of ethnic relations:* Advocates of this approach strive to integrate the analysis of the tourist host relationship into the wider field of ethnicity and ethnic relations (Pi-Sunyer 1977; Gamper 1981). Its major proponent is van den Berghe (1980) (cf also van den Berghe and Keyes 1984). This approach dovetails with some work on the impact of the production of ethnic arts for the tourist market on ethnic identities (Graburn 1976b: 23–30).

8 *Tourism as a form of neocolonialism:* The focus is on the role of tourism in creating dependencies between tourism-generating, "metropolitan" countries and tourism-receiving, "peripheral" nations that replicate colonial or "imperialist" forms of domination and structural underdevelopment. This approach was explicitly formulated in a paper by Nash (1977); Matthews (1978: 74–86) has discussed its various forms. The most ambitious empirical attempt to analyze tourism in these terms on a global scale in Høivik and Heiberg's (1980).

THE PRINCIPAL ISSUE AREAS IN THE SOCIOLOGY OF TOURISM

Sociological research on tourism falls naturally into four principal issue areas: the tourist, relations between tourists and locals, the structure and functioning of the tourist system, and the consequences of tourism.

The tourist

Research on the tourist is extremely varied, but the bulk of work in this area consists of purely empirical, "touristological" surveys and trend analyses that are oriented toward meeting the practical needs of governments and the tourist industry. They deal primarily with the demographic and socioeconomic characteristics of tourists (e.g. Burkart and Medlik 1974: 80–103); the frequency, purpose, length, and type of trip; and the nature of tourists' destinations and the kinds of activities undertaken. Though of rather limited sociological relevance in themselves, such data are import-ant resources for secondary analysis, enabling scholars to identify the major trends in modern tourism (e.g. Scheuch 1981).

International tourism became a major modern mass phenomenon after World War II when it came to embrace practically all social classes in industrialized Western societies (Scheuch 1981: 1095). This expansion was made possible by rising stand-ards of living and the shortening of the work year, which were by longer paid vaca-tions in the industrialized Western countries and a rapid improvement in the means of transportation (Dumazdier 1967: 129–30; Young 1973: 30; Scheuch 1981: 1094). To these factors the enhanced motivation to travel should also be added; it will be discussed below. The rate of expansion since World War II has been spectacular: in 1950 there were still only 25.3 million international tourists; in 1960, 75.3 million; in 1970, 169.0 million (Young 1973: 52); and in 1981, 291 million. Domestic tourism apparently grew at an even steeper rate and was estimated at 2.3 billion in 1981 (World Tourism Trends 1982: 1). The major destinations of international tourism are still North America and Europe. The share captured by other world regions, while still minuscule, is rapidly growing (Cleverdon 1979: 13).

On the whole, men travel more than women; older people travel somewhat less than younger and middle-aged ones; and the number of younger tourists is on the increase (Young 1973: 31; D. G. Pearce 1978: 4–7). Urban residents take far more yearly holiday trips than rural inhabitants (Dumazdier 1967: 124–61; Scheuch 1981: 1094).

A greater proportion of people in the higher income categories take yearly holiday trips, while those in the highest brackets take more than one trip a year on the aver-age (Newman 1973: 235). Despite the democratization of travel, significant class dif-ferences still exist in the industrialized Western countries, not only in the propensity to travel but also in the distance and type of destination, the organization of the trip, the motivations and traveling style, and the deeper cultural motifs informing tourism (Newman 1973; Gottlieb 1982: 165).

The sociopsychological study of tourism has only recently come to the attention of professional psychologists (P. L. Pearce 1982; Stringer 1984). Several topics have been closely examined: motivation (Crompton 1979; Dann 1981), the cultural or en-vironmental shock experienced at the destination (Cort and King 1979; P. L. Pearce 1981), decision-making (Myers and Moncrief 1978; V. L. Smith 1979), attitudes (Stoffle et al 1979, Farrell 1979), and satisfaction (Pizam et al 1978). Of particular relevance for our purposes is the recent reorientation in tourist motivation research.

Instead of conceiving of tourist motivation "as a simple short-term process assessed by measuring the immediate satisfactions and causes of travel behavior" (P. L. Pearce 1982: 51) as leisure researchers are prone to do, motivation for travel is now increasingly understood in terms of how it relates to the individual's long-term psychological needs and life-plans; intrinsic motives such as self-actualization seem to be particularly important. This approach accords with the work of those sociologists who view the tourist's motivations and desired experiences in the context of the basic structural themes and cultural symbols of modern society (Dann 1977; MacCannell 1973, 1976; Cohen 1979b; 1982c: 1–16).

MacCannell conceives of tourism as the modern equivalent of the religious pilgrimage: the two are homologous in that "both are quests for authentic experiences" (1973: 593). He argues that modern peoples' quest for authenticity is similar to the "concern for the sacred in primitive society" (MacCannell 1973: 590), and it is thus analogous to the religious quest for ultimate reality. Owing, however, to the shallowness and inauthenticity of modern life and the alienation of modern man, "reality and authenticity are thought to be elsewhere: in other historical periods and other cultures, in purer, simpler life-styles" (MacCannell 1976: 3). The search for authenticity thus induces moderns to become tourists. This seminal idea is combined with another one – namely, that structurally, *"sightseeing is a ritual performed to the differentiations of society"* (MacCannell 1973: 13; italics in original). The differentiations are symbolized in the variety of attractions, which are the modern equivalent of the undifferentiated totemic symbols of simpler societies. Although attractions are potential expressions of authenticity, not all of them are equally authentic (Schudson 1979: 1251). In fact, their authenticity is frequently staged by the hosts who thus surreptitiously subvert the tourist's endeavor. Caught in a staged "tourist space" from which there is no exit, modern mass tourists are denied access to the back regions of the host society where genuine authenticity can be found and are presented instead with "false backs." The unstated conclusion is that tourism is in fact a futile quest.

MacCannell's ideas have inspired much empirical work in recent years (e.g. Buck 1978a: 221–2; 1978b; Schmidt 1979; Papson 1981; Graburn 1983a). His approach, however, has also been criticized and modified along various lines (Cohen 1979b; Nash 1981; 462; Greenwood 1982; Schudson 1979: 1252–3).

Like Boorstin (1964: 77–117) before him, MacCannell talks of the tourist as a unitary role-type. There is overwhelming empirical evidence, however, that actual tourists differ considerably from one another in their motivations (Cohen 1979b), travelling styles, and activities, among other things. Various criteria have been proposed to classify tourists and tourism (see Cohen 1972, 1974; Noronha 1977: 6–9; V. L. Smith 1977b; Knox 1978: 3–5).

V. L. Smith's (1977b) typology is based on a combination of the number of tourists and their adaptation to local norms, while Cohen's (1972) typology of tourists roles is based on the extent of the tourist's exposure to the strangeness of the host environment as against his seclusion within the "environmental bubble" of his or her home environment that is supplied by the touristic establishment. Four types of tourists – the organized and the individual mass tourist, the explorer, and the drifter

– are distinguished. Much of the recent research on tourists can be classified using Smith's or Cohen's typologies. Most of the literature refers at least implicitly to the mass tourist (e.g. Boorstin 1964: 77–117; MacCannell 1976; Hiller 1976, 1977), often mistakingly assuming that he or she represents all tourists. Only a few studies, however, focus specifically on organized mass tourism (Nieto Piñeroba 1977: 150–1; U. Wagner 1977). Evans is one of the few who deals expressly with the explorer (1978: 48–50). Drifters or "travellers" have been the subject of several papers (e.g. Cohen 1973, 1982b; ten Have 1974; Vogt 1976). Only a few studies, however, have explicitly compared different types of tourists and their impact on their destinations (e.g. Cohen 1982b; Evans 1978: 48–51).

There are few detailed studies of tourist behavior. The most common kind is studies of vacationing behavior on beaches and in seaside resorts (Edgerton 1979; Laurent 1973; U. Wagner 1977; Cohen 1982b) that very strongly bring out the antithetical character of the vacation as compared to ordinary life. In the former, normality is suspended and the individual is liberated from his or her ordinary preoccupations (Laurent 1973: 14, 18). Life at the beach is experienced as "out of time and place," as a relaxing, paradisiacal, or ludic existence (Cohen 1982b: 209; Laurent 1973: 179) that is separate from the ordinary life both of the tourist and even of the surrounding population. Cohen (1982b: 210) and Laurent (1973: 170–1) are less sanguine than U. Wagner (1977: 42–5), though, about the extent to which a Turnerian "communitas" actually emerges on the beaches; rather, in Edgerton's (1979) fitting phrase, tourists on beaches seem typically to be "alone together."

Tourists and locals

A large number of publications deal either primarily or incidentally with relations between locals and tourists or "Hosts and Guests," in the somewhat ironic title of V. L. Smith's (1977c) book on the subject. Few studies deal specifically, however, with the nature and dynamics of the tourist–local relationship, which has three principal dimensions: people's interactions, perceptions and attitudes [see reviews by P. L. Pearce (1982: 68–96) and by Knox (1978)]. Sutton initiated the analysis of the distinct character of the tourist–local interaction and characterized it as a "series of encounters [between] visitors who are on the move to enjoy themselves . . . and hosts who are relatively stationary and who have the function of catering to these visitors' needs and wishes" (1967: 220). Such encounters are essentially transitory, nonrepetitive and asymmetrical; the participants are oriented toward achieving immediate gratification rather than toward maintaining a continuous relationship.

These basic traits of the "encounter" have been further amplified in later research. Owing to the transitory and nonrepetitive nature of the relationship, the participants do not have to take account of the effects their present actions will have on the relationship in the future; hence, there is neither a felt necessity nor an opportunity to create mutual trust. Consequently, such "relationships are particularly open to deceit, exploitation and mistrust, since both tourists and natives can easily escape the consequences of hostility and dishonesty" (van den Berghe 1980: 388). The asymmetry of the relationship and the quest for immediate gratification compound these possibilities. Sutton focused mainly on the asymmetry of knowledge where the host has an advantage over the visitor, which accounts for tourists' alleged "gullibility" (e.g. Mitford 1959: 6; Boorstin 1964: 107). But other asymmetries also exist. For example, the meaning of the encounter is different for each of the participants: tourism means work for most locals, leisure for the visitors, and this situation creates misunderstandings and conflicts of interest (V. L. Smith 1977a: 59; Nieto Piñeroba 1977: 149).

The tourist–local relationship is, to varying degrees, embedded in and regulated by two sociocultural systems: a native system, which is invaded by tourism, and the emergent tourist system itself. The principal evolutionary dynamics of the relationship consist of a transition from the former to the latter. Studies of this evolution usually present the process as a commercialization or "commoditization" (Greenwood 1977) of hospitality. Tourists are initially treated as part of the traditional guest–host relationship (Pi-Sunyer 1977: 150–1), but as their numbers increase, they become less and less welcome (Cohen 1982a: 248). Pressures then build up that transform the guest–host relationship that is based on customary, but neither precise nor obligatory, reciprocity into a commercial one that is based on remuneration. This transformation involves incorporating hospitality – an area that many societies view as founded on values that are the very opposite of economic ones – into the economic domain. Therefore, it is frequently a slow and tortuous process (Mansur 1972: 65; Cohen 1982a: 246–9).

As tourism moves out of the realm of native hospitality, it often passes through an anomic stage during which locals develop what Sutton (1967: 221) has termed a predatory orientation toward tourists. They strive to extract as much gain as possible from each encounter, irrespective of the long-term consequences that such conduct may have on the tourist flow. During this stage, which is often marked by considerable hostility to tourists (P. L. Pearce 1982: 83–5), a significant increase in tourist-oriented discrimination, deviance, and petty crime takes place (e.g. Nieto Piñeroba 1977: 149; Pi-Sunyer 1977: 154–5; Cohen 1982b: 219–21). Such occurrences, however, are detrimental to the long-term development of tourism, and they give rise to efforts – on the part of either tourist entrepreneurs or the authorities – to create and institutionalize a professionalized tourist system.

The principal motive of professionalization is to preserve and enhance the area's reputation and thereby ensure the long-term benefits of a continuous and growing flow of tourists. Though economically motivated, a professionalized local–tourist relationship does not take on the character of a wholly depersonalized, neutral economic exchange. Rather, it becomes professionally "staged" in MacCannell's sense, with the locals "playing the natives" and the tourist establishment's personnel correctly providing a competently "personalized" service. Professionalization thus consists of the effort to surmount the potential conflict between the economic and the social components of the service role. While this conflict is never completely resolved (Shamir 1978), professionalization may prevent or attenuate host hostility (de Kadt 1979: 58–9); more often, however, it merely becomes an outer veneer of exaggerated servility, and considerable host hostility lingers on beneath it (e.g. Fukunaga 1975: 61, 226).

The attitudes and mutual perceptions of tourists and locals have been studied primarily from the locals' perspective; there is little reliable information on the impact of touring on the tourist (P. L. Pearce 1982: 85). In the past, advocates of tourism claimed that it improves international understanding (e.g. Waters 1966: 101–11) while their critics denied this (e.g. Joerges and Karsten 1978: 6), but both claims remain largely unsubstantiated. P. L. Pearce suggests, on the basis of the meager evidence available, that "tourists do develop, albeit marginally, more positive attitudes to their hosts as a consequence of their travelling" (1982: 92); but he also thinks that "holiday experiences tend to confirm preexisting attitudes" (1982: 92) – negative as well as positive ones.

Turning now to the locals' perspective, we should note that tourists are virtually never the first strangers to penetrate even the more isolated societies of the world. They are usually preceded by conquerors, administrators, traders, missionaries, adventurers, anthropologists, etc (cf, for example, V. L. Smith 1977a: 52–3). Such prior contacts condition the locals' initial perceptions of and attitudes towards tourists who may be classified in traditional terms as friends or enemies (Dress 1979: 129) or even as foreign "soldiers" (Leach 1973: 357). According to Pi-Sunyer, insofar as they are accepted as guests, they are at first treated as individuals in a personalized relationship (1977: 150–1). With the advent of widespread tourism, however, locals become incapable of relating to each visitor individually and tend to create an "ethnic typology." As tourism develops further, mass tourists may become separated in the locals' consciousness from normal humankind and debarred of their "essential individuality and human qualities" (Pi-Sunyer 1977: 155). This perception legitimizes exploitative behavior (see also P. L. Pearce 1982: 84; Cohen 1982b: 220).

Doxey (1976) proposed a general evolutionary model of change in locals' attitudes toward tourists consisting of four stages: euphoria, apathy, annoyance, and antagonism. A positive attitude toward tourism may indeed accompany the initial stages of its development (e.g. Belisle and Hoy 1980; J. A. Pearce 1980), but euphoria does not always mark the beginning of tourism, especially when it is imposed from the outside (Fukunaga 1975: 209; Blakeway 1980: 79–80)

Antagonism is certainly found in many touristic areas (Noronha 1977: 43–6; Knox 1978: 14–20), particularly where tourist densities have increased rapidly (Greenwood 1972: 90; V. L. Smith 1977a: 57), and the tourist industry has exacerbated the socioeconomic and cultural differences between locals and tourists or engendered competition over scarce local resources (Jordan 1980: 43; Kent 1975; Knox 1978: 14–5). It may also be stimulated or reinforced by the conduct of tourists, especially in situations of considerable asymmetry where there is little segregation (P. L. Pearce 1982: 84–5; V. L. Smith 1977a: 59). Tourists also frequently serve as the concrete focus of a more general resentment toward white foreigners found, for example, in some ex-colonies (Young 1973: 141; Knox 1978: 17; de Kadt 1979: 59–61). Nonetheless, antagonism toward tourists is not a necessary or ubiquitous consequence of mass tourism (Manning 1979: 173–5). Various factors, such as cultural similarities or the involvement of locals in tourism, may modify or improve local attitudes (Noronha 1977: 43; Knox 1978: 9). The local people also learn to cope with the foreigners and develop a tolerance toward their peculiar behavior (Stott 1978: 82). Mature touristic areas such as Switzerland where tourism is a highly professionalized occupational area may thus be marked by an absence of both host hostility and genuine human contact between locals and visitors (cf MacCannell 1976: 106).

The development and structure of the tourist system

Modern tourism is an ecological, economic, and political system that is complex and global. As it matures, it attains a degree of separation from the rest of society (Cohen 1972: 171–3). The system is marked by a centrifugal tendency (Christaller 1955: 5–6) as it constantly expands into new areas, whether in a spontaneous "organic" pattern as a result of some inner impetus or in a sponsored, induced form through the efforts of the national authorities or large-scale developers (Cohen 1972: 198; 1979c: 24).

The core of the global tourist system is located in the major tourism-generating countries (Williams and Zelinsky 1970); its modern roots reach back to the Grand Tour (Brodsky-Porges 1981), which provided the geographical backbone from which the system expanded into more and more peripheral areas. It is presently penetrating the most remote and hitherto unaccessible areas of the Third World and the polar regions (e.g. Leach 1973; Cohen 1979a; Reich 1980). Speculation on tourism in space has already begun (e.g. Kaufmann 1983).

Socioeconomically, the system hinges on a group of national and increasingly transnational corporate actors and governmental and intergovernmental agencies, such as airlines; travel companies, travel agencies, and tour operators; hotel chains; international travel organizations (e.g. International Association of Travel Agents [IATA] and IUOTO); and various governmental and intergovernmental organizations (Matthews 1978; Young 1973; Cleverdon 1979; Dunning and McQueen 1982). Studies of the major corporate actors on the global scene reveal extensive metropolitan domination of the tourist industry (Matthews 1978: 43). The tourist industry is thus becoming internationalized (Lanfant 1980: 23; see also Dunning and McQueen 1982). The structure of the tourist industry on the global level has important repercussions at the national and local levels in the host countries. These effects are the principal preoccupation of those who study tourism from the perspective of dependency theory (e.g. Pérez 1973; Wood 1979, 1981; Britton 1982; for a critique of this approach, see Matthews 1978: 79–81, 85–6).

Sociologists and anthropologists have studied the dynamics of the tourist system mainly on the regional and local levels. The "genetic approach" was pioneered by Forster who drew attention to the processual nature of tourism, which "creates a type of 'cumulative causation', and ultimately a new economic base" (1964: 218) as it penetrates a new area. Greenwood's (1972) study of Fuenterrabia is the most influential publication in this field.

Building on Greenwood's work, Noronha (1977: 17–27) elaborated a general model of the development of tourism that consists of three stages: I. discovery; II. local response and initiative; and III. institutionalization. The model is based on the assumption that tourism in a newly discovered destination initially develops spontaneously and is based on local initiatives. Later on, however, as local resources prove insufficient to support further growth, the "wider political authorities and economic blocks intervene" until control eventually passes into the hands of outsiders during Stage III. In the process, craft tourism changes to industrial tourism (Pi-Sunyer 1977: 13–18; Rodenburg 1980) as facilities become bigger and are upgraded to international standards. The general implication of the model is that, as the industry develops, locals lose control and their relative share in the total benefits from tourism gradually declines (Rodenburg 1980: 186; but see Jenkins 1982). Peck and Lepie (1977: 160–1), however, argue that there is not merely one but several types of dynamics of development; their typology resembles Cohen's (1979c: 24; 1982b; 1983b) distinction between those local tourist systems that grow organically and those whose initial growth is induced from the outside.

The organic model is paralleled in the geographer's concept of the "resort cycle" (Stansfield 1978) or the "tourist area cycle" (Butler 1980). In his sophisticated model, Butler differentiates among five developmental stages: (a) evolution, (b) involvement, (c) development and consolidation, and (d) stagnation, which (e) either leads to decline or is transformed by rejuvenation. The last is well illustrated by the attempts to revitalize Atlantic City, primarily through legalized gambling (Stansfield 1978: 49–50). Hovinen (1982) has applied a modified version of this model to Lancaster County, Pennsylvania. It appears to describe adequately the evolution of many older resorts such as Nice (Nash 1979) or Queenstown, New Zealand (D. G. Pearce 1980), that grew organically, but it probably is not applicable to those newer resorts in Third World countries whose growth has been extraneously induced.

The impact of tourism

The impact of tourism is by far the most intensively researched issue area within the sociology of tourism. The great bulk of the impact studies focus on the host community or society; the effect on the tourists' country of origin is neglected.

Most authors distinguish between the socioeconomic and sociocultural effects of tourism (UNESCO 1976). Noronha (1977: 5–77) and Cleverdon (1979) have provided the most comprehensive surveys of the range of socioeconomic impacts that tourism has under different conditions. The socioeconomic studies cover primarily eight major topics: foreign exchange, income, employment, prices, the distribution of benefits, ownership and control, development, and government revenue. There is considerable agreement on the impact tourism has on them. It is well established that tourism generates foreign exchange (Gray 1982: 29–32; Varley 1978: 37; Wall and Ali 1977: 5–6), income for the host country (Cleverdon 1979: 32–6), and employment for the local population (e.g. Noronha 1977: 52–60; Cleverdon 1979: 39–42; de Kadt 1979: 35–44). Tourism often becomes an important source of governmental revenue as well (Cleverdon 1979: 45–8), which may be one of the reasons why many governments are eager to encourage its rapid development. Nonetheless, the positive economic effects of tourism frequently fall significantly short of expectations or predictions.

In addition, tourism generates or reinforces inflationary tendencies by putting pressure upon resources whose supply is inelastic – particularly some types of food (Urbanowicz 1977: 88; Cohen 1982b: 218) and land (e.g. Noronha 1979: 188). Thus, while tourism frequently benefits those locals who are directly involved in it, it may cause hardships for the rest of the population.

The development of a tourist industry often involves the penetration of outsiders (e.g. Cohen 1983b; Noronha 1979: 188) and both national and foreign outside financial interests (Cleverdon 1979: 20–2, 56–67; Noronha 1977: 19–20; de Kadt 1979: 28–32). This process frequently leads to a loss of local control over the industry (Greenwood 1972; Rodenburg 1980: 184).

Beyond these points of general agreement, the findings vary a great deal. Tourism has the most serious dislocating effects (Forster 1964: 219) and yields the smallest relative benefits for locals when large-scale, high-standard facilities are rapidly introduced by outside developers into an otherwise poorly developed area; dependency, rather than development, then results (Cleverdon 1979: 49–50; Hiller 1979; Geshekter 1978). Under such conditions, the disproportionate growth of the tourist sector fails to engender linkages with other sectors, particularly with agriculture (e.g. Elkan 1975: 129; Wilson 1979: 235 Cleverdon 1979: 43); rather, it causes dislocations, thus institutionalizing structural underdevelopment.

Where small-scale, locally owned, lower-standard, "craft" tourism is slowly introduced into a less-developed context, gross earnings may be smaller, but a greater percentage will be locally retained and there will be fewer disruptive effects (Rodenburg 1980). There is a better chance that linkages with the local economy will be established (e.g. Hermans 1981). The impetus such tourism provides may not suffice to stimulate sustained local development (Cohen 1982b: 224; Jenkins 1982: 235), however, in the absence of sufficient local capital and technical and entrepreneurial resources.

The sociocultural impacts of tourism are numerous and varied, but most of them can be classified under one of ten major topics: community involvement in wider frameworks, the nature of interpersonal relations, the bases of social organization, the rhythm of social life, migration, the division of labor, stratification, the distribution of power, deviance, and customs and the arts. There is a broad agreement among scholars on the findings about most of these topics.

Under tourism the local community becomes increasingly involved in the wider national and international systems, with a concomitant loss of local autonomy; the community's welfare comes to depend more and more upon external factors (such as changing fashions and worldwide prosperity or recession) over which it has no control (Greenwood 1972: 90).

On the level of local interpersonal relations, tourism tends to loosen diffuse solidarities and increase individualization (e.g. Stott 1978: 81) and creates stress and conflicts (Redclift 1973: 7–8; Boissevain 1977: 530; Andronicou 1979: 248–9); these in turn generate pressures for a greater formalization of local life. But under some circumstances, especially among marginal ecological or ethnic groups, it also produces a reaction in the opposite direction – i.e. a strengthening of group solidarity in the face of the intruding foreigners (e.g. Boissevain 1977: 530–2; Reynoso y Valle and de Regt 1979: 133).

Tourism's major impact on the bases of social organization, particularly in simple and traditional societies, consists of an expansion of the economic domain: some areas of life that were not primarily regulated by economic criteria become commercialized or "commoditized" (Greenwood 1977). Moreover, considerations of economic gain take a more prominent place in locals' attitudes and relationships – not only in their dealings with tourists, but also among themselves.

Many researchers have noted the impact that tourism has on the rhythm of social life. Tourism is a highly seasonal activity that drastically affects the traditional way of life in agricultural communities (e.g. Clarke 1981: 453–5; Greenwood 1972; Jordan 1980). It also changes the daily division of time between work and leisure for employees in the industry (e.g. Boissevain 1979: 87–8), which may, in turn, affect family life.

Tourism creates new employment opportunities in the host area and hence influences migration patterns in two principal directions: it helps the community retain members who would otherwise migrate away, particularly unemployed or underemployed youths in economically marginal areas such as islands or mountains; but it also attracts outsiders who are searching for work or economic opportunity and who often come from other branches of the economy, particularly agriculture (e.g. Noronha 1977: 54–5, 67; de Kadt 1979: 35–6, 43; Cohen 1983b). Thus, in mature tourist areas, tourism spurs urbanization (e.g. Rambaud 1967; McKean 1976a: 138; Preau 1982).

One of the most ubiquitously noted effects of tourism is its impact on the division of labor, particularly between the sexes. By creating new kinds of employment, tourism draws into the labor force parts of the local population previously outside it – specifically, young women who now find employment either in tourist services, such as hotels (e.g. Noronha 1977: 65; de Kadt 1979: 43–4); in the production of crafts and souvenirs for the market (e.g. Boissevain 1979: 83–4; Swain 1977); or in tourism-oriented prostitution. This change, in turn, affects not only the division of labor within the household but also the status of women *vis-à-vis* their families and husbands, and the control of parents over children. It occasionally leads to increased conflict and deviance within the family (e.g. Noronha 1977: 65–6; Boissevain and Serracino-Inglott 1979: 275).

The impact of tourism on stratification has been noted by many researchers (e.g. de Kadt 1979: 47–9), but the issues involved have not always been analytically distinguished. Tourism certainly promotes a change in the criteria of stratification (e.g. Stott 1978: 81): by placing greater emphasis on the economic domain, it enhances the value of money as a criterion of stratification vs more traditional criteria such as a person's origin or status-honor. It thus tends to effect a transformation of the existing stratificational system (Greenwood 1972: 89; Reynoso y Valle and de Regt 1979: 133). Moreover, even when its consequences are less profound, it creates new social strata, particularly middle classes (de Kadt 1979: 47–8).

The revaluation of local resources because of the new uses to which they are put as a result of tourism may produce fortuitous changes in the standing of some individuals, including local elites – e.g. poor or hitherto unused land may suddenly acquire considerable value (e.g. Cohen 1982b: 215; Clarke 1981: 458; Noronha 1979: 198). As a rule, however, the new tourist entrepreneurs do not come from the established local elites but are members of the urban middle classes (de Kadt 1979: 48–9; Cohen 1983b).

The most general impact that tourism has on stratification is that it augments social disparities and hence widens the span of the local stratificational system (Cleverdon 1979: 44; de Kadt 1979: 48; Boissevain 1977: 129; for an exception, see Stott 1978: 81). This change reflects both the increased division of labor engendered by tourism and the unequal distribution of benefits that usually accompany it.

Tourism is not a particularly effective mechanism of social mobility: while some individuals may greatly benefit from it (Greenwood 1972: 89; de Kadt 1979: 49), rank and file employees of the industry have limited chances for advancement, owing to the peculiar employment structure; it has a broad base of unskilled and semiskilled workers and narrow upper echelons. Moreover, in poorly developed areas these echelons tend to be occupied by outsiders (Noronha 1977: 55–7), to the detriment of local employees. Tourism does, however, encourage new economic activities in ancillary and complementary services and thus indirectly creates new opportunities for economic mobility among the locals (Cohen 1982b; Wahnschafft 1982: 435–6).

The specific political consequences of tourism have only received scant attention (but see Matthews 1983). It appears that tourism gives rise to new kinds of political interests and leads to a pluralization of local power structures by creating new centers of power, new political offices, and new types of leaders who often compete with the traditional leadership. The frequent result is increased community conflict around novel issues (Nuñez 1963: 351; Redclift 1973: 7–8; Boissevain and Serracino-Inglott 1979: 275–6).

The argument that tourism encourages deviance of various sorts has frequently been made (Nicholls 1976). While various kinds of tourism-oriented deviance such as theft, begging (Noronha 1979: 193; Cohen 1983c), prostitution (e.g. Jones 1978; Cohen 1982d; Wahnschafft 1982: 436–7), and fraud (e.g. Loeb 1977) have been reported, the role of tourism in the etiology of such activities appears to have been much exaggerated, particularly in the case of prostitution.

The impact of tourism on customs and the arts has been extensively examined (Graburn 1976c), but it can only be reviewed briefly here. Customs and the arts are frequently drawn into the economic domain or "commoditized" (Greenwood 1977) as resources to encourage tourism. While the fact that this occurs is commonly accepted, the question of their transformation and debasement through tourism is still hotly debated. "Commoditization" does not, in itself, necessarily change customs or

63

the arts – indeed, in some instances it may conserve them in the interests of tourism (e.g. Aspelin 1977; Wilson 1979: 230).

In most cases, customs and the arts have, in fact, undergone changes as they have been addressed to a new "external" public (Graburn 1976b: 8; McKean 1976b: 242–3) that does not share the cultural background, language, and values of the traditional, "internal" public. Dances and rituals have been shortened or embellished, and folk customs or arts altered, faked, and occasionally invented for the benefit of tourists (e.g. Boorstin 1964: 108; Graburn 1976b: 19–20).

Tourism has often been presented as a major debaser and destroyer of customs and the arts that leads to the emergence of a "phony-folk-culture" (Forster 1964: 226) and to the mass production of cheap, artless souvenirs and fake "airport art" (Schadler 1979: 147–8) adapted to tourists' expectations (Boorstin 1964: 106–7). While such phenomena are indeed quite widespread, there are, however, other developments that the culture critics have overlooked. In particular, there are instances where tourism furthered the survival of an otherwise moribund folk art (e.g. Andronicou 1979: 252–3; Boissevain 1977: 532–4; Cohen 1983a) or stimulated the development of new arts or styles, occasionally of considerable artistic merit (e.g. Graburn 1976a; Cohen 1983a).

In conclusion, rather than looking at transformations engendered by tourism in customs and the arts as mere aberrations, it is more useful to approach them as another, albeit accelerated, stage in the continuous process of cultural change. It presently gives rise to a variety of "transitional arts" created for the tourist market and meriting attention on their own terms as genuinely new artistic creations.

CONCLUSION

Mainstream sociology has only recently discovered tourism as a field of systematic inquiry, but many sociologists still view it with suspicion or even disdain. While this may in part reflect the commonsense view of tourism as a frivolous, superficial activity unworthy of serious investigation, it also certainly reflects the fact that the study of tourism has not been well integrated into mainstream sociology. This situation has been only partly remedied by recent work. While a variety of often intriguing conceptual and theoretical approaches for studying the complex and manifold touristic phenomena have emerged, none has yet withstood rigorous empirical testing; while field-studies have proliferated, many lack an explicit, theoretical orientation and hence contribute little to theory building. It is hoped that this review helps to bring theory and empirical research closer together and to codify the field, as well as to further recognition of it as a legitimate and significant sociological specialty.

LITERATURE CITED

Andronicou, A. 1979. Tourism in Cyprus. See de Kadt 1979, pp. 237–64

Aspelin, P. L. 1977. The anthropological analysis of tourism: Indirect tourism and the political economy of the Mamainde of Mato Grosso, Brazil. *Ann. Tourism Res.* 9(3):135–60

Belisle, F. J., Hoy, D. R. 1980. The perceived impact of tourism by residents: A case study in Santa María, Colombia. *Ann. Tourism Res.* 7(1):83–101

Blakeway, M. 1980. The dilemmas of paradise. *PHP* Sept.: 73–84

Bodio, L. 1899. Sul movimento dei foresteri in Italia e sul dinero chi vi spendono. *G. Econ.* 15:54–61

Boissevain, J. 1977. Tourism and development in Malta. *Dev. and Change* 8(4):528–38

Boissevain, J. 1979. The impact of tourism on a dependent island: Gozo, Malta. *Ann. Tourism Res.* 6(1):76–90

Boissevain, J., Serracino-Inglott, P. 1979. Tourism in Malta. See de Kadt, 1979, pp. 265–84

Boorstin, D. J. 1964. *The Image: A Guide to Pseudo-Events in America.* New York: Harper & Row

Britton, S. G. 1982. The political economy of tourism in the Third World. *Ann. Tourism Res.* 9(3):331–58

Brodsky-Porges, E. 1981. The Grand Tour: Travel as an educational device, 1600–1800. *Ann. Tourism Res.* 8(2):171–86

Buck, R. C. 1978a. Boundary maintenance revisited – Tourist experience in an Old Order Amish community. *Rural Sociol.* 43(2):221–34

Buck, R. C. 1978b. From work to play: Some observations on a popular nostalgic theme. *J. Am. Cult.* 1(3):543–53

Burkart, A. J., Medlik, S. 1974. *Tourism: Past, Present and Future.* London: Heinemann

Butler, R. W. 1980. The concept of a tourist area cycle of evolution: Implications for management of resources. *Can. Geogr.* 24(1):5–12

Christaller, W. 1955. Beiträge zu einer Geographie des Fremdenverkers. *Erdkunde* 9(1):1–19

Clarke, A. 1981. Coastal development in France: Tourism as a tool for regional development. *Ann. Tourism Res.* 8(3):447–61

Cleverdon, R. 1979. *The Economic and Social Impact of International Tourism in Developing Countries, Spec. Rep. No. 6.* London: Econ. Intell. Unit

Cohen, E. 1972. Toward a sociology of international tourism. *Soc. Res.* 39(1):164–82

Cohen, E. 1973. Nomads from affluence: Notes on the phenomenon of drifter-tourism. *Int. J. Comp. Sociol.* 14(1/2):89–103

Cohen, E. 1974. Who is a tourist? A conceptual clarification. *Sociol. Rev.* 22(4):527–55

Cohen, E. 1979a. The impact of tourism on the Hill Tribes of northern Thailand. *Int. Asienforum* 10(1/2):5–38

Cohen, E. 1979b. A phenomenology of tourist experiences. *Sociology* 13:179–201

Cohen, E. 1979c. Rethinking the sociology of tourism. *Ann. Tourism Res.* 6(1):18–35

Cohen, E., ed. 1979d. Sociology of tourism. *Ann. Tourism Res.* 6(1–2):18–194 (Spec. issue)

Cohen, E. 1982a. Jungle guides in northern Thailand: The dynamics of a marginal occupational role. *Sociol. Rev.* 30(2):234–66

Cohen, E. 1982b. Marginal paradises: Bungalow tourism on the islands of southern Thailand. *Ann. Tourism Res.* 9(2):189–228

Cohen, E. 1982c. The Pacific Islands from utopian myth to consumer product: The disenchantment of paradise. *Cah. Tourisme, Ser. B* 27:1–34

Cohen, E. 1982d. Thai girls and farang men: The edge of ambiguity. *Ann. Tourism Res.* 9(3):403–28

Cohen, E. 1983a. The dynamics of commercialized arts: The Meo and Yao of northern Thailand. *J. Natl. Res. Counc. Thailand* 15(1): Pt. II, 1–34

Cohen, E. 1983b. Insiders and outsiders: The dynamics of development of bungalow tourism on the islands of southern Thailand. *Hum. Organ.* 27(2):227–51

Cohen, E. 1983c. Hill Tribe tourism. In *Highlanders of Thailand*, ed. W. Bhruksasri, J. McKinnon. Kuala Lumpur: Oxford Univ. Press

Cohen, E. 1984. Pilgrimage and tourism: Convergence and divergence. In *Journeys to Sacred Places*, ed. E. A. Morinis. In press

Cort, D. A., King, M. 1979. Some correlates of culture shock among American tourists in Africa. *Int. J. Intercult. Relat.* 3(2):211–24

Crompton, J. 1979. Motivations for pleasure vacation. *Ann. Tourism Res.* 6(4):408–24

Dann, G. M. S. 1977. Anomie, ego enhancement and tourism. *Ann. Tourism Res.* 4(2):184–94

Dann, G. M. S. 1981. Tourism motivation: An appraisal. *Ann. Tourism Res.* 8(2):187–219

de Kadt, E. 1979. *Tourism – Passport to Development?* New York: Oxford Univ. Press

Doxey, G. V. 1976. A causation theory of visitor–resident irritants: Methodology and research inferences. In *The Impact of Tourism Proc. 6th Ann. Conf. Travel Res. Assoc. San Diego California*, pp. 195–8

Dress, G. 1979. *Wirtschafts- und sozialgeographische Aspekte des Tourisms in Entwicklungsländern.* München: Florentz

Dumazdier, J. 1967. *Toward a Society of Leisure.* New York: Free Press

Dunning, J., McQueen, M. 1982. Multinational corporations in the international hotel industry. *Ann. Tourism Res.* 9(1):69–90

Edgerton, R. B. 1979. *Alone Together: Social Order on an Urban Beach.* Berkeley, CA: Univ. California Press

Elkan, W. 1975. The relation between tourism and employment in Kenya and Tanzania. *J. Dev. Stud.* 11(2):123–30

Evans, N. H. 1978. Tourism and cross-cultural communication. See Smith 1978a, pp. 41–53

Farrell, B., ed. 1977. *The Social and Economic Impact of Tourism on Pacific Communities.* Santa Cruz, CA: Cent. South Pac. Stud., Univ. California

Farrell, B. 1979. Tourism's human conflicts: Cases from the Pacific. *Ann. Tourism Res.* 6(2):122–36

Finney, B. R., Watson, K. M. A., eds. 1975. *A New Kind of Sugar: Tourism in the Pacific.* Honolulu: East-West Cent.

Forster, J. 1964. The sociological consequences of tourism. *Int. J. Comp. Sociol.* 5(2):217–27

Fukunaga, L. 1975. A new sun in North Kohala: The socio-economic impact of tourism and resort development on a rural community in Hawaii. See Finney and Watson 1975, pp. 199–227

Gamper, J. A. 1981. Tourism in Austria: A case study of the influence of tourism on ethnic relations. *Ann. Tourism Res.* 8(3): 432–46

Geshekter, Ch. L. 1978. International tourism and African underdevelopment: Some reflections on Kenya. See Smith 1978b, pp. 57–88

Gottlieb, A. 1982. Americans' vacations. *Ann. Tourism Res.* 9: 165–87

Graburn, N. H. H. 1976a. Eskimo art: The eastern Canadian Arctic. See Graburn 1976c, pp. 39–55

Graburn, N. H. H. 1976b. Introduction: Arts of the Fourth World. See Graburn 1976c, pp. 1–32

Graburn, N. H. H., ed. 1976c. *Ethnic and Tourist Arts: Cultural Expressions From the Fourth World.* Berkeley, CA: Univ. California Press

Graburn, N. H. H. 1977. Tourism: The sacred journey. See Smith 1977c, pp. 17–32

Graburn, N. H. H. 1983a. To pray, play and pay: The cultural structure of Japanese domestic tourism. *Cah. Tourisme, Ser. B* 25: 1–89

Graburn, N. H. H., ed. 1983b. Anthropology of tourism. *Ann. Tourism Res.* 10(1): 1–189 (Spec. issue)

Gray, H. P. 1982. The economics of international tourism. *Ann. Tourism Res.* 9(1): 1–125. (Spec. issue)

Greenwood, D. J. 1972. Tourism as an agent of change: A Spanish Basque case. *Ethnology* 11(1): 80–91

Greenwood, D. J. 1977. Culture by the pound: An anthropological perspective on tourism as cultural commoditization. See Smith 1977c, pp. 129–38

Greenwood, D. J. 1982. Cultural "authenticity." *Cult. Survival Q.* 6(3): 27–8

Hermans, D. 1981. The encounter of agriculture and tourism: A Catalan case. *Ann. Tourism Res.* 8(1): 462–79

Hiller, H. L. 1976. Escapism, penetration and response: Industrial tourism in the Caribbean. *Caribb. Stud.* 16(2): 92–116

Hiller, H. 1977. Industrialism, tourism, island nations and changing values. See Farrell 1977, pp. 115–21

Hiller, H. L. 1979. Tourism. Development or dependence? In *The Restless Caribbean*, ed. R. Millet, W. M. Hill, pp. 51–61. New York: Praeger

Høivik, T., Heiberg, T. 1980. Centre–periphery tourism and self-reliance. *Int. Soc. Sci. J.* 32(1): 69–98

Homberg, E. 1978. Reisen – zwischen Kritik und Analyse; zum Stand der Tourismusforschung. *Z. Kult.* 28(3): 4–10

Hovinen, G. R. 1982. Visitor cycles: Outlook for tourism in Lancaster county. *Ann. Tourism Res.* 9: 565–83

International Union of Official Travel Organizations. (IUOTO). 1963: *The United Nations' Conference on International Travel and Tourism.* Geneva: IUOTO

Jenkins, C. L. 1982. The effects of scale in tourism projects in developing countries. *Ann. Tourism Res.* 9(2): 229–49

Joerges, B., Karsten, D. 1978. Tourismus und Kulturwandel. *Z. Kult.* 28(3): 4–10

Jones, D. R. W. 1978 *Prostitution and Tourism.* Presented at PEACESAT Conf. Impact Tourism Dev. Pac. Sess. 4, April 19, 1978. Suva: Univ. South Pac. Ext. Serv. (Mimeo)

Jordan, J. W. 1980. The summer people and the natives: Some effects of tourism in a Vermont vacation village. *Ann. Tourism Res.* 7(1): 34–55

Kaufmann, W. F. III. 1983. Tourism in the twenty-first century. *Sci. Dig.* 91(4): 52–64

Kent, N. 1975. A new kind of sugar. See Finney and Watson 1975, pp. 169–98

Knebel, H. J. 1960. *Soziologische Strukturwandlungen im modernen Tourismus.* Stuttgart: Enke

Knox, J. M. 1978. *Resident–visitor interaction: A review of the literature and general policy alternatives.* See Jones 1978

Lanfant, M. F. 1980. Introduction: Tourism in the process of internationalization. *Int. Soc. Sci. J.* 32(1): 14–43

Laurent, A. 1973. *Libérer les vacances?* Paris: Seuil

Leach, J. W. 1973. Making the best of tourism: The Trobriand case. In *Priorities in Melanesian Development,* ed. R. J. May, pp. 357–61. Canberra: Aust. Natl. Univ.

Leiper, N. 1979. The framework of tourism: Towards a definition of tourism, tourist and the tourist industry. *Ann. Tourism Res.* 6(4): 390–407

Lengyel, P., ed. 1980. The anatomy of tourism. *Int. Soc. Sci. J.* 32(1): 1–150 (Spec. issue)

Loeb, L. D. 1977. Creating antiques for fun and profit: Encounters between Iranian Jewish merchants and touring coreligionists. See Smith 1977c, pp. 185–92

MacCannell, D. 1973. Staged authenticity: Arrangements of social space in tourist settings. *Am. J. Sociol.* 79(3): 589–603

MacCannell, D. 1976. *The Tourist: A New Theory of the Leisure Class.* New York: Schocken

Manning, F. E. 1979. Tourism and Bermuda's black clubs: A case of cultural revitalization. See de Kadt 1979, pp. 157–76

Mansur, F. 1972. *Bodrum: A Town in the Aegean.* Leiden, The Netherlands: Brill

Matthews, H. G. 1978. *International Tourism – A Political and Social Analysis.* Cambridge, MA: Schenkman

Matthews, H., ed. 1983. Political Science and Tourism. *Ann. Tourism Res.* 10(3) (Spec. issue)

McKean, Ph. F. 1976a. Interaction between tourists and Balinese: An anthropological analysis of partial equivalence structures. *Majalah Ilmu-Ilmu Sosial Indones.* 3: 135–46

McKean, Ph. F. 1976b. Tourism, culture change and culture conservation in Bali. In *Changing Identities in Modern Southeast Asia,* ed. D. J. Banks, pp. 237–48. The Hague: Mouton

Mitford, N. 1959. The tourist. *Encounter* 13(4): 3–7

Myers, P. B., Moncrief, L. W. 1978. Differential leisure travel decision-making between spouses. *Ann. Tourism Res.* 5(1): 157–65

Nash, D. 1977. Tourism as a form of imperialism. See Smith 1977c, pp. 33–47

Nash, D. 1979. The rise and fall of an aristocratic tourist culture – Nice, 1763–1936. *Ann. Tourism Res.* 6(1):61–75

Nash, D. 1981. Tourism as an anthropological subject. *Curr. Anthropol.* 22(5):461–81

Nettekoven, L. 1979. Mechanisms of intercultural interaction. See de Kadt 1979, pp. 135–45

Newman, B. 1973. Holidays and social class. In *Leisure and Society in Britain*, ed. M. A. Smith *et al*, pp. 230–40. London: Allen Lane

Nicholls, L. L. 1976. Tourism and crime. *Ann. Tourism Res.* 3:176–82

Nieto Piñeroba, J. A. 1977. Turistas y nativos: el caso de Formentera. *Rev. Esp. Opin. Pública* 47:147–67

Noronha, R. 1977. Social and cultural dimensions of tourism: A review of the literature in English. Washington, DC: World Bank (Draft)

Noronha, R. 1979. Paradise reviewed: Tourism in Bali. See de Kadt 1979, pp. 177–204

Norval, A. J. 1936. *The Tourist Industry: A National and International Survey*. London: Pitman

Nuñez, Th. A. 1963. Tourism, tradition and acculturation: *Weekendismo* in a Mexican village. *Ethnology* 2(3):347–52

Ogilvie, F. W. 1933. *The Tourist Movement: An Economic Study*. London: Staples

Papson, S. 1981. Spuriousness and tourism: Politics of two Canadian provincial governments. *Ann. Tourism Res.* 8(2):220–35

Pearce, D. G. 1978. Demographic variations in international tourism. *Rev. Tourisme* 33(1):4–9

Pearce, D. G. 1980. Tourism and regional development: A genetic approach. *Ann. Tourism Res.* 7(1):69–82

Pearce, J. A. 1980. Host community acceptance of foreign tourists: Strategic considerations. *Ann. Tourism Res.* 7(2):224–33

Pearce, P. L. 1981. "Environmental shock": A study of tourists' reactions to two tropical islands. *J. Appl. Soc. Psychol.* 11(3):268–80

Pearce, P. L. 1982. *The Social Psychology of Tourist Behaviour*. New York: Pergamon

Peck, J. G., Lepie, A. Sh. 1977. Tourism and development in three North Carolina towns. See Smith 1977c, pp. 159–72

Pérez, L. A. 1973. Aspects of underdevelopment: Tourism in the West Indies. *Sci. and Soc.* 37(4):473–80

Pi-Sunyer, O. 1977. Through native eyes: Tourists and tourism in a Catalan maritime community. See Smith 1977c, pp. 149–55

Pizam, A., Neumann, Y., Reichel, A. 1978. Dimensions of tourist satisfaction with a destination area. *Ann. Tourism Res.* 5(3):314–22

Preau, P. 1982. Tourisme et urbanisation en montagne: le cas de la Savoie. *Rev. Géogr. Alp.* 70(1/2):137–51

Rambaud, P. 1967. Tourisme et urbanisation des campagnes. *Sociol. Rur.* 7:311–34

Redclift, M. 1973. The effects of socioeconomic changes in a Spanish Pueblo on community cohesion. *Sociol. Rur.* 13(1):1–14

Reich, R. J. 1980. The development of Antarctic tourism. *Polar Rec.* 20(126):203–14

Reynoso y Valle, A., de Regt, J. P. 1979. Growing pains: Planned tourism development in Ixtapa-Zihuatanejo. See de Kadt 1979, pp. 111–34

Rodenburg, E. E. 1980. The effects of scale in economic development: Tourism in Bali. *Ann. Tourism Res.* 7(2):177–96

Schadler, K. F. 1979. African arts and crafts in a world of changing values. See de Kadt 1979, pp. 146–56

Scheuch, E. K. 1981. Tourismus. In *Die Psychologie des 20 Jahrhunderts*, 13:1089–114. München: Kindler

Schmidt, C. J. 1979. The guided tour: Insulated adventure. *Urb. Life* 7(4):441–67

Schudson, M. S. 1979. On tourism and modern culture. *Am. J. Sociol.* 84(5):1249–58

Shamir, B. 1978. Between bureaucracy and hospitality – Some organizational characteristics of hotels. *J. Mgmt. Stud.* 15(3):285–307

Smith, V. L. 1977a. Eskimo tourism: Micromodels and marginal man. See Smith 1977c, pp. 51–70

Smith, V. L. 1977b. Introduction. See Smith 1977c, pp. 1–14

Smith, V. L., ed. 1977c. *Hosts and Guests*. Philadelphia: Univ. Pennsylvania Press

Smith, V. L., ed. 1978a. *Tourism and Behavior, Studies in Third World Societies, No. 5.* Williamsburg, VA: Dept. Anthropol., Coll. William and Mary

Smith, V. L. ed. 1978b. *Tourism and Economic Change, Studies in Third World Societies, No. 6.* Williamsburg, VA: Dept. Anthropol., Coll. William and Mary

Smith, V. L. 1979. Women: The taste makers in tourism. *Ann. Tourism Res.* 6(1):49–60

Smith, V. L., ed. 1980. Tourism and development: Anthropological perspectives. *Ann. Tourism Res.* 7(1):11–119 (Spec. issue)

Stansfield, C. 1978. Atlantic City and the resort cycle. *Ann. Tourism Res.* 5(2):238–51

Stoffle, R. W., Last, C. A., Evans, M. J. 1979. Reservation-based tourism: Implications of tourist attitudes for Native American economic development. *Hum. Organ.* 38(3):300–6

Stott, M. A. 1978. Tourism in Mykonos: Some social and cultural responses. *Mediterr. Stud.* 1(2):72–90

Stringer, P., ed. 1984. Social psychology and tourism. *Ann. Tourism Res.* 11(1) (Spec. issue)

Sutton, W. A. 1967. Travel and understanding: Notes on the social structure of touring. *Int. J. Comp. Sociol.* 8(2):218–23

Swain, M. B. 1977. Cuna women and ethnic tourism: A way to persist and an avenue to change. See Smith 1977c, pp. 71–82

Taylor, A. E. 1932. "Tourism" – The business of organized hospitality. *Commer. Rep.* 35:188–90

ten Have, P. 1974. The counter culture on the march: A field study of youth tourists in Amsterdam. *Mens Maatsch.* 49(3):297–315

Tourismus und Kulturwandel. 1978. Tourismus und Kulturwandel. *Z. Kult.* 28(3):1–119 (Spec. issue)

Turner, L., Ash, J. 1975. *The Golden Hordes: International Tourism and the Pleasure Periphery.* London: Constable

UNESCO. 1976. The effects of tourism on socio-cultural values. *Ann. Tourism Res.* 4(2): 78–105

Urbanowicz, Ch. F. 1977. Tourism in Tonga: Troubled times. See Smith 1977c, pp. 83–92

van den Berghe, P. 1980. Tourism as ethnic relations: A case study of Cuzco, Peru. *Ethn. Racial Stud.* 3(4): 375–92

van den Berghe, P., Keyes, Ch. F., eds. 1984. Tourism and ethnicity. *Ann. Tourism Res.* 11(3) In press (Spec. issue)

Varley, R. C. G. 1978. *Tourism in Fiji: Some Economic and Social Problems, Bangor Occas. Pap. Econ. No. 1.* Bangor: Univ. Wales

Vogt, J. W. 1976. Wandering: Youth and travel behavior. *Ann. Tourism Res.* 4(1): 25–41

von Wiese, L. 1930. Fremdenverkehr als zwischenmenschliche Beziehung. *Arch. Fremdenverkehr* 1(1)

Wagner, U. 1977. Out of time and place: Mass tourism and charter trips. *Ethnos* 42(1/2): 38–52

Wahnschafft, R. 1982. Formal and informal tourism sectors: A case study of Pattaya, Thailand. *Ann. Tourism Res.* 9(3): 429–52

Wall, G., Ali, I. M. 1977. The impact of tourism in Trinidad and Tobago. *Ann. Tourism Res.* 5: 43–9

Waters, S. R. 1966. The American tourist. *Ann. Am. Acad. Polit. Soc. Sci.* 368: 109–18

Williams, A. V., Zelinsky, W. 1970. Some patterns in international tourist flows. *Econ. Geogr.* 46(4): 549–67

Wilson, D. 1979. The early effects of tourism on the Seychelles. See de Kadt 1979, pp. 205–36

Wood, R. E. 1979. Tourism and underdevelopment in Southeast Asia. *J. Contemp. Asia* 8(4): 274–87

Wood, R. E. 1981. The economics of tourism. *Southeast Asia Chron.* 78: 2–9

World Tourism Trends. 1982. World tourism trends in 1981. *Int. Tourism Q.* 1:1

Young, G. 1973. *Tourism – Blessing or Blight?* Harmondsworth: Penguin

Part II

THE TOURISM SYSTEM AND THE INDIVIDUAL

4

MOTIVATION AND ANTICIPATION IN POST-INDUSTRIAL TOURISM*

Giuli Liebman Parrinello

INTRODUCTION

The importance of motivation in tourism is quite obvious. It acts as a trigger that sets off all the events involved in travel. In other words, it represents the whys and the wherefores of travel in general, or of a specific choice in particular. There have been a number of works on tourist motivation, but it is only in the last 10 years that Dann (1981, 1983), Pearce (1982), and Stringer and Pearce (1984) have all argued, though admittedly from different standpoints, in favor of an interdisciplinary and pluralistic approach in this field. Both Pearce (1982) and Dann (1977, 1981, 1983) have proposed broadening the question of motivation to a more thorough methodological and epistemological study.

It must be remembered that the most recent psychological studies on motivation suggest considering the issue in a much more complex framework than that of individual psychology, thus including other traditional areas of psychological investigation, as well as social psychology (Graumann 1981; Schmalt 1986). In recent years, environmental psychology, which is by definition interdisciplinary, has also made some very useful contributions, and acts as a necessary frame of reference (Fridgen 1984).

One of the characteristics of post-industrial societies (Bell 1973) is that generally more than 50% of their population practice tourism (usually defined in statistical data as at least four or five days away from one's usual place of residence). This is, for example, the case in most Western European countries (Krippendorf 1987; Scheuch 1981). These countries are also characterized by an increasingly fast, intensive, and capillary-like process of information passed on by both individual and social experience and the traditional and new forms of mass media. It is possible to perceive a trend towards generalized tourist behavior at an international level, though the pattern of the motivational process maintains its own specific cultural identity.

Many scholars now accept that, conceptually, tourism must be considered a process taking place in a cycle of various phases (for the different points of view see Guzman 1986). Both psychologists and sociologists of tourism have stressed the

* Reprinted with kind permission of Elsevier Science Ltd, Pergamon Imprint, Oxford, England, from *Annals of Tourism Research*, 1993, 20: 233–49.

complexity of and length of time involved in the psychic processes (Cohen 1984; Pearce 1982). Fridgen (1984) adopts the conceptual five-phase framework first proposed by Clawson and Knetsch (1966) for recreation and applies it to tourist experience. These five phases are anticipation, travel to the destination, on-site behavior, return travel, and recollection. Anticipation must first be considered within the context of the whole trip and then of each successive phase of the trip. Special attention must be given to its relationship with the last phase that follows the return home (Chon 1990; Fridgen 1984; Mercer 1971) for the all-important feedback processes that link the first and the last phases of a trip. Then it must also be related to the concept of satisfaction (Pizam, Neumann and Reichel 1978, 1979).

In post-industrial societies, anticipation has become an increasingly significant phase. It is like traveling in time within a particular socio-cultural context of everyday life (Berger and Luckmann 1966; Thurn 1980), but without any actual move taking place in space. It still needs to be seen how motivation enters the first phase of anticipation, the pre-trip, which is a stage of planning, of collecting information and making decisions about the trip. Today motivation is becoming more and more complex from the cognitive point of view, in so far as the actual decision-making of the tourist is a consequence of a much wider experience and of a much greater quantity of information available. It has, in fact, become such a complicated process of elaboration that perhaps there may be doubts as to whether it is possible to talk about motivation in traditional terms.

Although this paper does not deal specifically with the tourist image which has been the subject of many studies, the image can, nevertheless, be considered as strictly connected with, or even as the "goal" of, motivation and anticipation. The consequences of the process of diffusing and acquiring information will most probably spill over on to the image itself.

This paper focuses on motivation in relation to the first phase of the tourist trip – its anticipation – within the framework of ordinary everyday life in post-industrial societies.

THEORETICAL AND METHODOLOGICAL PERSPECTIVES

At the outset, it is important to outline theories and orientations in the study of motivation and their applications to the study of tourist motivation. In psychology, the term "motivation" is at times given a very broad definition. It has become a "super-concept" and it is so pervasive that it would be almost impossible to say which processes taking place in the body or in the personality are not connected with it (Thomae 1965a).

According to De Charms and Muir's (1978) schematic classification, the origins of motivational theories can be traced back to Freud, Hull, and Lewin, as representatives of the psychodynamic theory, of the stimulus–response learning theory, and of cognitive psychology, though obviously more subtle distinctions should be made.

In fact, this classification can also be limited to only two groups (Thomae 1988): the homeostatic theories (including behavioral theories ranging from psychoanalysis to neobehaviorism) and cognitive theories (or activation, exploratory behavior, and competence). Text books, such as those by Thomae (1965b), Weiner (1985), or the more sophisticated one by Madsen (1974), provide a complete and well-constructed picture of motivation studies in the various American and European schools.

Tourism motivation involves all the components of personality in both the cognitive and affective dimensions over a long period. Therefore, it cannot be satisfactorily accounted for by the simple behavioristic approach of stimulus–response, as often used in experiments on the behavior of animals.

From the 1950s onward, even though behavioristic modules continued to influence studies, the introduction of the concept of a constantly active organism at a neurophysiological level led to more attention being paid to motivation with the gradual inclusion and acceptance of cognitive processes. Berlyne (1960, 1978) has focused on the exploratory and epistemological type of behavior, which is peculiar to humans and is sparked off by curiosity and their need for information about the world. Echoes of it can be found in the more updated version of intrinsic motivation (Deci 1975) that emphasizes competence and self-determination and is often adopted for tourist and leisure motivation, which stress an optimum level of activation between boredom and stress (Hartmann 1979). Iso-Ahola (1980, 1982) and Mannel and Iso-Ahola (1987) take the conclusions on leisure motivation as their starting point and, following a dialectical method of analysis, deal with the individual and social motivation of escaping and seeking by comparing them with life situations that are either over- or understimulating.

In respect to tourism motivation, Dann's push and pull subdivisions clearly placed "anomie" and "ego-enhancement" in the former category as early as 1977. Since then, scholars seemed to have reached some agreement on a multivariate formulation, on the basis of 5–8 motivations (Crompton 1979; Kaspar 1972; McIntosh 1977; Opaschowski 1977).

It is, in fact, this need for a more complete view which explains the continued success of Maslow's so-called humanistic psychology in studies on tourism. This is based on the premise of homeostasis which is the reconstruction of a basic equilibrium. It is here that Maslow (1954) classifies five needs into an interesting hierarchy ranging from biological needs to the more complex need of self-realization.

Freud's dynamic theory also seemed to go some way toward explaining tourist motivation (Grinstein 1955). But apart from Freud, it is the practical application of his psychoanalysis which forms Berne's transactional analysis. This is particularly relevant to tourism in all its three parts, with Dann arguing in favor of the motivational needs for tourist travel (Dann 1989; Mayo and Jarvis 1981).

The most recent motivational theories are founded on very complex interactive models, which are based on personal and situational factors represented in cognitive and emotional terms, with a wider perspective given to the interpretation of a simple segment of behavior (Graumann 1981; Schmalt 1986). In spite of the differences that persist between American and European schools of psychology, in the field of dynamic cognitive studies, the gap is gradually narrowing between their very different positions on, for example, the question of cognitive expectations (Atkinson

1983), the heuristic value of the dynamic model (Heckhausen 1980), and the connection between action and the "goal." The time dimension is implicit in a plan of action (already in Kelly's 1955 personal construct theory). This is, however, made explicit by some authors (Heckhausen 1967; Nuttin 1980), especially in the case of achievement motivation. Time anticipation and planning can then be considered corollaries of the dynamic motivational theories.

According to von Cranach and Valach (1984), scholars of the theory of action, a goal-directed action is social in motivation in any case, because of the control society exercises over individuals and vice versa. It can even be claimed that, since human behavior is by its very nature social, the use of the adjective is superfluous (Reykowski 1982). It is obvious that when motivational psychology takes on a dynamic interrelational approach, the issue is automatically opened out to the social dimension and, therefore, to a sociological perspective, as soon as the "subject" is replaced by the social "actor" (Cohen 1984; Dann 1981).

In the European school of social psychology, the question of motivation is not presented as such: traces of it can be found in "attitudes," in so far as they are dispositions that influence behavior (Jaspars 1978), and implicitly in Moscovici's (1984) "social representations," which have a symbolic and communicative value, but are also connected with psychic mechanisms.

Recent developments in environmental psychology have led to a broadening from a behavioristic conception to an anthropological–cultural view (Altman, Rapoport, and Wohlwill 1980; Stokols 1987) of the interactional nature of the relationship between the individual and the environment (Stokols 1977, 1978). Credit for this goes to the notions of behavior setting (Barker 1968; Wicker 1979) and of meaning (Rapoport 1982). Since these notions concur with ideas originating from social psychology, they are considered valid instruments for investigation into tourist motivation (Pearce 1982).

Dann, Nash and Pearce (1988) suggest a diagram with four quadrants resulting from the relationship between methodological sophistication and theoretical awareness for the methodology to be used in tourism studies. The dangers that may arise from a lack of attention to theoretical aspects leading to almost obsessive forms of methodological sophistication are pointed out in the proposed third quadrant.

The difficulties that emerge in a complete motivational study may appear insurmountable: the study should follow a typical group of tourists through all the five phases, from anticipation of the trip itself, right through to recollection, which may cover a number of years. Although some checks can be made, there always exists the problem of *post hoc* motivation, which represents a rationalization of a previous emotional motivation at a later date (Pearce 1982).

The field of motivational research ranges from large-scale market research to small studies carried out on atypical groups. Examples of studies and their relative methodology (in particular the Canadian Motivation to Travel and Vacation Study) are given by Pearce (1982). A section of the German Reiseanalyse presents a multi-motivational multivariate study, which is preceded and backed by very precise theoretical research (Studienkreis für Tourismus 1969, 1981). However, some criticisms have been made about its intentional ideological bias (Prahl and Steinecke 1979). Even though Krippendorf (1987) is aware of the limitations of this kind of study, he stresses the unique contribution made by the German Reiseanalyse, because it gives scholars the opportunity to follow developments in motivational studies from 1970 onwards.

Today motivational studies have become remarkably sophisticated, thanks to the use of the computer. Even the use of path analysis can be misleading (Dann, Nash and Pearce 1988). Potter and Coshall (1988) suggest a non-parametric hand-operable method. However, factor analyses (generally cluster analyses) also raise the problem of the cognitive dimension, which lies at the base of all items to be examined in interviews (Hartmann 1981). One must also bear in mind the objections made by Jaspars and Fraser (1984) to the usual attitude scales, which presuppose exactly the same cognitive approach to reality (i.e. the same culture). Furthermore, the culture-specific nature of language must be remembered. The problem of motivation among non-travelers also arises (Haukeland 1990).

The interdisciplinary and eclectic qualitative methodology proposed by Cohen (1979) has played a very important role. He himself points out the difficulties involved in its application (Cohen 1988), but there are signs of the beginning of a methodology that brings together the social science disciplines in studying tourism. The anthropological–cultural approach has generally not been applied to complex societies (however, see Gottlieb 1982) and quite rightly Smith (1979) and Nash (1981) stress the need for a cultural and cross-cultural investigation.

A motivational study can start off from an everyday ordinary life context (Berger and Luckmann 1966; Schütz 1971), which offers the advantage of being rich in implications and open to a number of methodological instruments. In this context, tourism emerges as a "counter-everyday" non-ordinary experience (Krippendorf 1987), while the tourist always carries his or her ordinary experience as a part of his or her cultural baggage (Opaschowski 1977).

Push motivations are generally accepted as the dominant factors. It is the ordinary world which provides the push for motivation. It is, at the same time, the context in which anticipation is experienced in various degrees, and in which it must necessarily be measured. At this point, three basic social sciences come into play: sociology, cultural anthropology, and (social) psychology. Social representations are formed here. Concepts like status and role, built environment, crowding, and stress must also be considered in this context. It is indeed in this context that researchers, influenced by a certain disciplinary bias, draw up the drafts of their questionnaires.

The ordinary or everyday context has some culturally specific factors, which can vary noticeably even between post-industrial countries like Germany, the United States, or Japan. It would not be altogether impossible to carry out a motivational study in five or more big cities in different post-industrial countries. The study could be organized by a small group of scholars of tourism representing the countries involved. It would be carried out on the same type of sample group using the same methods, which would have to be discussed, evaluated, and decided upon in deference to the cultural differences (at a semantic level, too) of the different countries.

Two big cities in developing countries could also act as a means of comparison. A possible study of the environment, a preliminary discussion about methodology, a careful choice of the sample group, and a cross-cultural comparative approach would all help guarantee the validity of the investigation, despite the necessary limited statistical basis.

POST-INDUSTRIAL SOCIETIES AND TOURISM

As Bell (1973) has pointed out, post-industrial societies are characterized by a dominant service sector (the tertiary sector, but also the advanced tertiary, quaternary, and quinary sectors) and by the production of knowledge (Touraine 1969). Technology, information, the speed of change, and a projection toward the future are all commonly accepted as features of post-industrial societies.

As defined by Kahn and Wiener (1968), Bell (1973), and Touraine (1969, 1977), several characteristics of post-industrial societies – but only those that differ from mass-consumer societies and are particularly relevant to tourism – may be identified. One, there is a continual increase in the amount of free time, but also the inclusion of free time in the main economic sector. Although social time has been reorganized (Rezsohazy 1986; Zoll 1988), there is more freedom of choice, different attitudes to work and free time, including tourism (quantitative factors of budget-time and qualitative factors of evaluation). Two, decentralization of production, and gradual growth of tertiary and quaternary sectors, are present. This includes the emptying of inner-city areas and their growth of significance for group rituals in free time and for culture tourism. Three, mobility has become the key to the working of the system (Touraine 1969) and the presence of conditions that make spatial mobility more intensive, frenetic, and non-stop (Knebel 1960). Four, characteristics of post-industrial societies include ecological threats and awareness, the rediscovery of "nature," and the increasing importance given to places and forms of tourism outside the traditional tourist circuit. Five, growing stress is placed on the quality of new forms of social needs, such as friendship and community life (Heller 1975, 1978). The final characteristic is the distribution of information through telematic and television links in today's global village (McLuhan 1965), including decentralization prospects.

Against this background, which is also characterized by a quick pace of change (Toffler 1970), tourism is developing into a series of "tourisms." These follow along more personalized lines and adapt to the new awareness of nature and self, as well as the search for new forms of subjectivity and old and new forms of culture (CENSIS 1983). It is perhaps within this context that Opaschowski (1977) draws up a list of eight needs that are typically post-industrial: recreation, compensation, education (knowledge), contemplation, communication, integration, participation, and enculturation.

A look at the European situation

At first glance, Europe seems to offer a uniform picture. According to a survey (Commission des Communautés Européennes 1986), 56% of all adult Europeans went on at least one holiday lasting four or more days in 1985. However, this percentage conceals differences within the European community, which vary from the 65% for the Dutch, 61% for the British, 60% for the Germans, and 57% for the Italians, to the 44% for the Spaniards and 41% for the Belgians. Since the rate of departures is supposed to be linked to the GNP, some countries appear to have a higher rate than the economic indicator alone would suggest, while the case of Belgium, with a high level of income, stands out as an exception with one of the lowest rates in Europe. Those who stayed at home mostly claimed they did so for lack of economic means.

The Reiseanalyse produces some very significant data about the former West Germany, which has one of the highest rates in Europe. It has already been mentioned that this survey is particularly interesting for the regularity with which it is carried out each year, still using the same quantitative and qualitative method since it was introduced in 1970 on a wide sample group of about 6,000 people above the age of 14. It only considers holidays lasting more than five days. In 1987, 64.6% of the population in the Bundesrepublik went on holiday, in 1988, 64.9%, and in 1989, 66.8%, thus showing a continuous increase in the percentage of travelers (Lohmann and Besel 1990).

There are several conclusions, both general and specific, to be drawn from the data. At a general level, repeated movements of people from all over the country mean a direct acquisition of information through personal experience. This is clear even without breaking down the data for domestic and international tourism, the latter being particularly high in the case of Germany. If figures are added for those who actually go on a number of holidays throughout the year, and those who go to health resorts, plus those on business trips, which are deliberately not included in the survey, as well as short holidays or weekend breaks, then the amount of information acquired will inevitably grow quite noticeably. Tourist experience is linked to the frequency of trips and proof of this can be seen in the enormous increase in trips over the last 15 years. In the 1989 survey, 57.9% Austrians, 45.3% Italians, and 38.3% Spaniards (as opposed to 36.6%, 22.1%, and 13.3% in 1974) visited a foreign country.

The Reiseanalyse also offers some interesting indications about the sources of information used, which can be of particular interest for the purposes of this paper. The 1989 questionnaire, offering the chance of a multiple answer, showed an average figure of 1.58 sources of information per tourist. The results put personal conversations at the head of the list, with conversations with relatives and friends accounting for 33.3%, and information from travel agencies and tourist bodies 17%. Personal experience acquired through owning a holiday flat or villa in a holiday spot represented 38.7%, while the total amount given by catalogues and brochures added up to 37.6%. The so-called "neutral" information (tourist guides, holiday reports) covered 26.2% and advertising in newspapers and magazines, on radio and television, and at fairs only 5.4%.

Even if one considers the last three years, the "word-of-mouth" publicity still proves to be well established, with a very high rate of 33.3% in 1989 compared with 37.7% in 1987.

There appears to be a tendency toward a fall in the number of information sources used, from the 1.78% per tourist in 1987 to 1.58% in 1989. Italians also show this tendency – 24.4% want to know everything about the holiday beforehand, 33.9% only about the important aspects of it, 21.9% prefer to embark on an adventure, while 15.2% trust their habits, and only 0.5% resort to advertising (CENSIS 1988). This can be interpreted perhaps as a persisting attitude that has been reinforced by a new cognitive stimulus.

Anticipation and motivation

Schade and Hahn (1969) have already described the psychic journey of a holiday, which is, in fact, much longer than the holiday itself, and begins with planning and ends with recollection and re-elaboration. Following along similar lines Hahn and Hartmann (1973) later decided to swim against the current and, probably basing their arguments on Thomae (1965a, 1965b), minimized the absolute importance of decision-making. They started by examining processes involving all the components of personality and considered this psychic journey more as a kind of progressive drifting from one phase to another. They marked out the various phases from the first stimulus to the reinforcement of the actual decision and to the preparation, and paid particular attention to the phase of time anticipation and the value of satisfaction that it can give.

Dann concentrated on fantasy and pointed out 10 aspects of it, linking it to push factors (1976, 1977), while Rubenstein (1980) investigated fantasy by pointing out five types of motivation. One step further on from fantasy is reverie or daydreaming, an aspect which Hahn and Hartmann (1973) discussed, especially in connection with women's tourism. This type of dream can be considered as a dream of compensation, as Bloch defined it (Blöcker 1979), or as a kind of break in everyday life, rather like Csikszentmihalyi's microflow (1975).

The question of time anticipation has also been touched upon by Dann (1981) on the basis of Schütz's phenomenology (1971), which makes a distinction between "because-of" motivations and "in-order-to" motivations that are directed towards the future. Tourist motivation can, therefore, be seen as strictly linked with anticipation insofar as it is the moving from one area of meaning to another (i.e. from work time to free time). Even though Schütz (1971) considered work time as the starting point, there must be some doubts nowadays about the absolute parameters used in this analysis. There is still the belief among some people that the values of the Protestant Ethic are very widespread even today (Mayo and Jarvis 1981) and that they are still handed down from one generation to another in the middle class, at least insofar as work and free time are concerned (Furnham 1984). However, some scholars of tourism, such as Krippendorf (1987), maintain that a new conception of free time and tourism is emerging, which builds up the unit of the individual without the escapism of holidays.

To proceed with this discussion, it is important to see how the motivation process is intertwined with the phase of anticipation in post-industrial societies. These societies are literally saturated with tourist culture, which originates from increasingly widespread and intensive channels of information, so that motivation and anticipation are now packed with meanings that would have been impossible to note 15 years ago.

In this type of society, not only the individual level of income and of education, but also people's cultural attitudes (which depend on how long the country has enjoyed a certain level of prosperity and on the degree of participation in tourism among different social classes), are important indicators for tourism and contribute to a kind of tourist maturity. It is almost as if a social representation is formed, in the sense Moscovici (1984) gives the term, a kind of environment in relation to the individual and the group. The symbolic value of this representation, which is relatively new, derives from a specific and unique context.

If motivation is considered at an individual psychological level, then it can be seen how cognitive elements determined by past experience intervene. The notion of mobility through travel is becoming increasingly important, especially as the tendency to have more than one annual holiday noticeably increases the tourist's background knowledge. These past experiences can emerge as an attitude toward a particular disposition to act (Mayo and Jarvis 1981). The process of feedback from the last phase of the journey after the return home, when affective and emotional factors emerge alongside cognitive factors, is very important. It is not necessarily a passive process, but it can set off the so-called "vicarious exploration" (Fridgen 1984), which can find new stimuli in the environment. Even in the psychological context developed by Sessa (1987), both the affective motivational and cognitive factors are present.

Since the hypothesis of an isolated subject has been dismissed, much of motivational research now seriously considers the part played by women or the whole family in decision-making (Jenkins 1978; Smith 1979; van Raaij and Francken 1984). As early as 1973, Hahn and Hartmann pointed out that decisions about a holiday are not simply taken at an individual level, and stressed the importance of the family or the group of holiday companions who would undertake the trip together.

As far as the question of social motivation is concerned (rather than simply refer to generally accepted concepts, such as the traditional concept of role with the basic membership or reference groups), it is more interesting to see how the motivational process of tourism actually manifests itself in the ordinary world. The distinctive characteristics of this world, which is more limited in its outlook (Thurn 1980), are social interaction and language. It is also regulated by a conventional social time (Berger and Luckmann 1966) which determines seasons for tourism promotional campaigns. These satisfy many deep-felt needs in a consumer-oriented market where tourism is sold as a commodity directly or indirectly in, for example, big department stores, with weeks dedicated to the sale of products from distant countries. The much sought after tourist destinations can also be advertised through competitions or holiday prizes.

While tourism advertising and tourism image have been studied, mostly in reference to Maslow's (1954) universal hierarchy of needs, much less attention has been paid to the indirect effect of advertising those products representing the tourist's "props" (Schober 1981) which induce a general "push" motivation. These products include sun creams, sunglasses, bathing suits, anticellulitis creams, and cameras, and more. They make up the essential tourism "accessories," even though the potential tourists may never go away.

Specialized organizations, such as travel agencies, are now in a position to activate and stimulate motivation not only with the use of more refined photography, but also with the increasingly frequent use of videos and films. Here the potential tourists have the chance to "sail" through the space quickly, to convert space into time, and to also concentrate space in time, thus accelerating traditional perceptions or even turning them upside down completely (Bloch 1959). Since people have developed a kind of defensive discriminatory behavior toward the traditional mass media, the new media, thanks to their adaptability, can be much more effectively used whenever required (Winterhoff-Spurk and Vitouch 1988) to anticipate or to even repeat the tourist experience, as, for example, the tourist videocassettes which are now available at many bookshops, among others.

The great importance of "word-of-mouth" publicity has already been pointed out. It represents much more than just a very high percentage of tourist advertising (Smith 1979). It must, in fact, be attributed to the real social culture of tourism. Dundler (1985) estimated that in Germany there are as many as 14 million opinion leaders, they are young, of above average education, and willing to share their touristic experiences with relatives, friends, and acquaintances. Once again, an abstract concept of information is not sufficient to explain this because these opinion leaders make impartial comments, which are made and received as such. The friendly atmosphere of trust in which these ideas are passed on, often backed up by pictorial evidence (postcards, photographs, and films), is a vital part of this phenomenon.

It is a well-established fact in industrial and post-industrial countries that the consumption of tourism follows its own peculiar pattern, which is quite different from those for other consumer goods. At a certain level of discretionary income, there are non-economic variables that intervene. In Germany, only 33% admit that they base their choice of destination on economic reasons (Hahn and Hartmann 1973) and in Italy the figure is even lower – only 11.5% (CENSIS 1988). Tourist expenditure comes third in the list of consumption items in both Germany (Scheuch 1981) and Italy (CENSIS 1983).

Furthermore, unlike other consumer goods, tourism finds an audience receptive to information and advertising through the year (Datzer 1981). This can be interpreted as latent interest and intrinsic motivation (Wärneryd 1980), and can also be explained by the atmosphere of tourist culture. Therefore, tourism is not simply sold like any other commodity, but it permeates imperceptibly throughout all aspects of everyday life, where the distinction between work time and free time tends to become less and less marked within the new organization of social time. It involves a process of enculturation that appears in entertainment, recreation, food, clothing, and language. It often presupposes a type of culture in which tourism is connected with learning foreign languages or a local craft.

In addition, the salvaging of the inner-city areas not only encourages that type of urban domestic tourism which is continuously growing, but also gives an anticipation of international tourism as a kind of a group ritual during which tourists think about (and/or recollect) and also prepare themselves for future trips. As a result of the continual demands being made on cognitive factors (intertwined in any case with emotional factors), it becomes extremely difficult within this framework to identify the exact moment when the motivational process is triggered off.

CONCLUSIONS

The significance of the "cognitive revolution" in tourism has already been discussed. From the socio-psychological point of view, it has been observed how tourist experience is accumulated both during a person's lifetime and through the number and frequency of holidays in the course of the year. From a sociological and anthropological point of view, it has been shown how, in post-industrial societies, a social culture of tourism has developed which continually transmits stimuli and can be referred to as a "social representation." This paper's discussion of a comparative motivational study concerning different post-industrial countries with their culturally specific factors would fit into this framework.

If considered from a semiotic point of view, tourism is immersed in a universe of signs, which also spill over into working hours, thus creating a continuity between work and leisure. This relationship is typical of post-industrial societies. One can also foresee that the cognitive factors will inevitably have debilitating effects on the tourist image, especially on the "image imaginée" (Lanquar 1990), considering the equation, which can easily be represented as: more information = impossibility of dream.

Another important aspect that has not yet been considered is the development of tourist anticipation technology. New forms of media have already been mentioned. Certainly videocassettes and cameras at the ready offer a number of occasions for anticipation and instigation that still have to be explored. Today they still constitute a novelty for the adult tourist, but they will represent an obvious advantage for the tourist in the making.

The field of virtual reality will also be of great importance. The "virtuality" machine involves a full immersion in computerized programs that will provide contact in real time (i.e. the answers of the operator will interact with the human "proprioceptor system"). A number of people may even share the same experience. The machine, which for all intents and purposes is a journey in itself, offers a wide range of getaway places and tourist settings. This type of anticipatory experience with its plausible interactive relationship could reinforce the *déjà vu* feeling even further and perhaps condition the psychology of perception in a way which one cannot yet foresee.

It is possible, therefore, to imagine a kind of inversion of motivation from dream to reality with a subsequent increase in cultural motivation, which would be a natural result of the strong need for "inculturation" felt in post-industrial societies.

However, the simplest of all types of motivation – i.e. recreation – would inevitably continue to exist and perhaps become even more important.

But the knowledge that there is "nowhere else" may have new and unforeseen consequences. It could lead to a kind of "anticognitivehood," which stops an individual from acquiring sources of information in order to keep at least some idea of novelty about his or her holiday, as can be perceived in the German and Italian cases mentioned in this paper. It is as if the tourist of the future will want in some way to safeguard an element of surprise in his or her own personal itinerary, even if the destination no longer holds any secrets for him or her.

Another line of development comes from the awareness of the environment. European "soft" tourism leads to international tourism by starting off not just with the usual kind of domestic tourism, but also with the educational exploration of one's own region or city. In fact, this tourism can take the tourist, after a very short trip, back to his or her own doorstep. In this way, true tourist motivation and anticipation are transformed deliberately into a negation of tourism itself (Jungk 1990). Another area of development could be in a kind of anticipation that expects a disappointing holiday (large crowds, spoiled environment, etc.), so that the tourist opts for a substitutive holiday or abandons the idea completely.

Any decision-changing necessarily involves an anticipatory process, which has apparently not yet been taken into consideration and studied. This hypothesis, therefore, implies that even the simplest decision-making would prove to be a much more complex process, made up of a number of anticipations.

Since motivational research is necessarily becoming more and more complicated, one could well witness a greater division between true scientific motivational research and research carried out simply for market analysis. Furthermore, the question will probably arise whether it is still justified to take tourist motivation *tout court* as an object of study for more fruitful results. Perhaps it can and should be studied as tourist anticipation within a narrower framework but along wider interdisciplinary lines of inquiry.

REFERENCES

Altman, I., A. Rapoport, and J. F. Wohlwill, eds. 1980 *Human Behavior and Environment: Advances in Theory and Research.* (Vol. IV): *Environment and Culture.* New York: Plenum Press.

Atkinson, J. W. 1983 *Personality, Motivation and Action. Selected Papers.* New York: Praeger.

Barker, R. G. 1968 *Ecological Psychology: Concepts and Methods for Studying the Environment of Human Behavior.* Stanford: Stanford University Press.

Bell, D. 1973 *The Coming of the Post-Industrial Society: A Venture in Social Forecasting.* New York: Basic Books.

Berger, P. L., and T. Luckmann 1966 *The Social Construction of Reality.* Garden City, NY: Doubleday.

Berlyne, D. E. 1960 *Conflict, Arousal and Curiosity.* New York: McGraw-Hill.

—— 1978 Struktur und Motivation. In *Die Psychologie des 20. Jahrhunderts,* Vol. VII: *Piaget und die Folgen,* G. Steiner, ed. Zürich: Kindler.

Bloch, E. 1959 *Das Prinzip Hoffnung.* Frankfurt am Main: Suhrkamp.

Blöcker, M. 1979 Der Tagtraum. In *Die Psychologie des 20. Jahrhunderts*, Vol. XV: *Transzendenz, Imagination und Kreativität*, G. Condran, ed. Zürich: Kindler.

CENSIS 1983 Come cambia il turismo. Quindicinale di note e commenti 29: 3–48.

—— 1988 La domanda turistica degli italiani. Tipologie dei comportamenti di vacanza. CENSIS Turismo/2.

Chon, Kye-Sung 1990 The Role of Destination Image in Tourism. A Review and Discussion. *The Tourist Review* 45: 2–9.

Clawson, M., and J. L. Knetsch 1966 *Economics of Outdoor Recreation*. Baltimore, MD: Johns Hopkins.

Cohen, E. 1979 Rethinking the Sociology of Tourism. *Annals of Tourism Research* 6: 18–35.

—— 1984 The Sociology of Tourism: Approaches, Issues, and Findings. *Annual Review of Sociology* 10: 373–92.

—— 1988 Traditions in the Qualitative Sociology of Tourism. *Annals of Tourism Research* 15: 29–36.

Commission des Communautés Européennes 1986 *Les Européens et les Vacances*. Bruxelles: Commission des Communautés Européenes.

von Cranach, M., and L. Valach 1984 The Social Dimension of Goal-Directed Action. In *The Social Dimension: European Developments in Social Psychology* (Vol. 1), T. Tajfel, ed., pp. 285–299. Cambridge: Cambridge University Press.

Crompton, J. 1979 Motivations for Pleasure Vacations. *Annals of Tourism Research* 6: 408–424.

Csikszentmihalyi, M. 1975 *Beyond Boredom and Anxiety. The Experience of Play in Work and Games*. San Francisco: Jossey-Bass.

Dann, G. M. S. 1976 The Holiday was Simply Fantastic. *The Tourist Review* 31: 19–23.

—— 1977 Anomie, Ego-Enhancement and Tourism. *Annals of Tourism Research* 4: 184–194.

—— 1981 Tourist Motivation: An Appraisal. *Annals of Tourism Research* 8: 187–219.

—— 1983 Comment on Iso-Ahola's "Toward a Social Psychological Theory of Tourism Motivation." *Annals of Tourism Research* 10: 273–276.

—— 1989 The Tourist as Child. *Cahiers du Tourisme*. Aix-en-Provence: CHET.

Dann, G., D. Nash, and P. Pearce 1988 Methodology in Tourism Research. *Annals of Tourism Research* 15: 1–28.

Datzer, R. 1981 Ein Ueberblick über Ansätze der psychologischen und sozialpsychologischen Tourismusforschung. In *Reisemotive-Länderimages-Urlaubsverhalten*. Starnberg: Studienkreis für Tourismus.

De Charms, R., and M. S. Muir 1978 Motivation: Social Approaches. *Annual Review of Psychology* 29: 91–113.

Deci, E. L. 1975 *Intrinsic Motivation*. New York: Plenum Press.

Dundler, F. 1985 *Die Meinungsmacher reisen selber gern. Wie 14 Millionen Opinion Leaders die übrigen Touristen beeinflussen*. Starnberg: Studienkreis für Tourismus.

Fridgen, J. D. 1984 Environmental Psychology and Tourism. *Annals of Tourism Research* 11: 19–39.

Furnham, A. 1984 The Protestant Work Ethic: A Review of the Psychological Literature. *European Journal of Social Psychology* 14: 87–104.

Gottlieb, A. 1982 American's Vacation. *Annals of Tourism Research* 9: 165–187.

Graumann, C. F. 1981 *Motivation*. Wiesbaden: Akad. Verlagsges.

Grinstein, A. 1955 Vacations: A Psycho-analytic Study. *The International Journal of Psychoanalysis* 36: 177–186.

Guzman, L. F. J. 1986 Teoria turistica. Un enfoque integral del hecho social. Bogotá: Universidad Externado de Colombia.

Hahn, H., and K. D. Hartmann 1973 *Reiseinformation, Reiseentscheidung, Reisevorbereitung*. Starnberg: Studienkreis für Tourismus.

Hartmann, K. D. 1979 Psychologie des Reisens. In *Reisen und Tourismus*, N. Hinske and M. J. Müller, eds., pp. 15–21. Trier Beiträge: Universität Trier.

——— 1981 Einige Gedanken zum "Theoriedefizit" der Tourismusforschung. In *Reisemotive-Länderimages-Urlaubsverhalten*. Starnberg: Studienkreis für Tourismus.

Haukeland, J. V. 1990 Non-Travelers: The Flip Side of Motivation. *Annals of Tourism Research* 17: 172–184.

Heckhausen, H. 1967 *The Anatomy of Achievements Motivation*. New York: Academic Press.

——— 1980 *Motivation und Handeln. Lehrbuch der Motivationspsychologie*. New York: Springer.

Heller, A. 1975 *Sociologia della vita quotidiana*. Roma: Editori Riuniti.

——— 1978 *La teoria, la prassi e i bisogni*. Roma: Savelli.

Iso-Ahola, S. E. 1980 Toward a Dialectical Social Psychology of Leisure and Recreation. In *Social Psychological Perspectives on Leisure and Recreation*, S. E. Iso-Ahola, ed., pp. 19–37. Springfield, IL: Charles C. Thomas.

——— 1982 Toward a Social Psychological Theory of Tourism Motivation: A Rejoinder. *Annals of Tourism Research* 9: 256–262.

Jaspars, J. M. F. 1978 Nature and Measure of Attitudes. In *Introducing Social Psychology*, H. Tajfel and C. Frazer, eds. Harmondsworth: Penguin.

Jaspars, J., and C. Fraser 1984 Attitudes and Social Representations. In *Social Representations*, R. M. Farr and S. Moscovici, eds., pp. 101–123. Cambridge: Cambridge University Press.

Jenkins, R. L. 1978 Family Vacations Decision-Making. *Journal of Travel Research* 16: 2–7.

Jungk, R. 1990 Wenn einer keine Reise tut. . . . *Natur* 3 (Dossier):54.

Kahn, H., and A. J. Wiener 1968 *The Year 2000*. New York: Macmillan.

Kaspar, C. 1972 *Die Fremdenverkehrslehre im Grundriß*. Bern: Haupt.

Kelly, G. A. 1955 *The Psychology of Personal Constructs* (2 Vols). New York: Norton.

Knebel, H. J. 1960 *Soziologische Strukturwandlungen im modernen Tourismus*. Stuttgart: Enke.

Krippendorf, J. 1987 *The Holiday Makers: Understanding the Impact of Leisure and Travel*. London: Heinemann.

Lanquar, R. 1990 *Sociologie du tourisme et des voyages*. Paris: Presses Universitaires de France.

Lohmann, M., and K. Besel 1990 Urlaubsreisen 1989. *Kurzfassung der Reiseanalyse 1989*. Starnberg: Studienkreis für Tourismus.

Madsen, K. B. 1974 *Modern Theories of Motivation*. Copenhagen: Munksgaard.

Mannel, R. C., and S. E. Iso-Ahola 1987 Psychological Nature of Leisure and Tourism Experience. *Annals of Tourism Research* 14: 314–331.

Maslow, A. H. 1954 *Motivation and Personality*. New York: Harper and Brothers.

Mayo, E. J., and L. P. Jarvis 1981 *The Psychology of Leisure Travel: Effective Marketing and Selling of Travel Services*. Boston: CBI.

McIntosh, R. W. 1977 *Tourism: Principles, Practices, Philosophies*. Columbus, OH: Grid.

McLuhan, H. M. 1965 *Understanding Media: The Extensions of Man*. New York: McGraw-Hill.

Mercer, D. 1971 The Role of Perception in the Recreational Experience: A Review and Discussion. *Journal of Leisure Research* 3: 261–276.

Moscovici, S. 1984 The Phenomenon of Social Representations. In *Social Representations*, R. M. Farr and S. Moscovici, eds. Cambridge: Cambridge University Press.

Nash, D. 1981 Tourism as an Anthropological Subject. *Current Anthropology* 22: 461–481.

Nuttin, J. 1980 *Théorie de la motivation humaine. Du besoin au projet d'action*. Paris: Presses Universitaires de France.

Opaschowski, H. W. 1977 Urlaub: Der Alltag reist mit. *Psychologie heute* 6: 18–21.

Pearce, P. L. 1982 *The Social Psychology of Tourist Behavior*. Oxford: Pergamon.

Pizam, A., Y. Neumann, and A. Reichel 1978 Dimensions of Tourist Satisfaction with a Destination Area. *Annals of Tourism Research* 5: 314–322.

——— 1979 Tourist Satisfaction. Uses and Misuses. *Annals of Tourism Research* 6: 195–197.

Potter, R. B., and J. Coshall 1988 Sociopsychological Methods for Tourism Research. *Annals of Tourism Research* 15: 63–75.

Prahl, H. W., and A. Steinecke 1979 *Der Millionen-Urlaub. Von der Bildungsreise zur totalen Freizeit*. Darmstadt: Luchterhand.

Rapoport, A. 1982 *The Meaning of the Built Environment*. Beverly Hills, CA: Sage.

Rezsohazy, R. 1986 Recent Social Developments and Changes in Attitudes to Time. *International Social Science Journal* 38: 33–47.

Reykowski, J. 1982 Social Motivation. *Annual Review of Psychology* 33: 123–154.

Rubenstein, C. 1980 Vacations, Expectations, Satisfactions, Frustrations, Fantasies. *Psychology Today* (May): 62–76.

Schade, B., and H. Hahn 1969 Psychologie und Fremdenverkehr. In *Wissenschaftliche Aspekte des Fremdenverkehrs. Raum und Fremdenverkehr 1*, pp. 35–53. Hannover: Gebr. Jänecke Verlag.

Scheuch, E. K. 1981 Tourismus. In *Die Psychologie des 20. Jahrhunderts*, Vol. XIII, F. Stoll, ed., pp. 1089–1114. Zürich: Kindler.

Schmalt, H. D. 1986 *Motivationspsychologie*. Stuttgart: Kohlhammer.

Schober, R. 1981 Empfehlungen zur Tourismusforschung. In *Reisemotive-Länderimages-Urlaubsverhalten. Neue Ergebnisse der psychologischen Tourismusforschung*. Starnberg: Studienkreis für Tourismus.

Schütz, A. 1971 *Gesammelte Aufsätze*, Vol. I: *Das Problem der sozialen Wirklichkeit*, A. Gurvitsch, ed. Den Haag: Martinus Nijhoff.

Sessa, A. 1987 *Elementi di Sociologia e Psicologia del Turismo*. Roma: CLITT.

Smith, V. 1979 Women: The Taste-Makers in Tourism. *Annals of Tourism Research* 6: 49–60.

Stokols, D. 1977 Origins and Directions of Environment. Behavioral Research. In *Perspectives on Environment and Behavior: Theory, Research and Application*, D. Stokols, ed., pp. 5–36. New York: Plenum Press.

—— 1978 Environmental Psychology. *Annual Review of Psychology* 29: 253–295.

—— 1987 Conceptual Strategies of Environmental Psychology. In *Handbook of Environmental Psychology*, Vol. 1, D. Stokols and I. Altman, eds. New York: Wiley.

Stringer, P., and P. L. Pearce 1984 Towards a Symbiosis of Social Psychology and Tourism Studies. *Annals of Tourism Research* 11: 5–17.

Studienkreis für Tourismus 1969 *Motive-Meinungen-Verhaltensweisen*. Starnberg: Studienkreis für Tourismus.

—— 1981 *Reisemotive-Länderimages-Urlaubsverhalten. Neue Ergebnisse der psychologischen Tourismusforschung*. Studienkreis für Tourismus.

Thomae, H. 1965a Die Bedeutungen des Motivationsbegriffes. Zur Charakteristik des Motivationsgeschehens. In *Allgemeine Psychologie*, Vol. II: *Motivation*, H. Thomae, ed., pp. 3–122. Göttingen: Verlag für Psychologie.

—— 1965b. Einführung. In *Die Motivation menschlichen Handelns*, H. Thomae, ed. Köln-Berlin: Kiwi.

—— 1988 Motivation. In *Handwörterbuch der Psychologie*, R. Asanger and G. Wenninger, eds., pp. 463–467. München-Weinheim: Psychologie Verlags Union.

Thurn, H. P. 1980 *Der Mensch im Alltag. Grundrisse einer Anthropologie des Alltagslebens*. Stuttgart: Enke.

Toffler, A. 1970 *Future Shock*. London: Pan Books.

Touraine, A. 1969 *La société post-industrielle*. Paris: Denöel.

—— 1977 *The Self-Production of Society*. Chicago: University of Chicago Press.

van Raaij, W. F., and D. A. Francken 1984 Vacation Decisions, Activities, and Satisfactions. *Annals of Tourism Research* 11: 101–112.

Wärneryd, K. E. 1980 The Limits of Public Consumer Information. *Zeitschrift für Verbraucherpolitik* 4: 127–141.

Weiner, B. 1985 *Human Motivation*. New York/Berlin: Springer Verlag.

Wicker, A. 1979 *An Introduction to Ecological Psychology*. Monterey: Brooks.

Winterhoff-Spurk, P., and P. Vitouch 1988 Mediale Kommunikation. In *Empirische Medienpsychologie*, J. Groebel and P. Winterhoff-Spurk, eds. München-Weinheim: Psychologie Verlags Union.

Zoll, R. 1988 *Zerstörung und Wiederaneignung von Zeit*. Frankfurt am Main: Suhrkamp.

5

A PHENOMENOLOGY OF TOURIST EXPERIENCES*

Erik Cohen[1]

INTRODUCTION

What is the nature of the tourist experience? Is it a trivial, superficial, frivolous pursuit of vicarious, contrived experiences, a 'pseudo-event' as Boorstin (1964: 77–117) would have it, or is it an earnest quest for the authentic, the pilgrimage of modern man, as MacCannell (1973: 593) believes it to be?

Tourists are often seen as 'travellers for pleasure';[2] however, though sufficient for some purposes, this is a very superficial view of the tourist. The more precise quality and meaning of the touristic experience have seldom been given serious consideration, either in theoretical analysis or in empirical research. Not that we lack controversy – indeed, recently, the nature and meaning of tourism in modern society became the subject of a lively polemic among sociologists and social critics. In one camp of the polemic we find those, like Boorstin (1964) and lately Turner and Ash (1975), for whom tourism is essentially an aberration, a symptom of the *malaise* of the age. Boorstin bemoans the disappearance of the traveller of old, who was in search of authentic experiences, and despises the shallow modern mass tourist, savouring 'pseudo-events'. The opposing, newer camp is represented primarily by MacCannell; he criticizes the critics, claiming that '. . . Boorstin only expresses a long-standing touristic attitude, a pronounced dislike . . . for other tourists, an attitude that turns man against man in a they-are-the-tourists-I-am-not equation' (MacCannell, 1973: 602). He argues that Boorstin's approach, '. . . is so prevalent, in fact (among the tourists themselves as well as among travel writers), that it is a part of the problem of mass tourism, not an analytical reflection on it' (MacCannell, 1973: 600). As in every polemic, however, the protagonists of the opposing views tend to overstate their case. Thus MacCannell, claiming to confute Boorstin's view with empirical evidence, states that 'None of the accounts in my collection (of observations of tourists) support Boorstin's contention that tourists want superficial, contrived experiences. Rather, tourists demand authenticity, just as Boorstin does' (*ibid*: 600). But, MacCannell himself is very selective in the choice of his observations: his accounts are mostly of young, 'post-modern' (Kavolis, 1970) tourists; Boorstin's thesis may well find more support in a different sample, composed primarily of sedate,

* Reproduced with permission of B.S.A. Publications Ltd, from *Sociology*, 1979, 13: 179–201.

middle-class, middle-aged tourists. Hence, even if one admits that Boorstin's claims may be too extreme and that some tourists may indeed be in search of 'authenticity', it nevertheless appears too far-fetched to accept MacCannell's argument that all tourists single-mindedly pursue 'real', authentic experiences, but are denied them by the machinations of a tourist establishment which presents them with staged tourist settings and 'false backs'. The conflict between these contrasting conceptions of tourists remains thus unresolved, as the proponents of each claim to describe '*the* tourist' as a general type, while implicitly or explicitly denying the adequacy of the alternative conception.

In my view, neither of the opposing conceptions is universally valid, though each has contributed valuable insights into the motives, behaviour and experiences of *some* tourists. Different kinds of people may desire different modes of touristic experiences; hence '*the* tourist' does not exist as a type. The important point, however, is not merely to prove that both conceptions enjoy some empirical support, though neither is absolutely correct; rather it is to account for the differences within a more general theoretical framework, through which they will be related to, and in turn illuminated by, some broader views of the relationship of modern people to their society and culture. In this paper I shall attempt to do so by examining the place and significance of tourism in a modern man's life; I shall argue that these are derived from his total world-view, and depend especially on the question of whether or not he adheres to a 'centre', and on the location of this 'centre' in relation to the society in which he lives. Phenomenologically distinct modes of touristic experiences are related to different types of relationships which obtain between a person and a variety of 'centres'.

TOURISM AND THE QUEST FOR THE CENTRE

The concept of the 'centre' entered sociological discourse in several overlapping, but not identical fashions. M. Eliade (1971: 12–17) pointed out that every religious 'cosmos' possesses a centre; this is '. . . pre-eminently the zone of the sacred, the zone of absolute reality' (*ibid*: 17). In traditional cosmological images, it is the point where the *axis mundi* penetrates the earthly sphere, '. . . the meeting point of heaven, earth and hell' (*ibid*: 12).

However, the centre is not necessarily geographically central to the life-space of the community of believers; indeed, as Victor Turner has pointed out, its ex-centric location may be meaningful in that it gives direction and structure to the pilgrimage as a sacred journey of spiritual ascension to 'The Center Out There' (Turner, 1973). The 'centre', however, should not be conceived in narrowly religious terms. E. Shils (1975) has argued that every society possesses a 'centre', which is the charismatic nexus of its supreme, ultimate moral values. While Shils does not deal explicitly with the location of the symbolic bearers of the charismatic 'centre', there is little doubt that he considers the locus of its paramount symbols, e.g. the monarch or the crown (Shils and Young, 1953), to be ordinarily within the geographical confines of the society. Shils' concept of the centre was further developed by S. N. Eisenstadt (1968) who distinguishes between multiple 'centres', e.g. political, religious or cultural; in modern society these centres do not necessarily overlap, and their paramount symbols may be differentially located. The individual's 'spiritual' centre, whether religious or cultural, i.e. the centre which for the individual symbolizes ultimate meanings, is the one with which we are concerned in this paper.

Structural–functionalist theory, particularly in the Parsonian variety, assumes as a matter of course that the spiritual centre of the modern man will be normally located within the confines of his society – he will 'conform' with this society's ultimate values. Such conformity may indeed generate tensions and dissatisfactions. These, however, will be taken care of by the mechanisms of 'pattern maintenance' and 'tension management'. The latter will include various types of leisure and recreational activity in which the individual finds release and relief. Such activities take place in segregated settings, which are not part of 'real' life; in Schutz's phenomenological terminology, they may be called 'finite provinces of meaning' (Berger and Luckman, 1966: 39). Though consisting of activities representing a reversal of those demanded by the central value-nexus (e.g. 'play' as against 'work'), they are 'functional' in relieving the tension built up in the individual and hence reinforce, in the long run, his allegiance to the 'centre'.[3] The individual may need relief from tension, created by the values, but he is not fundamentally alienated from them. Tourism, in the Parsonian scheme, is a recreational activity *par excellence*: it is a form of temporary getaway from one's centre, but in relation to the individual's biography, his life-plan and aspirations, it remains of peripheral significance. Indeed, in terms of a functional theory of leisure, tourism only remains functional so long as it does not become central to the individual's life-plan and aspirations – since only so long will it regulate his tensions and dissatisfactions, refreshing and restoring him, without destroying his motivation to perform the tasks of his everyday life. This means that tourism is essentially a temporary reversal of everyday activities – it is a no-work, no-care, no-thrift situation; but it is in itself devoid of deeper meaning: it is a 'vacation', i.e. 'vacant' time. If tourism became central, the individual would become 'deviant', he would be seen as 'retreating', opting-out, or escaping the duties imposed upon him by his society.

The assumption that modern man is normally a conformist and that he will hence generally adhere to the centre of 'his' society is, to say the least, simplistic. Many moderns are alienated from their society. What about the 'spiritual' centre of such alienated people? Several alternatives can be discerned: (a) some may be so completely alienated as not to look for any centre at all, i.e. not to seek any ultimate locus of meaning; (b) some, aware of what to them looks an irretrievable loss of their centre, seek to experience vicariously the authentic participation in the centre of others, who are as yet less modern and less, in E. Heller's (1961) term, 'disinherited'; (c) some, particularly those whom Kavolis (1970) described as 'post-modern', often possess a 'decentralized personality', and equivocate between different centres, almost turning the quest into the purpose of their life; (d) finally, some may find that their spiritual centre lies somewhere else, in another society or culture than their own. I argue that within the context of each of these possible types of attitude to the centre, tourism will be endowed with a different significance. In the following I shall develop a phenomenology of modes of touristic experiences and relate them to these alternative forms of relationship between a modern person and various 'centres'.

THE MODES OF TOURIST EXPERIENCES

Travelling for pleasure (as opposed to necessity) beyond the boundaries of one's life-space assumes that there is some experience available 'out there', which cannot be found within the life-space,[4] and which makes travel worthwhile. A man who finds relief from tensions within his life-space, or does not perceive outside its boundaries any attractions the desire for which he cannot also fulfil at home, will not travel for pleasure.

Risking some over-simplification, I argue that primitive society usually entertained an image of a limited 'cosmos', ideally co-terminous with its life-space, surrounded by a dangerous and threatening chaos. Insofar as the sacred centre was geographically located within the life-space, primitive man had no reason or desire to venture beyond its boundaries. It is only when a powerful mythological imagery locates the 'real' centre in another place, beyond the limits of the empirical world, a 'paradise' beyond the surrounding chaos, that 'paradisiac cults' terminating in large scale voyages develop (Eliade, 1969: 88–111). This is the original, archaic pilgrimage, the quest for the mythical land of pristine existence, of no evil or suffering, the primaeval centre from which man originally emerged, but eventually lost.[5] The pilgrimage later on becomes the dominant form of non-instrumental travelling in traditional and particularly peasant societies (Turner, 1973). However, the traditional pilgrimage differs from the archaic in that the pilgrim's goal, the centre, is located within his 'world', but beyond the boundaries of the immediate life-space; this contingency is predicated upon a separation between the limited life-space and his 'world': the image of the latter is vastly expanded and embraces a large number of life-spaces of individual communities or societies. Thus, Jerusalem becomes the centre of the Jewish and Christian 'world', Mecca that of the Muslim 'world'. Traditional pilgrimage is essentially a movement from the prophane periphery towards the sacred centre of the religious 'cosmos'.

Modern mass tourism, however, is predicated upon a different development: the gradual abandonment of the traditional, sacred image of the cosmos, and the awakening of interest in the culture, social life and natural environment of others. In its extreme form, modern tourism involves a *generalized* interest in or appreciation of that which is different, strange or novel in comparison with what the traveller is acquainted with in his cultural world (Cohen, 1974: 533, 1972: 165). Hence, it leads to a movement *away* from the spiritual, cultural or even religious centre of one's 'world', into its periphery, towards the centres of other cultures and societies.

Pilgrimages and modern tourism are thus predicated on different social conceptions of space and contrary views concerning the kind of destinations worth visiting and of their location in the socially constructed space; hence they involve movement in opposite directions: in pilgrimage from the periphery towards the cultural centre, in modern tourism, away from the cultural centre into the periphery.

These differences notwithstanding, the *roles* of pilgrim and tourist are often combined, particularly in the modern world (Dupont, 1973, Cohen, 1974: 542). The fusion of the role does not, however, mean a fusion of the divergent cognitive structures. MacCannell, who views the tourist as a modern pilgrim (1973: 593), does not expressly discuss the problem of the cognitive structure of the tourist's 'world', in contrast to that of the pilgrim.

Here I shall develop a phenomenological typology of tourist experiences by analysing the different meanings which interest in and appreciation of the culture, social life and natural environment of others has for the individual traveller. The degree to which his journey represents a 'quest for the centre', and the nature of that centre, will be at the heart of this analysis. The typology, in turn, relates to different points of continuum of privately constructed 'worlds' of individual travellers (not necessarily identical with those prevalent in their culture), ranging between the opposite poles of the conception of space characteristic of modern tourism on the one hand and that of the pilgrimage on the other. I have distinguished five main modes of touristic experiences:

1 The Recreational Mode
2 The Diversionary Mode
3 The Experiential Mode
4 The Experimental Mode
5 The Existential Mode.

These modes are ranked here so that they span the spectrum between the experience of the tourist as the traveller in pursuit of 'mere' pleasure in the strange and the novel, to that of the modern pilgrim in quest of meaning at somebody else's centre. Let us now discuss each in some detail.

The recreational mode

This is the mode of touristic experiences which a structural–functionalist analysis of society would lead us to expect as typical for modern man. The trip as a recreational experience is a form of entertainment akin in nature to other forms of entertainment such as the cinema, theatre, or television. The tourist 'enjoys' his trip, because it restores his physical and mental powers and endows him with a general sense of well-being. As the term 'recreation' indicates, even this mode of tourist experience is ultimately and distantly related to and derived from the religious voyage to the sacred, life-endowing centre, which rejuvenates and 're-creates'.[6] Indeed, one can follow the process of 'secularization' of tourism historically, e.g. in the change from 'thermalists', whose belief in the healing properties of thermal springs was ultimately grounded in mythological images of springs as 'centres' from which supernatural powers penetrate the empirical world, to tourists, who 'take the waters' primarily as a form of high-class socializing (Lowenthal, 1962). Though the belief in the recuperative or restorative power of the tourist trip is preserved, it is a secular, rational belief in the value of leisure activities, change of climate, rest, etc.

While the traditional pilgrim is newly born or 're-created' at the centre, the tourist is merely 'recreated'. In the recreational tourist trip, the intent and meaning of the religious voyage is secularized: it loses its deeper, spiritual content. Though the tourist may find his experiences on the trip 'interesting', they are not personally significant. He does not have a deep commitment to travel as a means of self-realization or self-expansion. Like other forms of mass-entertainment, recreational tourism appears from the perspective of 'high' culture as a shallow, superficial, trivial and often frivolous activity, and is ridiculed as such by Boorstin and other cultural critics. A correlate of this view is that the tourist travelling in that mode appears often to be gullible to the extreme (Mitford, 1959), easy to be taken in by blatantly inauthentic or outrightly contrived, commercialized displays of the culture, customs, crafts and even landscapes of the host society. His apparent gullibility, however, ought not to be ascribed solely to his ignorance; rather, he does not really desire or care for the authentic (Huetz de Lemps, 1964: 23); he is 'no stickler for authenticity' (Desai, 1974: 4). Since he seeks recreation, he is quite eager to accept the make-believe and not to question its authenticity; after all one does not need to be convinced of the authenticity of a TV play or a motion picture in order to enjoy it as a recreative, entertaining or relaxing experience.

The recreation-seeking tourist, hence, thrives on what Boorstin (1964) calls 'pseudo-events'. But the depth of contempt in which he is held on that account by intellectuals and 'serious' travellers is misplaced: the tourist gets what he really wants – the pleasure of entertainment, for which authenticity is largely irrelevant. Such recreation-oriented tourists should be looked upon not as shallow, easily gullible simpletons who believe any contraption to be 'real', or as stooges of a prevaricating tourist establishment, but rather as persons who attend a performance or participate in a game; the enjoyability of the occasion is contingent on their willingness to accept the make-believe or half-seriously to delude themselves. In a sense, they are accomplices of the tourist establishment in the production of their own deception.[7] Recreation-oriented tourists like the audience of a play can completely legitimately enjoy themselves despite, or even – as in the case of some of the more outlandish performances of local custom – because of, the fact that the experience is not 'real'; the real thing may be too terrifying or revolting to be enjoyable. For the recreation-seeking tourist, the people and landscapes he sees and experiences are not part of his 'real' world; like other recreational settings, they are 'finite provinces of meaning' separate from reality, though this is not explicitly admitted by either the tourists or the staff of tourist establishments. Indeed, tourists as well as staff may be mutually aware of the fact that each is playing a role in order to upkeep an inauthentic, indeed artificial, but nevertheless enjoyable, 'construction of (touristic) reality'. If this is openly admitted, the tourist situation would be homologous to that of mass-entertainment. The distinguishing trait of the tourist situation, however, is that such an admission would spoil the game.

Tourism as recreation is, in itself, not a 'serious business'; rather it is an 'idle pleasure' (Lowenthal, 1962: 124), and as such had a hard time in gaining recognition as a legitimate reason for travelling. It achieved such legitimation, indeed, not because it is enjoyable in itself, but rather on the strength of its recuperative powers, as a mechanism which recharges the batteries of weary modern man (Glasser, 1975: 19–20), refreshes and restitutes him so he is able again to return to the wear and tear of 'serious' living. Such tourism serves as a 'pressure-valve' for modern man. When he cannot take the pressures of daily living any more, he goes on a vacation. If he overdoes it, or fails to return to serious living, his behaviour becomes 'dysfunctional', in its extreme anomic escapism. But ordinarily it is 'functional' because it manages the tensions generated by modern society and hence helps to preserve the adherence of the individual to it – in a similar way in which the Carnival (e.g. Baroja, 1965: 23–4) and other forms of legitimate debauchery, normatively circumscribed in time and place, served as a 'pressure-valve' of traditional Christian society. In the functionalist view, recreational tourism is chiefly caused by the 'push' of the tourist's own society, not by the particular 'pull' of any place beyond its boundaries. The recreational tourist is primarily 'getting away'. Hence, he is often equanimous as to the choice of possible destinations for his 'holiday', thus providing the advertisement industry with plentiful opportunities to tilt his decision in a variety of competing directions.

Though not serious business in itself, recreation, then, performs a serious 'function' – it restitutes the individual to his society and its values, which, despite the pressures they generate, constitute the centre of his world. Insofar as he is aware of this function and values it, it becomes in an oblique sense the meaning of his trip. If it were not for the pressures generated in his daily life at home, or if the pressures were resolved by alternative mechanisms, as e.g. they are in traditional societies, he might find no need to travel; he would stay at home. Here we have one of the main reasons for the tremendous upsurge of tourism in modern, and particularly in urban, society (Dumazdier, 1967: 125–6): this society generates pressure which it has few means to resolve; peasants, even in modern societies, travel little.

The diversionary mode

Recreational tourism is a movement away from the centre, which serves eventually to reinforce the adherence to the centre. Hence, it may possess a meaning for the person oriented to that centre.

As we pointed out above, however, modern men are often alienated from the centre of their society or culture. Some of them may not be seeking alternative centres: their life, strictly speaking, is 'meaningless', but they are not looking for meaning, whether in their own society or elsewhere. For such people, travelling in the mode just described loses it recreational significance: it becomes purely diversionary – a mere escape from the boredom and meaningless of routine, everyday existence, into the forgetfulness of a vacation, which may heal the body and soothe the spirit, but does not 'recreate' – i.e. it does not re-establish adherence to a meaningful centre, but only makes alienation endurable. Diversionary tourism is then, in terms of what Glasser calls the 'Therapy School' of sociology of leisure, '. . . a healing balm for the robots . . . It accepts that for most people work will always be emotionally uncommitting and therefore unrewarding, and that they are condemned to seek in their leisure temporary oblivion and comfort for abraded nerve endings . . . the Therapy School . . . [puts] emphasis on immediate diversion . . .' (Glasser, 1975: 21).

The diversionary mode of tourist experience, hence, is similar to the recreational, except that it is not 'meaningful', even in an oblique sense. It is the meaningless pleasure of a centre-less person.

The recreational and diversionary modes of touristic experience have been the target of the savage criticism of tourism by culture critics such as Boorstin (1964) and Turner and Ash (1975). They are apparently characteristic of most mass tourists from modern, industrial urban societies. On this point I tend to agree with Boorstin, rather than with MacCannell. Even then, however, an interesting question remains unresolved: which one of these two modes is the prevalent one? One cannot approach this question without first taking a stand on that most basic problem: how deeply is modern man alienated? Even the critics of tourism may not be unanimous on this question. Hence, even the criticisms may differ: if modern man is conceived of as adhering to a central nexus of 'Western values', his prevailing mode of travel is recreational; he may then be criticized for his narrow 'parochialism', his lack of readiness to relate to the values of others except in a superficial, casual manner. If modern man is conceived of as alienated, then his prevailing mode of travel is diversionary; tourism is then criticized primarily as a sympton of the general *malaise* of modern society.

The two modes of tourism discussed above, however, do not exhaust the field; some tourists, primarily the minority of 'post-modern', and other, non-institutionalized types of tourists (Cohen, 1972), indeed derive a deeper meaning from their travels, of the kind MacCannell finds characteristic of tourists in general. The remaining three modes of touristic experience represent different levels of depth of meaning which tourism may possess for the individual.

Experiential mode

The recreational tourist adheres to the centre of his society or culture; the diversionary tourist moves in a centre-less space. But what happens when the disenchanted or alienated individuals become growingly aware of their state of alienation, and the meaninglessness and fatuity of their daily life, as many younger members of the middle classes in the 'post-modern' society have become?

One direction which their search for meaning might take is the attempt to transform their society through revolution; another, less radical alternative is to look for meaning in the life of others – tourism (MacCannell, 1976: 3).

The renewed quest for meaning, outside the confines of one's own society, is commenced, in whatever embryonic, unarticulated form, by the search for 'experiences':[8] the striving of people who have lost their own centre and are unable to lead an authentic life at home to recapture meaning by a vicarious, essentially aesthetic, experience of the authenticity of the life of others (MacCannell, 1973). This mode of tourism we shall call 'experiential'.

The 'experiential' mode characterizes the tourist as he emerges from MacCannell's description. If Boorstin is among the most outspoken critics of recreational and *a forteriori* diversionary tourism, which in his view encompass all modern tourism, MacCannell attempts to endow tourism with a new dignity by claiming that it is a modern form of the essentially religious quest for authenticity. But though he puts forward his view of the tourist against that of the 'intellectuals' (MacCannell, 1973: 598–601), implying that it holds for 'the tourist' in general, it is clear that his claim is based on a view of modern man who, alienated from the spiritual centre of his own society, actively, though perhaps inarticulately, searches for a new meaning. Indeed, MacCannell argues that 'The concern of moderns for the shallowness of their lives and inauthenticity of their [everyday] experiences parallels concern for the sacred in primitive society' (MacCannell, 1973: 589–90). Unlike in situations where such shallowness engenders a desire for an internal spiritual revolution, the modern tourist turns elsewhere for authenticity: 'The more the individual sinks into everyday life, the more he is reminded of reality and authenticity elsewhere' (MacCannell, 1976: 160). MacCannell claims that 'Pretension and tackiness generate the belief that somewhere, only not right here, not right now, perhaps just over there someplace, in another country, in another life-style, in another social class, perhaps, there is *genuine* society' (MacCannell, 1976: 159). Therefore 'Authentic experiences are believed to be available only to those moderns who try to break the bonds of their everyday existence and begin to "live"' (MacCannell, 1976: 159). The search for authentic experiences is essentially a religious quest: therefore it follows that '. . . tourism absorbs some of the social functions of religion in the modern world' (MacCannell, 1973: 589). However, since 'Touristic consciousness is motivated by the desire for authentic experience . . .' (*ibid*: 597), rather than trivial ones, the chief problem facing the tourist becomes '. . . to tell for sure if the experience is authentic or not' (*ibid*: 597). As against Boorstin and others who maintain that the tourist is content with contrived experiences, or is a mere superficial stooge, MacCannell endeavours to prove that the tourist is in fact a serious victim of a sophisticated deception: the tourist establishment 'stages authenticity', so that tourists are misled to believe that they succeeded in breaking through the contrived 'front' of the inauthentic, and have penetrated into the authentic 'back' regions of the host society, while in fact they were only presented with 'false backs', staged by the tourist establishment, or, in Carter's (1971) term, 'fenced in'. The problem is not the cultural shallowness of the tourists but the sophisticated machinations of the tourist establishment. However, though critical of the tourist establishment as the progenitor of a 'false (touristic) consciousness' (MacCannell, 1973: 589), MacCannell is nevertheless convinced of the 'functional' importance of tourism. Indeed, in an admittedly Durkheimian mode, he claims that tourism '. . . is a form of ritual respect for society' (MacCannell, 1973: 589) and hence, apparently, reinforces social solidarity. But he probably means 'Society' in general (and not necessarily the one of which the tourist is a member), since it was precisely the inauthenticity of life in his own society, coupled with the '. . . reminder (through the availability of souvenirs) of reality and authenticity elsewhere' (MacCannell, 1976: 160) and the '. . . availability of authentic experiences at other times and in other places' (*ibid*: 148), which motivated the tourist for his quest in the first place. MacCannell likens tourism to the religious pilgrimage: 'The motive behind a pilgrimage is similar to that behind a tour: both are quests for authentic experiences' (MacCannell, 1973: 593). But, the similarity he points out notwithstanding, there are some important, and to my mind crucial, differences: first, the pilgrim always undertakes his journey to the spiritual centre of *his* religion, though that centre may be located far beyond the boundaries of his life-space or society. It is true that the tourist, too, may travel to the artistic, national, religious and other centres of his own society or culture and pay them 'ritual respect'. But one of the distinguishing characteristics of modern tourism is precisely the generalized interest in the environment, and the desire for experiences far beyond the limits of the traveller's own cultural realm; indeed, it is often the sheer strangeness and novelty of other landscapes, life-ways and cultures which chiefly attract the tourist (Cohen, 1972).

Secondly, in contrast to the pilgrim, the experience-oriented tourist, even if he observes the authentic life of others, remains aware of their 'otherness', which persists even after his visit; he is not 'converted' to their life, nor does he accept their authentic life-ways. The pilgrim senses spiritual kinship with even a geographically remote centre; the 'experiential' tourist remains a stranger even when living among the people whose 'authentic' life he observes, and learns to appreciate, aesthetically. The pilgrim's experience is 'existential': he participates in, partakes of and is united with his co-religionists in the *communitas* created by the sacredness of the centre (Turner, 1973). He is fully involved in and committed to the beliefs and values symbolized by the centre. MacCannell's tourist, however, experiences only vicariously the authenticity of the life of others, but does not appropriate it for himself. Hence, though his quest may be essentially religious, the actual experience is primarily aesthetic, owing to its vicarious nature. The aesthesis provoked by direct contact with the authenticity of others may reassure and uplift the tourist, but does not provide a new meaning and guidance to his life. This can best be seen where 'experiential' tourists observe pilgrims at a pilgrimage centre: the pilgrims experience the sacredness of the centre; the tourists may experience aesthetically the authenticity of the pilgrims' experience. The 'experiential' mode of tourism, though more profound than the 'recreational' or 'diversionary', does not generate 'real' religious experiences.

MacCannell provides the clues for an analysis of the search for new meaning through tourism. But his work falls short of accomplishing that task; an extension of his approach leads to the distinction of still more profound modes of touristic experiences, and to the eventual closure of the gap separating the mode of experience of the modern mass tourist from that of the traditional pilgrim.

Experimental mode

This mode of the touristic experience is characteristic of people who do not adhere any more to the spiritual centre of their own society, but engage in a quest for an alternative in many different directions. It is congenial to the more thoughtful among the disoriented post-modern travellers, particularly the more serious of the 'drifters' (Cohen, 1973), who, endowed with a 'decentralized personality' (Kavolis, 1970: 438–9) and lacking clearly defined priorities and ultimate commitments, are pre-disposed to try out alternative life-ways in their quest for meaning. Travel is not the only possible form of their quest; mysticism, drugs, etc., may serve as alternative paths to the same goal; indeed, Eliade considers that the internal and external quests for the centre are homologous (Eliade, 1971: 18). But for those who do travel in quest of an alternative spiritual centre, travel takes up a new and heightened significance. While the traveller in the 'experiential' mode derives enjoyment and reassurance from the fact that *others* live authentically, while he remains 'disinherited' (Heller, 1961) and content merely to observe the authentic life of others, the traveller in the 'experimental' mode engages in that authentic life, but refuses fully to commit himself to it; rather, he samples and compares the different alternatives, hoping eventually to discover one which will suit his particular needs and desires. In a sense, the 'experimental' tourist is in 'search of himself', insofar as in a trial and error process, he seeks to discover that form of life which elicits a resonance in himself; he is often not really aware of what he seeks, of his 'real' needs and desires. His is an essentially religious quest, but diffuse and without a clearly set goal.

Examples of such seekers who experiment with alternative life-ways abound among the younger, post-modern set of travellers: urban American, European or Australian youngsters who taste life in farming communities, the Israeli kibbutzim, the Indian aśramas, remote Pacific villages and hippie communes, engage in the experimental mode of tourism. An enlightening example is a short story, apparently written by a foreign student, in an Israeli student paper, entitled 'In search of in search of . . .' (Coven, 1971), which commences: 'I was in search of religion. I was in the depths, the bitter waters. No future, no meaning, loneliness, and boredom. I wanted religion, any religion' (*ibid.*: 22); after describing several attempts to find religion in different Christian and Jewish settings in Israel, the story ends inconclusively; the search goes on . . .

Indeed, in extreme cases the search itself may become a way of life, and the traveller an eternal seeker. Such may be the case with those 'drifters' who get accustomed to move steadily between different peoples and cultures, who through constant wandering completely lose the faculty of making choices, and are unable to commit themselves permanently to anything. If the 'seeker' attitude becomes habitual, it excludes the very possibility of that essentially religious 'leap of faith', which commitment to a new 'spiritual' centre consists of; the habitual seeker cannot be 'converted'.

Existential mode

If the preceding mode of touristic experience characterizes the 'seeker', the 'existential' mode in its extreme form is characteristic of the traveller who is fully committed to an 'elective' spiritual centre, i.e. one external to the mainstream of his native society and culture. The acceptance of such a centre comes phenomenologically closest to a religious conversion, to 'switching worlds', in Berger and Luckmann's (1966: 144) terminology, though the content of the symbols and values so accepted need not be 'religious' in the narrow sense of the term. The person who encounters in his visit to an Israeli kibbutz a full realization of his quest for human communion; the seeker who achieved enlightenment in an Indian aśrama; the traveller who finds in the life of a remote Pacific atoll the fulfilment of his cravings for simplicity and closeness to nature; all these are examples of 'existential' touristic experiences.

For the person attached to an 'elective' external centre, life away from it is, as it were, living in 'exile'; the only meaningful 'real' life is at the centre.[9] The experience of life at the centre during his visits sustains the traveller in his daily life in 'exile', in the same sense in which the pilgrim derives new spiritual strength, is 're-created', by his pilgrimage.

Those most deeply committed to a new 'spiritual' centre may attach themselves permanently to it and start a new life there by 'submitting'[10] themselves completely to the culture or society based on an orientation to that centre: they will desire to 'go native' and to become, respectively, Hindu recluses, Israeli kibbutz members, Pacific islanders, etc.

However, what makes 'existential' experiences a touristic phenomenon is the fact that there are many people – and their number is increasing in a growingly mobile world – who, for a variety of practical reasons, will not be able or willing to move permanently to their 'elective' centre, but will live in two worlds: the world of their everyday life, where they follow their practical pursuits, but which for them is devoid of deeper meaning; and the world of their 'elective' centre, to which they will depart on periodical pilgrimages to derive spiritual sustenance. Thus e.g. there are some non-Jewish tourists who every year return to live for a few months on a kibbutz, while spending the rest of the year in their home country.

The visit to his centre of the tourist travelling in the existential mode is phenomenologically analogous to a pilgrimage. Indeed, Turner (1973: 193–4) refers to the community of pilgrims as an 'existential *communitas*'. In terms of the relationship of their existential quest to the culture of their society of origin, traditional pilgrimage and 'existential' tourism represent two extreme configurations: the traditional religious pilgrimage is a sacred journey to a centre which, though geographically 'ex-centric', is still the centre of the pilgrim's religion; it is the charismatic centre from which the pilgrim's life derives meaning, the spiritual centre of his society. Hence, though living away from the centre, the pilgrim is not living in 'exile'. His world and daily abode is hallowed, or given meaning through the centre. The centre, however, is *given*; it is not elective, not a matter of choice.

The centre of the 'existential' tourist, however, is not the centre of his culture of origin; it is an 'elective' centre, one which he chose and 'converted' to. Hence, it is not only ex-centric to his daily abode, but beyond the boundaries of the world of his daily existence; it does not hallow his world; hence, he lives in 'exile'. His pilgrimage is not one from the mere periphery of a religious world towards its centre; it is a journey from chaos into another cosmos, from meaninglessness to authentic existence.

Between these two extremes, the pilgrimage to a traditionally given centre and to an 'elective' one, different intermediate types can be discerned. There exist other than purely *religious* traditional centres of pilgrimage – such as cultural, aesthetic (artistic or natural) or national ones. Visits to the great artistic centres of the past, the heritage of one's own culture, such as were included e.g. in the Grand Tour (Lambert (ed.), 1935, Trease, 1967), or any visit by people of 'Western' culture to the sites of classical antiquity, may take on the quality of *cultural* pilgrimages. Visits to the shrines of the civil religion (Bellah, 1967), such as the Capitol or the Lincoln Monument by U.S. citizens, or those of the official state religion, e.g. Lenin's Tomb by Soviet citizens (MacCannell, 1976: 85), are forms of *political* pilgrimage. A person's culture may include, in addition to the religious, any number of primary and secondary cultural, aesthetic and national centres, visits to which may be conducted in the existential mode of pilgrimages. Indeed, in the complexities of the modern world, the 'world' of any given culture and society is not clearly bounded; the cultural inheritance of one society is often appropriated by, and made part of, other cultures. Many Westerners consider the centres of the ancient Greek or Hebrew cultures as part of 'their' tradition. Hence, what is today an 'elective' centre of a few individuals, outside the confines of their culture of origin, may tomorrow be appropriated by that culture; centres are 'traditional' or 'elective' only relatively to a given point in history.

We spoke of the 'existential' tourist as one who adheres to an 'elective' centre. Such a centre may be completely extraneous to his culture of origin, the history of his society or his biography. But it may also be a traditional centre to which he, his forebears or his 'people' had been attached in the past, but become alienated from. In this case, the desire for a visit to such a centre derives from a desire to find one's spiritual roots. The visit takes on the quality of a home-coming to a historical home. Such travellers, so to speak, re-elect their traditional centre. This conception is perhaps most clearly articulated in the ideology of Zionism. The full realization of the Zionist ideal is 'aliyah', literally 'ascension', the essentially religious term used to describe the act of permanent migration of a Zionist Jew to Israel.

Many Zionists, however, though Israel is their centre, do not take the ultimate step of 'aliyah'. Their commitment to the 'centre' is expressed in a variety of less radical forms of behaviour, one of which is repeated sojourns in Israel, differing in content, frequency and length: periods of study and volunteer work on kibbutz settlements, yearly visits as private persons or in groups organized by different Zionist organizations, or eventual retirement to Israel, etc.[11] All of these are, in various degrees, forms of 'tourism' (Cohen, 1974). Particularly, those who return yearly for relatively short visits, for no other reason but to live for a while in Israel, exemplify the 'existential' mode of tourism, in the form of a renewed relationship to a historical centre.

It is interesting to note that recently, the motivation for 'existential' tourism to Israel has apparently widened to include not only Zionists in the narrow sense, but also Diaspora Jews who desire to taste 'genuine' Jewish communal life; the borderline between these and Jews who come for religious reasons, i.e. pilgrims in the narrower traditional religious sense, has thus become blurred. Even people who are not pilgrims in any sense may be overcome by an 'existential' experience at the centre. This comes through powerfully from a recent review of S. Bellow's book *To Jerusalem and Back*: 'The most saline of American writers finds himself unable to escape the tenebrous undertow of Jewish mysticism. "My inclination is to resist imagination when it operates in this way" he writes. "Yet I, too, feel that the light of Jerusalem has purifying powers and filters the blood and the thought. I don't forbid myself the reflection that light might be the outer garment of God" ' (*Time*, 1976: 62).

A craving for an existential experience at one's historical sources probably motivates many old-time immigrants – and their progeny – who travel from their country of abode to visit the 'old country', from which they or their parents once departed: e.g. the American Italians or Irish visiting Italy or Ireland, the Corsicans in France visiting Corsica, the American Chinese visiting pre-Communist mainland China, etc. Perhaps the most interesting recent example of the sudden awakening of such cravings among a long-exiled people is the renewed interest of American blacks in Africa as the land of their fathers (*Spiegel*, 1973). Though I have to add, a point to be discussed more fully below, that the mere desire for such an experience is not a guarantee for its fulfilment as many American blacks who visited Africa, and for that matter Jews who visited Israel, learned to their sorrow.

The various modes of tourist experience were here presented in an ascending order from the most 'superficial' one, motivated by the desire for mere 'pleasure', to that most 'profound', motivated by the quest for meaning. The modes were separated for analytic purposes; any individual tourist may experience several modes on a single trip; a change from one mode to another may also occur in the 'touristic biography' of any individual traveller. The mix of modes characteristic of different types of trips and the changes in the desired modes of experiences during a person's 'touristic biography' are empirical problems for further investigation.

One particular conceptual problem, however, remains to be clarified: the problem of 'multiple centres'. We have throughout proceeded on the tacit assumption that the individual adheres to only one principal 'spiritual' centre. If he is alienated from the centre of his society or culture, he may look for it elsewhere.

This, however, is an over-simplification, which needs two qualifications: first, some people, we may call them 'humanists', entertain extremely broad conceptions of 'their' culture and are willing to subsume under it everything or almost everything human, on the principle of Goethe's famous statement *Nichts Menschlichesist mir fern* ('Nothing human is alien to me'). For such people, there is no single principal 'spiritual' centre: every culture is a form in which the human spirit is manifested. They may thus travel in the experiential, or even existential, mode, without being alienated from their culture of origin; for them, the culture they happen to have been reared in is just one of the many equally valid cultures. The narrower the scope of cultures given equal status, the closer the 'humanist' comes to the 'cultural' tourist. The more important an external centre becomes relative to his culture of origin, the more closely he approximates the 'existential' tourist.

Secondly, there are people, we may call them 'dualists' or more broadly 'pluralists', who adhere simultaneously to two or more heterogeneous 'spiritual' centres, each giving rise to equally authentic, though different, forms of life. Such persons may feel equally at home in two or more 'worlds', and even enjoy 'existential' experiences from their sojourn at another centre or centres, without being alienated from their own. American Zionists e.g. may not necessarily feel in 'exile' in the United States, but may adhere simultaneously to the 'American Dream' and to Israel as the Zionist centre, and be equally committed to both.

'Humanists' and 'dualists' or 'pluralists' qualify the underlying hypothesis of this paper, that a person seeks and ultimately adheres to 'spiritual' centres of others only after he realizes the discomfort of his alienation from the centre of his own culture and society. They indicate the necessity for a more thorough phenomenological investigation of the variety of complex world-views which developed in the modern world, for the analysis of which Eliade's or Shils' basic models do not suffice any more.

CONCLUSIONS

The typology of modes of tourist experience presented above reconciles the opposing views of 'the tourist' in the current polemic on tourism and thereby prepares the way for a more systematic comparative study of touristic phenomena. Our discussion shows that, depending on the mode of the touristic experience, tourism spans the range of motivations between the desire for mere pleasure characteristic of the sphere of 'leisure' and the quest for meaning and authenticity, characteristic of the sphere of 'religion'; it can hence be approached from both the perspective of the 'sociology of leisure' as well as that of the 'sociology of religion'. But neither of these approaches will exhaust the whole phenomenon, owing to the differences in the modes of experiences desired by different tourists. The context within which the typology has been developed was borrowed from the sociology of religion: my point of departure was a tourist's fundamental world-view, and specifically, his adherence to or quest for a 'spiritual' centre. I assumed that different world-views are conducive to different modes of the touristic experience. In fact I tackled the same problem which MacCannell addressed himself to, but, instead of assuming that all tourists are 'pilgrims', I attempted to answer the question, under what conditions and in what sense tourism becomes a form of pilgrimage? It now remains to work out some of the implications of the typology developed in response to this question.

By claiming that tourists pursue different modes of experience, we did not imply that these are invariably realized in their trip. Two problems can be discerned here: first, from the viewpoint of the tourist, what are the chances of *realization* of the different modes of touristic experience? Second, from the point of view of the external observer, what are the possibilities of *falsification* of such experiences by the tourist establishment? Again, I raise questions which MacCannell has been concerned with, but my answers are somewhat different.

While MacCannell takes a lofty view of the desires of the tourists, and a pessimistic view of their realizability, I claim that the various modes of touristic experiences differ in the ease of their realization; generally speaking, the more 'profound' the mode of experience, the harder it becomes to realize it. The 'diversionary' mode is the easiest to realize: as with any kind of entertainment, it suffices if the travel experience has been pleasurable. The realization of the 'recreational' mode demands, in addition, that the experience perform a restorative function for the individual. Since the traveller in these two modes has no pretensions to authenticity, his experience cannot be falsified. He can achieve his aim even when he is fully aware that his experience was staged in a 'tourist space'. As in other forms of entertainment, there is no need fully to camouflage the staging. The art of the tourist 'producer' is to create in the tourist a semi-conscious illusion, and to engage his imagination until he is turned into a willing accomplice, rather than a stooge, of the game of touristic make-believe. The tourist and the touristic entrepreneur may agree that they deal in contrivances; indeed, the fact that these are contrivances often ensures their enjoyability. Insofar as much of what tourists around the world come in touch with in their sightseeing tours, e.g. on visits to 'native villages', or at performances of 'folkloristic dances and ceremonies', becomes explicitly defined as entertainment, rather than authentic culture, no falsification of the experience of the unpretentious 'diversionary' or 'recreational' tourist is involved.

The situation is completely different for tourists travelling in the other modes of touristic experience; for them, the authenticity of the experience is crucial for its meaning. This is true not only for the 'experiential' tourist, who is reassured by the authentic life of others, and for whom authenticity is obviously a *sine qua non* for the realization of his experience. It is equally true for the 'experimental' and 'experiential' tourist: one can hardly experiment with alternative ways of life if these are merely contrived for one's convenience, nor can one derive existential meaning from a 'spiritual centre' outside one's society or culture, if such a centre is only a chimera, advertised to lure tourists in quest of existential experiences. No wonder that MacCannell, who discusses mainly what we termed 'experiential' tourism, emphasizes that the tourist constantly faces the danger of a 'false (touristic) consciousness', by becoming the victim of the machinations of the tourist establishment, which presents him with a '. . . false back [which] is more insidious and dangerous than a false front; [hence] an inauthentic demystification of social life [of the hosts] is not merely a lie but a superlie, the kind that drips with sincerity' (MacCannell, 1973: 599). In MacCannell's view, the prevalent fate of tourists is to become entrapped in 'tourist space', never able to realize their craving for authenticity: '. . . there is no way out for them so long as they press their search for authenticity' (MacCannell, 1973: 601). This claim attains with MacCannell almost the status of a 'touristic condition' reflecting a generally absurd human condition captured in works of existentialist philosophers. If for Sartre, there is 'No Exit' from the human existence and no way to penetrate the subjectivity of others, for MacCannell there is no way for the tourist to penetrate the others' authenticity. Taken to its extreme, the quest of MacCannell's tourist, like that of Camus's or Sartre's heroes, is absurd.

I do not subscribe to this view and believe that at least some modern tourists, particularly the explorer and the original drifter (Cohen, 1972, 1973), are capable of penetrating beyond the staged 'tourist space' and its false backs and observe other people's life 'as it really is'. But this demands an effort and application, and a degree of sophistication, which most tourists do not possess. There is hence a high chance

105

that any of those tourists who desire authenticity will be misled by the tourist establishment, and their experience will be falsified; as long as they do not grasp the falsification, they may labour under the illusion that they have realized their aim; if and when they penetrate the deception, they will be both enlightened and disenchanted; their resentment will give rise to demands for 'honesty in tourism'.

The mechanisms which support the constitution of the touristic illusion and the processes of its denouement have yet to be studied in detail. Such a study would, in MacCannell's neo-Marxist terminology, represent the examination of the processes through which 'false (touristic) consciousness' is created and those through which '(touristic) class consciousness' emerges. MacCannell has done some pioneering work in this field, but much more systematic study is needed.

The tourist travelling in the experimental mode also faces the problem of authenticity. The danger of delusion will be less serious in his case, since his desire to experiment with other forms of life, and not just experience them, leads him off the beaten track and sharpens his critical faculties. Being inquisitive and uncommitted, he is tuned to discover deception. His major problem, however, is to achieve commitment to any of the life-ways with which he experiments. What originally appears as experimentation, with a view to an ultimate commitment to one of the alternatives, may turn into a predicament. An 'experimental' tourist with a decentralized personality may easily become an 'eternal seeker'. If false consciousness is the danger faced by the 'experiential' tourist, total disorientation, and ultimate alienation from all human society, is the threat to the 'experimental' tourist. The fate of some modern drifters strongly supports this argument.

The tourist travelling in the existential mode faces the most serious problem of realization. Commitment to and authenticity of the experience of the 'elective' centre are not enough; the ultimate problem is that of 'commensurability': is the 'true' life at the centre indeed commensurable to his high hopes and expectations? Does it enable the traveller to live authentically, to achieve self-realization? This is a problem which existential tourists share with pilgrims. The centre, of course, symbolizes an ideal. Ideals are not fully realizable, but can only be approached 'asymptotically'.[12] The geographical centre symbolizes the ideal one; between the two, however, there is necessarily a discrepancy: Jerusalem may be the Holy City, but ordinary human life in Jerusalem is far from holy. The pilgrim or the existential tourist 'ascends' spiritually to the ideal centre, but he necessarily arrives at the geographical one. How does he handle the discrepancy? For example: a person adhering to the ideal of voluntary collectivism may go to live on a kibbutz, as an 'elective' centre embodying his ideals; soon, however, he will realize that life on the kibbutz is far from ideal. He will thus encounter a discrepancy between the ideal conception and actual life, which, if not dealt with satisfactorily, may provoke a personal crisis of meaninglessness, futility and disenchantment.

I distinguish three kinds of 'existential' tourists in terms of the manner in which they deal with the perceived discrepancy:

(a) 'Realistic idealists', who are willing to concede that even the most ideal place, society or culture has shortcomings, and are thus able to achieve self-realization at the centre without deluding themselves as to its faultlessness. I suggest that these are often people who became committed to their 'elective' centre after a prolonged quest and experimentation, and are thus bereft of illusions.

(b) 'Starry-eyed idealists', those 'true believers' (Hoffer, 1952) who will see perfection in whatever they find at the centre and refuse to face the reality of life in it, inclusive of its shortcomings. From the point of view of the external observer, their self-realization will be based on self-delusion. I suggest that these are often people whose commitment to an 'elective' centre was a result of a sudden conversion, of a precipitous 'switching of worlds' in the certainty of discovery of a panacea.

(c) Finally, there are the 'critical idealists' who oscillate between a craving for the centre from afar, and a disenchantment when they visit it. They are attached to the ideal which the centre is meant to represent, but reject the reality they found at it. For these, the centre has meaning when they are remote, but tends to lose it when they approach it. Their attitude has been forcefully expressed by the Jewish writer Elie Wiesel, at a Conference on Jewish Intellectuals in New York in 1971: 'I am at home in Jerusalem when I am not there'.[13] I suggest that the 'critical idealists' tend to be people who adhered to the centre for a long time from afar, and for whom the trip was a realization of a long-cherished dream. They may preserve their dream, while denying the adequacy of its earthly embodiment, and advocating a reform of the actual centre to bring it closer to the ideal.

The problem of discrepancies, however, can be 'resolved' in another way – at the expense of the authenticity of the tourist's experience, i.e. by straightforward falsification. As demand for existential experiences increases, the tourist establishment and other bodies may set out to supply it. The existential mode of the tourist experience, based as it often is on a prior commitment, is particularly amenable to falsification. The tourist, expecting the ideal life at the centre, is easily taken in; he is helped, as it were, to become a 'starry-eyed idealist'. Like traditional pilgrimage centres, centres of 'existential' tourism are advertised and embellished; tours through 'existential tourist space', like traditional pilgrimages, are staged. New centres may even be straightforwardly invented. The purveyance of existential experiences becomes big business. Tourist-oriented centres of Eastern religion, catering for 'instant enlightenment', may be one example.[14] Another are the massive 'Zionist pilgrimages' staged by the Israeli governmental and national institutions, in which the visitors are brought to a pitch of Zionist ecstasy at the height of a well-planned and organized tour through staged 'Zionist tourist space'. The largest of these pilgrimages, equal in everything to its religious counterpart, was the massive United Jewish Appeal 'This Year in Jerusalem' tour of 1976, which brought several thousand people to the country and large contributions to the U.J.A. The study of staging the 'existential' touristic sites and tours, such as the U.J.A. pilgrimages, is just commencing, but promises rich and interesting data for comparison with traditional religious pilgrimages.[15]

One last word on the relationship between the modes of touristic experiences and the problem of strangeness. It is generally assumed that tourists, when leaving their familiar environment, expose themselves to increasing degrees of strangeness, against which the more routine, less adventurous mass tourists are protected by an 'ecological bubble of their home environment' (Cohen, 1972: 171), so as not to suffer a disorienting culture shock which would spoil the pleasure of their trip. This argument is based on a tacit assumption that the tourist, adhering to the 'spiritual centre' of his own society or culture, prefers its life-ways and thought-patterns, and feels threatened and incommoded when presented with the different, unfamiliar ones of the host country. Strangeness, however, may be not only a threat, but also a lure and challenge (Cohen, in preparation (a)). This seems particularly true for those travellers for whom the above assumption does not hold and who have either lost their 'centre' and travel in the experiential or experimental mode, or adhere to a new 'elective' one outside their society (existential mode). Such travellers may well desire exposure to strangeness and not shun it, but rather seek to 'submit' to it. Unlike the mass tourist, they will not suffer from a culture shock when exposed to the host environment, but may rather experience what Meintel (1973: 52) calls a 'reverse culture shock' upon return home. Talking of the personal experience of (particularly postmodern) anthropologists, Meintel observes: 'Desirable values . . . which were not experienced before and which may have been attained as a stranger in a foreign setting may appear unrealizable in the home situation. Nash attributes the fact that "many anthropologists come alive only when a field trip is in prospect for them" to the attractions of the stranger role (Nash, 1963: 163), but perhaps, desirable personal ends attained to a significant degree elsewhere are actually unattainable in the situations to which these individuals return' (Meintel, 1973: 53). Her observation may well apply to 'existential' tourists as well, provided that they succeeded in realizing the desired experiences. The problem of such travellers is, however, that being the most committed and nurturing the highest expectations, they may indeed experience a 'shock' upon arrival at their 'elective' centre – but not one emanating from the contrast between home and their 'elected' external centre, but rather from the fact that this 'centre' is too much like home and hence does not correspond to their idealized image.

The phenomenological analysis of tourist experiences in this paper has been highly speculative; contrary to other areas in the study of tourism, the in-depth study of tourist experiences is not yet much developed, though an endless number of surveys of tourist 'motivations' has been conducted. I hope that the conceptual framework and the typology here proposed will serve as the theoretical baseline for more profound, empirical studies of tourist experiences.[16]

NOTES

1 The collection of material on which this paper is based was facilitated by a grant of the Basic Research Unit of the Israel National Academy of Sciences and Humanities. Thanks are due to the Academy for its support and to Dr J. Dolgin and J. Michalowicz for their comments on an earlier draft of this paper.

2 Definitions of the concept 'tourist' abound in the literature. 'Travelling for pleasure' is the most commonly evoked dimension of the phenomenon; for additional dimensions necessary for a systematic definition of the tourist as a traveller role, see Cohen (1974). The present paper departs on a different track – it does not deal with the tourist's role, but with the precise nature of his supposedly 'pleasurable' experience.

3 Cf. e.g. Gross (1961: 5): 'In the area of tension management, the cathartic and restorative functions of leisure are pre-eminent, . . .'.

4 If the experience were available within the life-space, there would be no need to take the trouble to travel: cf. Stouffer (1950).

5 'Paradisiac cults' are predicated on the belief that paradise, i.e. the centre, is a place which can be approached by an actual voyage, though that voyage may include miraculous elements (e.g. men flying over the sea, Eliade, 1969: 101–4); if it is believed that the centre is located on a wholly different sphere, it will be approachable by a 'spiritual journey', such as that of the shaman (Rasmussen, 1972), in which a man is miraculously transported to other spheres without actual physical movement through empirical space.

6 This is evidenced by the recurrent use of paradisiac imagery in modern mass tourism (see e.g. Turner and Ash, 1975: 149 ff). But the 'paradise' these tourists seek is of a stereotyped, commercialized kind – it is an idyllic place equipped with all modern amenities. For a discussion of 'paradise' as a 'type of touristic community' see MacCannell (1976: 183). For an example of the process of debasement of the paradisiac image, see Cohen (in preparation (b)).

7 An excellent example, in which the game of make-believe has been brought almost to the level of a fine art is mass tourism in Hawaii. Thus Crampon describes a three-stage game through which the 'royal visitor to the Islands' (i.e. the tourist) becomes a Hawaiian; at the end of this process, the tourist comes to like Hawaii, since the Hawaiian kama'aina likes Hawaii. Crampon claims that 'Probably . . . this visitor is not "acting". He does like Hawaii. He is convinced that Hawaii is a Paradise' (Crampon, n.d.: 54). The game has terminated in successful self-delusion, with the full cooperation of the tourist.

8 For MacCannell's definition of 'experience' in the sense here used, see MacCannell, (1976: 23); for some concrete examples of touristic 'experiences' see *ibid*: 97.

9 This point is admirably illustrated in an anecdote told by Eliade of the famous German historian Th. Mommsen. After a lecture in which Mommsen gave by heart a detailed account of the topography of ancient Athens, a valet had to take him home, since '. . . the famous historian did not know how to go home alone. The greatest living authority on fifth-century Athens was completely lost in his own city of Wilhelminian Berlin' (Eliade, 1976: 19). Eliade continues: 'Mommsen admirably illustrates the existential meaning of "living in one's own world". His real world, the only one which was relevant and meaningful, was the classical Graeco-Roman world. For Mommsen, the world of the Greeks and Romans was not simple *history* . . .; it was *his* world – that place where he could move, think and enjoy the beatitude of being alive and creative . . . Like most creative scholars, he probably lived in two worlds: the universe of forms and values, to the understanding of which he dedicated his life and which corresponds somehow to the "cosmicized" and therefore "sacred" world of the primitives, and the everyday "profane" world into which he was "thrown" as Heidegger would say. Mommsen obviously felt detached from the profane, non-essential, and for him meaningless and ultimately chaotic space of modern Berlin'. (*ibid*: 19). While the historian Mommsen's 'real' world was remote in time, the existential tourist's real world is remote in space; but the cognitive structure of their respective worlds is otherwise identical.

10 On the concept of 'submission', as a voluntary form of transition from strangeness to familiarity, see Cohen (in preparation (a)).

11 I intend to deal in a separate paper with the different forms of temporary migration of

Jews to Israel which recently proliferated, and through which the boundary between Israeli Jews and Jews of the Diaspora became progressively blurred.

12 This idea has been mostly fully developed in the work of the philosopher E. Bloch; most pertinent for our purposes is his discussion of 'geographical utopias' (Bloch, 1959: 873–929). I am grateful to Dr. Paul Mendes-Flohr who introduced me to Bloch's ideas.

13 Reported to me by Paul Mendes-Flohr.

14 An excellent example is the Bhagwan Shree Rajneesh Ashram in Poona, visited primarily by Westerners. Rajneesh, who '. . . speeds up the usually slow Hindu attainment of meditation and bliss with a sort of pop-Hinduism . . .', argues that '"Westerners want things quickly, so we give it to them right away . . ."' (*Bangkok Post*, 1978: 7).

15 I am obliged for the information on the U.J.A. to Dr Janet O'Dea, who currently studies the U.J.A. 'pilgrimages' to Israel.

16 Accepted 10.1.78.

REFERENCES

Bangkok Post (1978). Sex Guru Challenges Desai, *Bangkok Post*, 23.8.1978, 7.

Baroja, E. J. C. (1965). *El Carnaval*, Madrid: Taurus.

Bellah, R. N. (1967). 'Civil Religion in America', *Daedalus*, Winter, 1967, 1–21.

Berger, P. and Luckmann, Th. (1966). *The Social Construction of Reality*, Harmondsworth: Penguin.

Bloch, E. (1959). *Das Prinzip Hoffnung*, Frankfurt am Main: Suhrkamp Verlag, Vol. II.

Boorstin, D. J. (1964). *The Image: A Guide to Pseudo-Events in America*, New York: Harper and Row.

Carter, J. (1971). 'Do Fence Them In!', *Pacific Islands Monthly*, 42(6), 49–53.

Cohen, E. (1972). 'Towards a Sociology of International Tourism', *Social Research*, 39(1), 164–89.

Cohen, E. (1973). 'Nomads from Affluence: Notes on the Phenomenon of Drifter-Tourism', *International Journal of Comparative Sociology*, 14(1–2), 89–103.

Cohen, E. (1974). 'Who Is a Tourist? A Conceptual Clarification', *Sociological Review*, 22(4), 527–55.

Cohen, E. (in preparation (a)). *Strangeness and Familiarity: The Sociology of Temporary Migration*.

Cohen, E. (in preparation (b)). The Pacific Islands from Utopia to Consumer Product: The Transformation of Paradise.

Coven, I. (1971). 'In search of in search of . . .', *Lilit*, No. 7, 22–3.

Crampon, L. J. (n.d.). 'The Impact of Aloha', in L. J. Crampon, *Tourist Development Notes*, Boulder, CO: Univ. of Colorado, Graduate School of Business Admin., Business Research Division, Vol. III, 51–60 (Mimeo).

Desai, A. V. (1974). 'Tourism – Economic Possibilities and Policies', in *Tourism in Fiji*, Suva: Univ. of the South Pacific, 1–12.

Dumazdier, J. (1967). *Toward a Society of Leisure*, New York: Free Press.

Dupont, G. (1973). 'Lourdes: Pilgrims or Tourists?', *Manchester Guardian Weekly*, 108(20), 10.5.1973, 16.

Eisenstadt, S. N. (1968). 'Transformation of Social, Political and Cultural Orders in Modernisation', in Eisenstadt, S. N. (ed.), *Comparative Perspectives on Social Change*, Boston: Little, Brown and Co., 256–79.

Eliade, M. (1969). *The Quest; History and Meaning in Religion*, Chicago and London: Univ. of Chicago Press.

Eliade, M. (1971). *The Myth of Eternal Return*, Princeton, NJ: Princeton Univ. Press.

Eliade, M. (1976). *Occultism, Witchcraft and Cultural Fashions*, Chicago: Univ. of Chicago Press.

Glasser, R. (1975). 'Life Force or Tranquilizer', *Society and Leisure*, 7(3), 17–26.

Gross, E. (1961). 'A Functional Approach to Leisure Analysis', *Social Problems*, 9(1), 2–8.

Heller, E. (1961). *The Disinherited Mind*, Harmondsworth: Penguin.

Hoffer, E. (1952). *The True Believer: Thoughts on the Nature of Mass Movements*, London: Secker and Warburg.

Huetz de Lemps. Ch. (1964). 'Le Tourisme dans l'Archipel des Hawaii', *Cahiers d'Outre-Mer*, 17(65), 9–57.

Kavolis, V. (1970). 'Post Modern Man: Psychocultural Responses to Social Trends', *Social Problems*, 17(4), 435–48.

Lambert, R. S. (ed) (1935). *Grand Tour: A Journey in the Tracks of the Age of Aristocracy*, London: Faber and Faber.

Lowenthal, D. (1962). 'Tourists and Thermalists', *Geographical Review*, 52(1), 124–7.

MacCannell, D. (1973). 'Staged Authenticity: Arrangements of Social Space in Tourist Settings', *American Journal of Sociology*, 79(3), 589–603.

MacCannell, D. (1976). *The Tourist: A New Theory of the Leisure Class*, New York: Schocken Books.

Meintel, D. A. (1973). 'Strangers, Homecomers and Ordinary Men', *Anthropological Quarterly*, 46(1), 47–58.

Mitford, N. (1959). 'The Tourist', *Encounter*, 13(4), 3–7.

Nash, D. (1963). 'The Ethnologist as Stranger: An Essay in the Sociology of Knowledge', *Southwestern Journal of Anthropology*, 19, 149–67.

Rasmussen, K. (1972). 'A Shaman's Journey to the Sea Spirit', in W. A. Lessa and E. Z. Vogt (eds), *Reader in Comparative Religion*, New York, Harper and Row, 388–91.

Shils, E. (1975). 'Center and Periphery', in E. Shils, *Center and Periphery: Essays in Macrosociology*, Chicago and London: Univ. of Chicago Press, 3–16.

Shils, E. and Young, M. (1953). 'The Meaning of the Coronation', *Sociological Review*, 1(2), 63–82.

Spiegel, (1973). 'Enge Bindung', *Der Spiegel*, 27(4), 22.1.1973, III.

Stouffer, S. A. (1950). 'Intervening Opportunities: A Theory Relating Mobility and Distance', *American Sociological Review*, 5, 845–67.

Time (1976). 'Review of S. Bellow's *To Jerusalem and Back*', *Time*, 8.11.1976, 62.

Trease, G. (1967). *The Grand Tour*, London: Heinemann.

Turner, L. and Ash, J. (1975). *The Golden Hordes*, London: Constable.

Turner, V. (1973). 'The Center Out There: The Pilgrim's Goal', *History of Religions*, 12(3), 191–230.

Part III

STRUCTURES OF SOCIAL INEQUALITY IN THE TOURISM SYSTEM

6

TOURISM, CULTURE AND SOCIAL INEQUALITY*

John Urry

INTRODUCTION

I have shown some of the connections between tourist practices and many other social phenomena. These are complex, partly because of the diverse nature of tourism and partly because other social phenomena increasingly involve elements of the tourist gaze. There is a universalisation of the tourist gaze in postmodern cultures – a universalisation which in Britain mainly takes the form of a vernacular and heritage reshaping of much of the urban and rural landscape.

None of the theories outlined are in themselves adequate to grasp the 'essence' of tourism, which is multi-faceted and particularly bound up with many other social and cultural elements in contemporary societies. It is inappropriate to think that it is possible to devise 'the theory of tourist behaviour'. Instead what is required is a range of concepts and arguments which capture both what is specific to tourism and what is common to tourist and certain non-tourist social practices. The concept of the tourist gaze attempts to do this. I categorised objects of the gaze in terms of romantic/collective, historical/modern, and authentic/inauthentic. Central to such objects is some notion of departure, particularly that there are distinct contrasts between what people routinely see and experience and what is extraordinary, the extraordinary sometimes taking the form of a liminal zone. The following are relevant to understanding the changing sociology of the tourist gaze: the social tone of different places; the globalisation and the universalisation of the tourist gaze; the processes of consuming tourist services; tourist meanings and signs; modernism and postmodernism; history, heritage and the vernacular; and post-tourism and play. Different gazes and hence different tourist practices are authorised in terms of a variety of discourses. These include education, as in the Grand Tour; enlightenment, as in much individual 'travel' and cultural tourism; health, as in tourism designed to 'restore' the individual to healthy functioning; group solidarity, as in many Japanese tourist practices; and play, as in the case of the post-tourist.

I shall now consider what is meant by the visual aspects of the gaze, by the idea of seeing and in turn being seen, and will bring together the various arguments on social

* Reprinted with permission of Sage Publications Ltd from John Urry (1991) *The Tourist Gaze: Leisure and Travel in Contemporary Societies*, Newbury Park, CA: Sage, pp. 135–56.

divisions and tourism, especially with regard to gender and ethnicity. I will reconsider the simulated character of the contemporary cultural experience, so-called 'hyper-reality' and the construction of themes and the notions of cultural and educational tourism – developments which bring us back to certain elements of the Grand Tour and which imply that further transformations are occurring in the complex relationships between work, leisure and holidays.

SEEING AND BEING SEEN

Mass tourism is a characteristic of modern societies. It could only develop when a variety of economic, urban, infrastructural and attitudinal changes had transformed the social experiences of large sections of the population of European societies during the course of the nineteenth century. The ways these changes worked themselves out in Britain were illustrated by analysing the causes and consequences of the growth of a new urban form, the seaside resort.

But there was one aspect of nineteenth-century developments which I did not describe in any detail. This concerns the emergence of relatively novel modes of visual perception which became part of the modern experience of living and visiting new urban centres, particularly the grand capital cities. Here I shall show the nature of this new mode of visual perception, the connections between it and the growth of the tourist gaze, and the centrality of photography to these processes. The immensely expanding popularity of photography in the later nineteenth century indicates the importance of these new forms of visual perception, and their role in structuring the tourist gaze that was emerging in this period. This new mode of visual experience has been eloquently characterised by Berman, who sees in the rebuilding of Paris during the Second Empire in the mid-nineteenth century the construction of the conditions for the quintessentially modern experience (see Berman, 1983: section 3). It is also one of the most celebrated of tourist gazes.

For Berman what is of central importance to Paris in this period is the reconstruction of urban space which permits new ways of seeing and being seen. This was engendered by the massive rebuilding of Paris by Haussmann, who blasted a vast network of new boulevards through the heart of the old medieval city. The rebuilding of Paris displaced 350,000 people; by 1870 one-fifth of the streets of central Paris were Haussmann's creation; and at the height of the reconstruction one in five of all workers in the capital was employed in construction (see Clark, 1984; 37).

The boulevards were central to this planned reconstruction – they were like arteries in a massive circulatory system, and were partly at least to be functional for rapid troop movements. However, they also restructured what could be seen or gazed upon. Haussmann's plan entailed the building of markets, bridges, parks, the Opera and other cultural palaces, with many located at the end of the various boulevards. Such boulevards came to structure the gaze, both of Parisians and later of visitors. For the first time in a major city people could see well into the distance and indeed see where they were going and where they had come from. Great sweeping vistas were designed so that each walk led to a dramatic climax. As Berman says: 'All these qualities helped to make Paris a uniquely enticing spectacle, a visual and sensual feast ... after centuries of life as a cluster of isolated cells, Paris was becoming a unified physical and human space' (1983: 151). Certain of these spectacular views have come to be signifiers of the entity 'Paris' (as opposed to the individual districts).

These boulevards brought enormous numbers of people together in ways that were relatively novel. The street level was lined with many small businesses, shops and especially cafés. The last of these have come to be known all over the world as signs of *la vie Parisienne*, particularly as generations of painters, writers and photographers have represented the patterns of life in and around them, beginning with the Impressionists in the 1860s (see Berman, 1983: 151; Clark, 1984).

In particular Berman talks of the way in which the boulevards and cafés created a new kind of space, especially one where lovers could be 'private in public', intimately together without being physically alone (1983: 152). Lovers caught up in the extraordinary movement of modern Paris in the 1860s and 1870s could experience their emotional commitment particularly intensely. It was the traffic of people and horses that transformed social experience in this modern urban area. Urban life was both exceptionally rich and full of possibilities; and at the same time it was dangerous and frightening. As Baudelaire wrote: 'I was crossing the boulevard, in a great hurry, in the midst of a moving chaos, with death galloping at me from every side' (Berman, 1983: 159). To be private in the midst of such danger and chaos created the perfect romantic setting of modern times; and millions of visitors have attempted to re-experience that particular quality among the boulevards and cafés of Paris.

This romantic experience could be felt especially intensely in front of the endless parades of strangers moving up and down the boulevards – it was those strangers they gazed upon and who in turn gazed at them. Part then of the gaze in the new modern city of Paris was of the multitude of passers-by, who both enhanced the lovers' vision of themselves and in turn provided an endlessly fascinating source of curiosity.

> They could weave veils of fantasy around the multitude of passers-by: who were these people, where did they come from and where were they going, what did they want, whom did they love? The more they saw of others and showed themselves to others – the more they participated in the extended 'family of eyes' – the richer became their vision of themselves.
>
> (Berman, 1983: 152)

Haussmann's reconstruction of Paris was not of course without its intense critics (see Clark, 1984: 41–50). It was very forcibly pointed out that demolishing the old *quartiers* meant that much of the working class was forced out of the centre of Paris, particularly because of the exceptionally high rents charged in the lavish apartment blocks that lined the new boulevards. Reconstruction therefore led to rapid residential segregation and to the worst signs of deprivation and squalor being removed from the gaze of richer Parisians and, later in the century, of visitors. Second, Paris was said to be increasingly a city of vice, vulgarity and display – ostentation not luxury, frippery not fashion, consumption not trade (see Clark, 1984: 46–7). It was a city of uncertainty in which there were too many surfaces, too few boundaries. It was the city of the *flâneur* or stroller. The anonymity of the crowd provided an asylum for those on the margins of society who were able to move about unnoticed, observing and being observed, but never really interacting with those encountered. The *flâneur* was the modern hero, able to travel, to arrive, to gaze, to move on, to be anonymous, to be in a liminal zone (see Benjamin, 1973; Wolff, 1985).

The *flâneur* was invariably male and this rendered invisible the different ways in which women were both more restricted to the private sphere and at the same time were coming to colonise other emerging public spheres in the late nineteenth century, especially the department store (see Wolff, 1985). The strolling *flâneur* was a forerunner of the twentieth-century tourist and in particular of *the* activity which has in a way become emblematic of the tourist: the democratised taking of photographs – of being seen and recorded and seeing others and recording them.

Sontag explicitly makes this link between the *flâneur* and photography. The latter:

> first comes into its own as an extension of the eye of the middle-class *flâneur*. . . . The photographer is an armed version of the solitary walker reconnoitering, stalking, cruising the urban inferno, the voyeuristic stroller who discovers the city as a landscape of voluptuous extremes. Adept of the joys of watching, connoisseur of empathy, the *flâneur* finds the world 'picturesque'.
>
> (1979: 55)

While the middle-class *flâneur* was attracted to the city's dark seamy corners, the twentieth-century photographer is attracted everywhere, to every possible object, event and person. And at the same time the photographer is also observed and photographed. One is both see-er and seen. To be a photographer in the twentieth century, and that is so much part of travel and tourism, is also to be seen and photographed.

There has been an enormous proliferation of photographic images since the invention of photography in 1839. Over that century and a half there has been an utter insatiability of the photographing eye, an insatiability that teaches new ways of looking at the world and new forms of authority for doing so. But it is crucial to understand that photography is a socially constructed way of seeing and recording. As such it has a number of central characteristics (on the following see Sontag, 1979; Berger, 1972; Barthes, 1981; Albers and James, 1988).

1 To photograph is in some way to appropriate the object being photographed. It is a power/knowledge relationship. To have visual knowledge of an object is in part to have power, even if only momentarily, over it. Photography tames the object of the gaze, the most striking examples being of exotic cultures. In the USA the railway companies did much to create 'Indian' attractions to be photographed, carefully selecting those tribes with a particularly 'picturesque and ancient' appearance (see Albers and James, 1988: 151).

2 Photography *seems* to be a means of transcribing reality. The images produced appear to be not statements about the world but pieces of it, or even miniature slices of reality. A photograph thus seems to furnish evidence that something did

indeed happen – that someone really was there or that the mountain actually was that large. It is thought that the camera does not lie.

3 Yet in fact photographs are the outcome of an active signifying practice in which those taking the photo select, structure and shape what is going to be taken. In particular there is the attempt to construct idealised images which beautify the object being photographed. Sontag summarises: 'the aestheticizing tendency of photography is such that the medium which conveys distress ends by neutralizing it' (1979: 109).

4 The power of the photograph thus stems from its ability to pass itself off as a miniaturisation of the real, without revealing either its constructed nature or its ideological content.

5 As everyone becomes a photographer so everyone also becomes an amateur semiotician. One learns that a thatched cottage with roses round the door represents 'ye olde England'; or that waves crashing on to rocks signifies 'wild, untamed nature'; or, especially, that a person with a camera draped around his/her neck is clearly a 'tourist'.

6 Photography involves the democratisation of all forms of human experience, both by turning everything into photographic images and by enabling anyone to photograph them. Photography is then part of the process of postmodernisation. Each thing or person photographed becomes equivalent to the other, equally interesting or uninteresting. Barthes notes that photography began with photographs of the notable and has ended up making notable whatever is photographed (1981: 34, and see Sontag, 1979: 111). Photography is a promiscuous way of seeing which cannot be limited to an elite, as art. Sontag talks of photography's 'zeal for debunking the high culture of the past . . . its conscientious courting of vulgarity . . . its skill in reconciling avant-garde ambitions with the rewards of commercialism . . . its transformation of art into cultural document' (1979: 131).

7 Photography gives shape to travel. It is the reason for stopping, to take (snap) a photograph, and then to move on. Photography involves obligations. People feel that they must not miss seeing particular scenes since otherwise the photo-opportunities will be missed. Tourist agencies spend much time indicating where photographs should be taken (so-called viewing points). Indeed much tourism becomes in effect a search for the photogenic; travel is a strategy for the accumulation of photographs. This seems particularly to appeal to those cultures with a very strong work ethic. Japanese, Americans and Germans all seem to 'have' to take photographs – it is a kind of leisure equivalent of the distorting obligations of a strong workplace culture (see Sontag, 1979).

8 Involved in much tourism is a kind of hermeneutic circle. What is sought for in a holiday is a set of photographic images, as seen in tour company brochures or on TV programmes. While the tourist is away, this then moves on to a tracking down and capturing of those images for oneself. And it ends up with travellers demonstrating that they really have been there by showing their version of the images that they had seen originally before they set off.

Photography is thus intimately bound up with the tourist gaze. Photographic images organise our anticipation or daydreaming about the places we might gaze upon. When we are away we record images of what we have gazed upon. And we partly choose where to go in order to capture places on film. The obtaining of photographic images in part organises our experience as tourists. And our memories of places are largely structured through photographic images and the mainly verbal text we weave around images when they are on show to others. The tourist gaze thus irreducibly involves the rapid circulation of photographic images.

SOCIAL DIVISIONS AND THE TOURIST GAZE

In much of the earlier discussion I emphasised the importance of social class divisions in structuring how tourist developments occurred in different ways in different places. For example, the respective social tone of different resorts and the patterns of landholding; the importance of the aristocratic connection in constructing the fashionability of certain places; the growth of the middle-class family holiday and the development of the bungalow as a specialised building form by the seaside; the importance of the 'romantic gaze' and its role in constructing nature as an absolutely central positional good; the character of the 'collective' gaze and the role of others like oneself in constituting the attraction of certain places; and the enhanced cultural capital of the service class and its impact in heightening the appeal of rural and industrial heritage, and of postmodernism.

I shall now broaden this discussion to consider some of the ways that generation, gender and ethnicity interconnect with class. These interconnections are important in forming the preferences that different social groupings develop about where to visit and in structuring the effects of such visits upon host populations. There are two key issues here: the social composition of fellow tourists and the social composition of those living in the places visited. These are important because of the way that most tourist practices involve movement into and through various sorts of public space – such as beaches, shops, restaurants, hotels, pump rooms, promenades, airports, swimming pools and squares. In such spaces people both gaze at and are gazed upon by others (and photograph and are photographed). Complex preferences have come to develop for the range of appropriate others that different social groups expect to look at in different places; and in turn different expectations are held by different social groups about who are appropriate others to gaze at oneself. Part of what is involved in tourism is the purchase of a particular social experience, and this depends upon a specifiable composition of the others with whom that experience is being shared in one way or another.

However, these preferences cannot be reduced solely to issues of social class. One different kind of argument developed has been with regard to the development of sex-tourism in certain South-East Asian societies. Here the combination of relations of gender and ethnic subordination had colluded to help construct very young Asian women as objects of a tourist/sexual gaze for male visitors from other societies which are in a sense ethnically dominant. The resulting tourist patterns cannot be analysed separately from relations of gender and racial subordination.

The importance of gender inequalities can be seen in another way. It is clear that in most societies men have enjoyed a higher standard of living than women. In Britain this has resulted from a privileged treatment in the household's distribution of food, heat and other material resources; and from the ability to escape the home to spend large amounts of leisure time in the 'masculine republic' of the pub, where at least 15 per cent of the household income was spent (see Hart, 1989, for an insightful analysis of the implications of this for politics). To the extent to which contemporary leisure patterns are more 'privatised' and shared, this may involve a reduced inequality of both household income and leisure time.

This relates in an important way to the development of holidays. Until the nineteenth century access to travel was largely the preserve of men. But this changed with the development of 'Victorian lady travellers', some of whom visited countries that were at the time considered 'uncivilised' and 'uncharted' (see Enloe, 1989: ch. 2). Other women took advantage of Cook's tours. As one woman wrote: 'We would venture anywhere with such a guide and guardian as Mr Cook' (quoted in Enloe, 1989: 29). From then onwards access to holidays has not been so unequally distributed as has access to some other forms of leisure, especially leisure time spent in the pub or bar. Working-class holidays to English seaside resorts were normally undertaken by couples. Moreover, the fact that such holidays developed first in industrial Lancashire was partly the result of high levels of female employment in the cotton textile industry, especially in weaving. This meant that household earning was higher than in other areas and women had more say over its distribution. So there is a sense that the development of the couple-based holiday has been a major mechanism for reducing some of the obvious inequalities of leisure access between the sexes and is something that has partly resulted from women's own economic activities.

The early forms of mass tourism were based around the heterosexual couple; during the course of the nineteenth century the holiday unit had increasingly come to be made up of such a couple plus their children. And by the inter-war period the family holiday had become very much child-centred. This was given a significant boost by the development of the holiday camp in the 1930s in which child-based activities were central. Their development was of benefit to women since it meant that a considerable amount of child care was undertaken by paid workers. The more recent growth of self-catering has moved in the opposite direction. Only a minority of holidays in Britain are now taken in serviced accommodation.

It is important to note how holiday-making could be described as overwhelmingly heterosexual, involving actual couples, with or without children, or potential couples. In the material produced by tour operators there are three predominant advertising images. These are the 'family holiday', that is a couple with two or three healthy school-age children; the 'romantic holiday', that is a heterosexual couple on their own gazing at the sunset (indeed the sunset is a signifier for romance); and the 'fun holiday', that is same-sex groups each looking for other-sex partners for 'fun'. There is also, as we have noted, the 'sex holiday' for men. It is well known that social groups that do not fall into any of these particular categories are poorly served by the tourist industry. Many criticisms have been made of how difficult holiday-making is for single people, sole-parent families, homosexual couples or groups, and those who are disabled.

Another social category is partly excluded from conventional holiday-making, namely black Britons. The advertising material produced by holiday companies shows that tourists are white; there are simply no black faces amongst the holiday-makers. Indeed if there are any non-white faces in the photographs it is presumed that they are the 'exotic natives' being gazed upon. The same process would seem to occur in those areas in Britain which attract large numbers of foreign tourists. If black or Asian people are seen there it would be presumed that they were visitors from overseas, or perhaps service workers, but not British residents themselves on holiday. The countryside particularly is constructed as 'white'.

An interesting question is the degree to which members of ethnic minorities do undertake western-type holidays. Clearly, relatively low incomes will make the rate of non-holiday-making higher than the UK average of 30 per cent (in any year). But it is also the case that aspects of the western holiday, in which one travels elsewhere simply because of the sun, hotel or scenery, form a cultural practice which must seem rather idiosyncratic at least to recent immigrants to Britain. The obsession of white people to darken their skins and thereby increase the possibilities of acquiring skin cancer must leave black and Asian people both excluded and perplexed. Many recent immigrants at least would consider that travel should have a more serious purpose than this: to look for work, to join the rest of one's family, or to visit relatives.

Many of the recent tourist developments are also likely to exclude many ethnic groups. On the one hand are the temples of consumerism, the new shopping centres, which I will discuss later. These clearly require considerable income for full participation; they also require a strong commitment to the values of a credit-based consumerism. This is certainly likely to make many citizens from Asia feel remarkably estranged. On the other hand, there is the heritage industry discussed in the previous chapter. It was noted that such a heritage is overwhelmingly populated by white faces. For example, in the new award-winning Maritime Museum in the Albert Dock in Liverpool there is a substantial section on the trade union history of the docks, but nothing on the complex patterns of ethnic subordination which has produced in Liverpool one of the most racist cities in Britain. Moreover, the first heritage attraction in Britain portraying black culture is now planned for Birkenhead, near Liverpool. Astonishingly, it will reconstruct life on board a ship built locally for the slavery-supporting Confederates in the American Civil War. One might also wonder whether part of the attraction of heritage for many white visitors is precisely the fact that it is seen as predominantly white – while many larger cities are disapprovingly viewed as having become 'multicultural'.

Ethnic groups are not unimportant in the British tourist industry, and in some respects play a key role. They are likely to be employed in those enterprises concerned with servicing visitors, especially in the major cities, apart incidentally from Liverpool. Almost 10 per cent of workers from Pakistan and Bangladesh, and 20 per cent of workers from the Mediterranean, the Middle East and Latin American countries, are employed in the hotel and catering industry (Bagguley, 1987). In some parts of Britain there is a remarkable concentration. In central London hotels, for example, 45 per cent of those employed are from 'black' ethnic groups. This is partly for the general reason that the majority of these jobs are badly paid, with poor conditions of employment and low levels of unionisation. White workers will tend to avoid such jobs; they are concentrated in other sectors of the economy where structural racial disadvantage discriminates against 'blacks'. The concentration of employment is also because a number of employers in hotel and catering in the past specifically re-cruited abroad, such as Grand Metropolitan. Such firms provided both transport and work permits for migrant workers. Although this does not now happen so clear-ly, there are still about 115,000 foreign nationals employed in the hotel and catering industry (Bagguley, 1987: 35).

In recent years certain ethnic groups have come to be constructed as part of the 'attraction' of some places. This is most common in the case of Asian rather than Afro-Caribbean groups. In Manchester this has occurred around its collection of outstanding restaurants in a small localised area, and resulted from the international-isation of British culinary taste in the postwar period (although this is less pro-nounced than in the USA; see Frieden and Sagalyn, 1989: 199–201), as well as from a cultural reinterpretation of racial difference. By the 1980s city planners were committed to a new vision of 'Chinatown', reconstructed and conserved as a now desirable object of the tourist gaze (see Anderson, 1988, on the similar process in Vancouver).

Bradford is a more distinctive example, since it had no tourist industry until 1980. It was a prime example of an industrial city of dark satanic mills. The setting up of a tourism initiative was self-consciously undertaken when it was realised that Bradford had a number of ingredients likely to appeal to the short-break holiday-maker. These were, apart from plenty of hotel beds, proximity to internationally renowned attractions such as Haworth and the Dales and Moors; a substantially intact industrial heritage of buildings, railways and canals derived from Bradford's status as 'Worstedopolis'; its location within the high-profile county of Yorkshire; and the existence of a large and vigorous Asian community which had generated a plethora of small enterprises.

The city council, moreover, realised that turning Bradford into a tourist mecca was itself a newsworthy item. They received considerable free publicity in the early 1980s as they launched their tourism initiative (such publicity being worth at least £250,000). One element in later campaigns has been to market the Asian community as a major visitor attraction. A separate booklet was produced entitled 'Flavours of Asia'. This details fifty 'Asian' restaurants, the largest Asian store in Europe, various curry tours, a dozen sari centres as well as a brief history of various Asian religions and of the patterns of immigration to Bradford. Consideration is now being given to developing a museum and a festival to complete the visitor's Asian experience in Bradford (see Davies, 1987).

Further analysis of this would need to explore the social effects for those of Asian origin of becoming constructed as an exotic object and whether this distorts patterns of economic and political development; and the effects on the white population of coming to view those of Asian origin as not so much threatening or even inferior but as exotic, as curiously different and possessing a rich and in part attractive culture. Such debates are developing in the context of a number of cultures taken to be exotically different. These issues arise not only in the case of those exceptionally inauthentic ethnic constructions but also where the representations are reasonably scholarly.

THEMES AND MALLS

If an ethnic group living in an area is sometimes constructed and presented as a visitor attraction, another way of expressing this is to say that ethnic difference, the 'exotic', may sometimes function as a 'theme'. In this section I shall discuss the themed character of urban and rural life directly, considering some aspects of recent theme parks before turning to the themed character of contemporary retailing, noting especially the features of the ubiquitous shopping mall.

There is an increasingly pervasive tendency to divide up Britain spatially. A series of new place names has been invented for the tourist. In the north of England there is 'Last of the Summer Wine Country', 'Emmerdale Farm Country', 'James Herriot Country', 'Robin Hood Country', 'Catherine Cookson Country', 'Brontë Country' and so on. Space is divided up in terms of signs which signify particular themes – but not themes that necessarily relate to actual historical or geographical processes. A similar process can be seen in Canada where the theme of 'Maritimicity' has been clearly developed since the 1920s as a result of the provincial state and private capital seeking to develop modern tourism in Nova Scotia. McKay describes it as 'a peculiar petit-bourgeois rhetoric of lobster pots, grizzled fishermen, wharves and schooners ... a Golden Age mythology in a region that has become economically dependent on tourism' (1988: 30). In particular Peggy's Cove has over the years become a purer and purer simulacrum, a copy of a prosperous and tranquil fishing village which never existed.

Even stranger is the case of the recently opened Granada Studios in Manchester. Part of the display consists of a mock-up of certain sets from the soap opera *Coronation Street*, including the famous Rover's Return public house. This is very popular with visitors, who are keen to photograph it. But as one commentator noted: 'when we develop our photos of that Rover's Return scenario we will consume a representation of a representation of a representation' (Goodwin, 1989). This set is part of the 'Coronation Street Experience' in which the Rover's Return is given a fictional history, starting in 1902.

Other recently developed themed attractions in Britain include the Jorvik Centre in York, the Chessington World of Adventures, the Camelot theme park in Lancashire, the American Adventure in the Peak District, Frontierland in Morecambe, the Oxford Story, the Crusades experience in Winchester ('history brought to life'), a planned Tudor theme park in Avebury, and the Pilgrim's Way in Canterbury. The last is described in the advertising material as 'a pilgrimage to the past'. However, the sense of history is exceptionally distorted. Faulks summarises one bizarre aspect: 'a man on children's television is the model for a dummy who is the adjunct to a non-existent scene in a mediaeval religious poem, none of whose words you hear' (Faulks, 1988).

Another distinctive example is to be found in Llandrindod Wells in Wales. Once a year most of the population dress up in Edwardian costume, but it has recently been suggested that the population could be dressed that way *for the entire year*. Thus the whole town and its population would be turned into a permanent Edwardian themed town. Already Visby in Sweden, an island in the Baltic, experiences a 'medieval week' when everyone dresses up in medieval costume.

This is, in Debord's terms, a 'society of the spectacle' (1983) or what Eco describes as 'travels in hyper-reality' (1986). In such themed areas the objects observed must seem real and absolutely authentic. Those responsible for Jorvik or the Oxford Story have attempted to make the experience authentic, through the use of smells as well as visual and aural simulation. The scenes are in a sense more real than the original, hyper-real in other words. Or at least the surfaces, as grasped through the immediate senses, are more real. Lowenthal notes that 'habituation to replicas tends to persuade us that antiquities should look complete and "new"' (1985: 293). The representations thus approximate more closely to our expectations of reality, of the signs that we carry around waiting to be instantiated: 'Disneyland tells us that faked nature corresponds much more to our daydream demands ... Disneyland tells us that technology can give us more reality than nature can' (Eco, 1986: 44).

This is currently being taken to the extreme in New Zealand. A popular nineteenth-century tourist attraction was a set of pink and white terraces rising up above Lake Rotomahana. These were destroyed by volcanic eruptions in 1886 although photographs of them have remained popular ever since. They are a well-known attraction even if they have not existed for a century. Now, however, there is a plan to recreate the physical attraction by running geothermal water over artificially built terraces in an entirely different location, but one close to existing tourist facilities. This set of what might be called themed terraces will look more authentic than the original which is only known about because of the hundred-year-old photographic images.

This technological ability to create new themes which appear more real than the original has now spread from tourist attractions *per se*, beginning with Disneyland, to shopping centres or malls. Some North American malls are now extraordinary tourist attractions in their own right and represent an exceptional degree of cultural de-differentiation, as previously discussed. Consider the following publicity material for the West Edmonton Mall:

Imagine visiting Disneyland, Malibu Beach, Bourbon Street, the San Diego Zoo, Rodeo Drive in Beverly Hills and Australia's Great Barrier Reef ... in one weekend – and under one roof.... Billed as the world's largest shopping complex of its kind, the Mall covers 110 acres and features 828 stores, 110 restaurants, 19 theatres . . . a five-acre water park with a glass dome that is over 19 storeys high. . . . Contemplate the Mall's indoor lake complete with four submarines from which you can view sharks, octopi, tropical marine life, and a replica of the Great Barrier Reef. . . . Fantasyland Hotel has given its rooms a variety of themes: one floor holds Classical Roman rooms, another '1001 Nights' Arabian rooms, another, Polynesian rooms. . . .

(Travel Alberta, undated)

The mall has been stunningly successful. In 1987 it attracted over 9 million tourists, making it the third most popular tourist attraction in North America after Walt Disney World and Disneyland. It represents a symbolic rejection of the normally understood world geography in which there are distant centres with Edmonton on the periphery. What is being asserted is a new collective sense of place based on transcending the geographical barrier of distance and of place. The real-space relations of the globe are thus replaced by imaginary-space relations (Shields, 1989: 153).

This has only been possible because of the pervasiveness of tourist signs, of the rapid circulation of photographic images in particular. It is this exchange of signs which makes possible the construction of a pastiche of themes, each of which seems more real than the original, particularly because of the way that shopping malls in general emphasise newness and cleanliness: 'It is a world where Spanish galleons sail up Main Street past Marks and Spencer to put in at "New Orleans", where everything is tame and happy shoppers mingle with smiling dolphins' (Shields, 1989: 154).

The closest to this phenomenon in Europe is the Metrocentre in Gateshead, interestingly again located in a place that has normally been considered peripheral to British and European life. It was constructed on derelict land which is part of an Enterprise Zone. The Centre was established by John Hall, a regional entrepreneur, and is now owned, somewhat improbably, by the Church Commissioners. It contains 3 miles of shopping malls with 300 shops, 40 restaurants, a 10-screen cinema, a bowling alley, an enormous fantasy kingdom of fairground rides and entertainments, a crèche, and three themed areas. These themes are 'Antique Village', with a phoney waterwheel and plastic ducks on the village pond; a 'Roman Forum', with areas on which to recline Roman-style; and a 'Mediterranean Village', with Italian, Greek and Lebanese restaurants lining a windingly quaint Mediterranean street. Shopping is clearly here only part of the appeal of the mall, which is as much concerned with leisure and with tourism. Within a few minutes' walk one can consume a range of tourist themes, can stroll gazing and being gazed upon as though 'on holiday', and can experience an enormous range of entertainment services (on malls in general, see Shields, 1989).

Malls represent membership of a community of consumers. To be in attendance at the 'court of commodities' is to assert one's existence and to be recognised as a citizen in contemporary society, that is as a consumer. However, the recent marketing philosophy for the 1980s has been to develop spectacles of 'diversity and market segmentation', although this is less clear in the case of mass middle-class malls such as the Metrocentre. The development of such differentiation in particular centres is 'because the display of difference will today increase a centre's "tourist" appeal to everyone else from elsewhere' (Morris, 1988: 205). The 'everyone else' means everyone within the same market segment. Trump Tower in New York is the ultimate upper-middle-class white shopping mall.

Developments of this sort also represent the changing nature of public space in contemporary societies. An increasingly central role is being played by privately owned and controlled consumption spaces. These involve high levels of surveillance where certain types of behaviour, clothing and comportment are expected, such as not sitting on the floor. The entrance and pathways of malls are 'policed' and undesirable categories of the population, such as the homeless or ethnic minorities, can be excluded. The Metrocentre boasts that it is the safest place in Britain to shop. There are some analogies between Bentham's panopticon prison and the visual and electronic surveillance found in these malls. They are also conspicuous for cleanliness and newness, with no space for untidy litter, the old, the shabby or the worn. Malls have to exude up-to-dateness and fashionability which is why they have to be regularly refurbished (see Fiske, 1989: 39–42).

It is suggested that malls attract their share of 'post-shoppers', people who play at being consumers in complex, self-conscious mockery. Users should not be seen simply as victims of consumerism, as 'credit card junkies', but also as being able to assert their independence from the mall developers. This is achieved by *flânerie*, by continuing to stroll, to gaze, and to be gazed upon: 'Their wandering footsteps, the modes of their crowd practice constitute that certain urban ambiance: a continuous reassertion of the rights and freedoms of the marketplace, the *communitas* of the carnival' (Shields, 1989: 161).

In an Australian study Pressdee showed that in spite of the control mechanisms in such malls 80 per cent of unemployed young people visited them at least once a week, and that more or less 100 per cent of young unemployed women were regular visitors (1986). Late-night shopping on Thursday was a particular occasion when the mall was invaded by young people with little intention to buy. The youths consumed images and space instead of commodities. Fiske talks of:

> a kind of sensuous consumption that did not create profits. The positive pleasure of parading up and down, of offending 'real' consumers and the gents of law and order, of asserting their difference within, and different use of, the cathedral of consumerism became an oppositional cultural practice.
>
> (1989: 17)

Fiske also points out the central importance of shops as public, or at least semi-public, spaces which are particularly attractive to women (1989; and see Morris, 1988). I noted earlier the importance of the nineteenth-century development of the department store in this respect, that it was both respectable and safe for unattached women. Zola described the department store as: 'a temple to women, making a legion of shop assistants burn incense before her' (quoted in Pevsner, 1970: 269).

The mall is somewhat similar, and indeed shopping is a sphere of social activity in which women are empowered. It links together the public and the domestic and involves activity in which women are permitted to demonstrate competence.

Finally, one should note a further setting for themed environments which have become particularly popular in the last decade or two: world fairs, which have developed into enormous international tourist attractions. For example, over 22 million people attended the 1986 Expo in Vancouver (see Ley and Olds, 1988, for further analysis). The development and popularity of these fairs represent the growing intrusion of leisure, tourism and the aesthetic into the urban landscape. They provide further examples of the de-differentiation of leisure, tourism, shopping, culture, education, eating, and so on.

Most of the 1988 Expo in Brisbane was organised around different national displays. There were well over fifty themed environments based on different national stereotypes, such as the British pub, American achievement in sport, the German beer garden and South Sea Islands' exotic dancing. Such themes were designed to demonstrate national pride in the cultural activities presumed specific to that country. Generally this pride was demonstrated either in repackaging aspects of that country's traditions and heritage or in demonstrating the high level of modern technology achieved. The display generally thought the most interesting (measured by the length of the queues to enter) was that of the Japanese, which combined both aspects, with, for example, high-technology representations of traditionally venerated dancing animals.

As with the Vancouver Expo, no single hegemonic set of messages was conveyed by the fair. Indeed they are such obviously postmodern phenomena that this would be difficult to achieve. Such fairs are, if anything, a kind of micro-version of international tourism. Rather than tourists having to travel worldwide to experience and gaze upon these different signs, they are conveniently brought together in one location, simply on a larger scale than the West Edmonton mall. Harvey says more generally: 'it is now possible to experience the world's geography vicariously, as a simulacrum' (1989: 300). This can be seen from the entertainment provided at such world fairs. At Vancouver there were 43,000 free on-site performances given by an incredible 80,000 performers (Ley and Olds, 1988: 203). Although there was high culture, including a presentation from La Scala to an audience of 40,000, most entertainment consisted of folk or popular forms, all in all a postmodern cultural pastiche, rather like the availability of cuisines from around the world in most major American cities (see Harvey, 1989: 300). Most performances were recognisably from a specific country and consisted of the sort of ethnic entertainment that is provided for tourists in each country that they visit. The difference here was that the visitors only had to walk from one tent or display to the next in order to gaze upon another cultural event signifying a further nation.

What people thus do in such a fair is to stroll, to be a *flâneur*, and what they stroll between are signs of different cultures. Therefore whether they are locals or visitors from outside the city hosting the fair, they are acting as tourists, gazing upon the signs of different cultures. People can do in an afternoon what otherwise takes a lifetime: gaze upon and collect signs of dozens of different cultures – the built environment, cultural artefacts, meals, and live, ethnic entertainment.

Many of the displays purport to be educational, and indeed groups of school-age children constitute a major category of visitors. And this is a further feature of the de-differentiation of the cultural spheres noted earlier. Education and entertainment are becoming merged, a process very much assisted by the increasingly central role of the visual media in both. Indeed some of the latest theme parks are involved in the provision of 'edu-tainment'. It is this topic, the relationships of tourism and education and culture, that I shall now consider.

EDUCATION AND ENTERTAINMENT

It was previously noted that the development of mass tourism was an aspect of the separation of work and leisure that characterised social development in the nineteenth century, as both work and leisure became characterised by increasing rationalisation. The emergence of mass tourism at the end of the century was a further aspect of this separation. Tourism came to be systematised and organised. It was the antithesis of work and of education and learning.

The main exception to these patterns had been in the Grand Tour, undertaken mainly by the sons of the wealthy. Such 'tourism' was not a leisure activity undertaken while away from work since those involved were not in work; and it did not involve an absence of learning and education – these were important elements of the tour.

I shall bring the argument full circle by suggesting that contemporary tourism is in part taking on some of these characteristics of the Grand Tour. This can be seen in the way that a substantial proportion of the population enjoys increased periods of time away from paid work. Often this is described as increased 'leisure' time. But this is a misleading term, since, especially for women, 'leisure' involves 'work'. However, there is for many people more time away from paid work, because of increased paid holidays for those in paid work, a rising proportion of the population who are in retirement, a high number of people unemployed or underemployed, and an increasing proportion of people in part-time work, particularly women. Work and non-work are more variable and flexible in comparison with the past, especially in the case of men. Holidays therefore do not have to offer such dramatic contrasts with paid work as previously. They need not only involve two weeks of 'seaside fun'.

Holidays are not so straightforwardly contrasted with education and learning as in the past. In a wide variety of ways much tourism is coming to be more closely interwoven with learning. I have already noted the increasing popularity of museums, the fascination with the lives of industrial workers in particular, and the popularity of hyper-real historical re-creations. Some developments are the increased desire to learn a new sport on holiday (such as skiing, watersports, hang-gliding); the development of arts and cultural tourism; the heightened attraction of unusual industrial sites, such as Sellafield nuclear reprocessing plant which now attracts 150,000 visitors a year; and the substantial increase in educational holidays (in schools, universities and hotels). Trusthouse Forte now offers a wide range of educational breaks at various hotels. Amongst the subjects that can be studied are arts and antiques, bridge, watercolour painting, archery, clay pigeon shooting, fly fishing, golf and pony-trekking.

One interesting tourist site symptomatic of these new developments is the Quarry Bank Mill at Styal in Cheshire. This water-powered cotton-spinning mill was built by Samuel Greg in 1784. Surrounding the mill were the buildings of an entire factory community, two chapels, a school, shop, houses for mill workers, and an apprentice house, all of which have remained physically well preserved. The museum was founded in 1976. It is described as 'a museum of the factory system', aiming to bring to life the role of the workforce, the Greg family and the circumstances which began the industrial revolution in the textile industry. The museum houses a number of displays on textile finishing and water power. Demonstrators, some dressed in appropriate clothing, show visitors how to spin cotton on a spinning jenny, how to hand-weave, how a carding machine operated, the workings of a weaving mule, and the domestic routines involved in cooking, cleaning and washing for the child workforce.

A great deal of research by professional historians has been undertaken to produce both the displays and the large number of supporting documents, given to or sold to visitors. Engineers have also been centrally involved in the development of the museum, in order to get the often derelict machinery reworking.

Half of the visitors to the site are parties of schoolchildren and so the staff of the museum are directly involved in education. The mill has produced a range of supporting material for such visitors, including a 'Resource and Document Pack'. Up to 100 guides are employed in explaining aspects of the mill's workings to such visitors. There are also a number of other educational activities undertaken by the museum. Courses run by the mill include weaving, spinning, patchwork and quilting, embroidery and lace, experimental textiles, fashion and clothing, design for textiles, dyeing and printing, and knitting.

At the same time the mill is making energetic efforts to attract the 'non-museum visiting public' by specifically increasing the entertainment elements of display. This is partly achieved by the use of people to demonstrate many of the processes and to interact in a role-playing way with the visitors. It is also assisted by organising a variety of special events: Mothering Sunday lunches, a tent-making project, St George's Day celebration, Spooky tours, Apprentices' Christmas and so on. However, there is an obvious danger of both being seen as over-commercial and of actually being over-commercial. The commercialisation of the mill is a reflection of the tendencies towards the postmodern museum.

The mill has had to grapple with the issue of authenticity. Although the building is genuine and has not been particularly cleaned up, the machinery does not of course stem from the eighteenth century. Some items had been in the mill since the nineteenth or early twentieth centuries, while quite a lot of it, including the immense waterwheel, has been imported from other, often derelict, industrial sites. The work on the machinery has involved using 'traditional' techniques which have had to be specially learnt. The mill does try to make explicit what is authentic, although this is not a straightforward exercise since what is thought to be authentic depends upon which particular period is being considered. Also, of course, existing 'authentic' factories contain machines from a variety of periods. What Quarry Bank Mill ultimately shows is that there is no simple 'authentic' reconstruction of history but that all involve various kinds of accommodation and reinterpretation.

Finally, the mill does not present an overly romanticised view of working-class life. There is plenty of evidence on the ill-health and squalor of much industrial work. However, the mill literature also draws considerable attention to the views of contemporaries which suggested that conditions in rural factory communities, such as Styal, were considerably better than those in the huge industrial cities, such as neighbouring Manchester and Salford. Thus there seem to have been lower levels of industrial unrest, although this could also be related to forms of surveillance and control available locally. It was also suggested by the curator that visitors would not necessarily return for further visits if an overly depressing account of factory life was represented. However, unlike some other industrial museums, Quarry Bank Mill is not a shrine to technology – if anything the textile machinery is likely to be regarded by visitors as noisy, dangerous and dirty.

The most surprising way in which the development of tourism is transforming the urban environment is in those places where a newly established cultural tourism has taken root. The best example of this in Britain is the transformation of Glasgow and its emergence as one of the centres of cultural tourism. It has been designated as the 'European City of Culture for 1990': 'Glasgow's regeneration has been largely arts-led, with the Mayfest and the opening of the Burrell collection all helping to change the city's image from a decaying industrial backwater to a dynamic growth area [attractive to tourists]' (quoted in McKellar, 1988: 14).

Two-thirds of visitors now consider that there is a wide variety of interesting museums and art galleries to visit in the city. And at least one-third think that there are so many cultural activities available that they wish they were able to stay longer. Fewer than one-fifth consider that Glasgow is the rough and depressing place to visit that once would have been the case (Myerscough, 1988: 88–9). Mysteriously but dramatically, Glasgow has become the kind of place that people now want to visit, to see and to be seen in. It has become a preferred object of the gaze of many tourists.

This transformation of Glasgow as an object of the gaze is the result of economic restructuring, social change, policy intervention and cultural re-evaluation. And part of that transformation has entailed tourism coming to be of central economic and social significance in western societies as the twenty-first century approaches. If Glasgow can be remade as a tourist attraction, one might wonder whether there are in fact any limits to the tourist, or post-tourist, gaze. And if there are few such limits what are the effects on societies whose built environment, conceptions of history, cultural symbols, social patterns and political processes can all be in part remade as objects of the gaze?

It is already clear that the trajectories of some countries are significantly determined by tourism, such as Spain, Austria, Greece, the West Indies and Bali. But what now is happening, as tourism develops into the largest industry worldwide, is that many, or even most, other countries will become engulfed by a tourist tidal wave. This is a wave that is not confined to particular places but where almost all spaces, histories and social activities can be materially and symbolically remade for the end-lessly devouring gaze, a gaze that in 1989 is partly responsible for bringing down some of the boundaries between East and West, as those in Eastern Europe have demanded the right to travel and to gaze upon those in the West. To return to Foucault, contemporary societies are developing less on the basis of surveillance and the normalisation of individuals, and more on the basis of the democratisation of the tourist gaze and the spectacle-isation of place.

REFERENCES

Albers, P. and James, W. (1988) 'Travel photography: a methodological approach', *Annals of Tourism Research* 15: 134–58.

Anderson, K. (1988) 'Cultural hegemony and the race-definition process in Chinatown, Vancouver: 1880–1980', *Environment and Planning D: Society and Space* 6: 127–49.

Bagguley, P. (1987) *Flexibility, Restructuring, and Gender: Changing Employment in Britain's Hotels*, Lancaster Regionalism Group Working Paper No. 24.

Barthes, R. (1981) *Camera Lucida*, New York: Hill & Wang.

Benjamin, W. (1973) 'The work of art in the age of mechanical reproduction', in W. Benjamin (ed.) *Illuminations*, pp. 219–54, London: Fontana.

Berger, J. (1972) *Ways of Escape*, Harmondsworth: Penguin.

Berman, M. (1983) *All That is Solid Melts into Air*, London: Verso.

Clark, T.J. (1984) *The Painting of Modern Life*, London: Thames & Hudson.

Davies, L. (1987) 'If you've got it, flaunt it', *Employment Gazette* April: 167–71.

Debord, G. (1983) *Society of the Spectacle*, Detroit, IL: Black & Red.

Eco, U. (1986) *Travels in Hyper-Reality*, London: Picador.

Enloe, C. (1989) *Bananas, Beaches and Bases*, London: Pandora.

Faulks, S. (1988) 'Disney comes to Chaucerland', *The Independent* June 11.

Fiske, J. (1989) *Reading the Popular*, Boston: Unwin Hyman.

Frieden, B. and Sagalyn, L. (1989) *Downtown, Inc.: How America Rebuilds Cities*, Cambridge, MA: MIT Press.

Goodwin, A. (1989) 'Nothing like the real thing', *New Statesman and Society* August 12.

Hart, N. (1989) 'Gender and the rise and fall of class politics', *New Left Review* 175: 19–47.

Harvey, D. (1989) *The Condition of Postmodernity*, Oxford: Blackwell.

Ley, D. and Olds, K. (1988) 'Landscape as spectacle: world's fairs and the culture of heroic consumption', *Environment and Planning D: Society and Space* 6: 191–212.

Lowenthal, D. (1985) *The Past is a Foreign Country*, Cambridge: Cambridge University Press.

McKay, I. (1988) 'Twilight at Peggy's Cove: towards a genealogy of "Maritimicity" in Nova Scotia', *Borderlines* Summer: 29–37.

McKellar, S. (1988) 'The enterprise of culture', *Local Work* June: 14–17.

Morris, M. (1988) 'Things to do with shopping centres', in S. Sheridan (ed.), *Grafts*, pp. 193–225, London: Verso.

Myerscough, J. (1988) *The Economic Importance of the Arts in Britain*, London: Policy Studies Institute.

Pevsner, N. (1970) *A History of Building Types*, London: Thames & Hudson.

Pressdee, M. (1986) 'Agony or ecstasy: broken traditions and new social state of working-class youth in Australia', Occasional papers, South Australia College of Adult Education, Magill, South Australia.

Shields, R. (1989) 'Social spatialization and the built environment: the West Edmonton Mall', *Environment and Planning D: Society and Space* 7: 147–64.

Sontag (1979) *On Photography*, Harmondsworth: Penguin.

Travel Alberta (n.d.) *West Edmonton Mall*, Edmonton: Alberta Tourism.

Wolff, J. (1985) 'The invisible flaneuse: women and the literature of modernity', *Theory, Culture and Society* 2: 37–48.

7

GENDER AND CLASS RELATIONS IN TOURISM EMPLOYMENT*

Michael Ireland

INTRODUCTION

This paper focuses on two areas neglected by those engaged in the academic study of tourism in the United Kingdom. These are the forms of interaction within families whose female members were or continue to be engaged in domestic employment in tourism, and the relations between visitors and local women. The article draws on material gathered during a period of anthropological fieldwork carried out in Sennen parish (West Cornwall, UK), between March 1981 and November 1982 (Figure 7.1) to explore these questions. In the intervening period, between the publication of this paper and the completion of the initial fieldwork, contact has been retained with the community.

The methodological approach adopted was the use of the extended case method and situational analysis (Van Velsen 1964: xxxiii–xxix). This has been complemented and supported with the use of historical and contemporary documents. The aim throughout the research was to ensure that the study, as far as possible, reflected the passage of real time, and, as such, documented a period in the history of Sennen, and its local resident and the transitory visitor populations. The role of women within the tourism economy of Sennen was seen as integral to the research.

In addressing the subject of female employment in tourism, the aim is to counter the criticism leveled at anthropology as being "an exotic kind of salvage operation" by drawing together the concerns of anthropology and feminism to show that "gender . . . is experienced everywhere through the specific mediation" of class and history (Moore 1989: 37, 39). Gender in this paper concerns the cultural and social construction of a person rather than biological differences. The focus is on women and on male–female relations in both private and public spheres, including both the difference and the exercise of power between the sexes (cf. Bell and Newby 1976; Caplan 1989: 206; Frankenberg 1976: 25).

The term class in this paper is intended to refer to a specific set of empirical relationships investigated during the fieldwork (Crompton 1989: 567; Urry 1990: 140). In

* Reprinted with kind permission of Elsevier Science Ltd, Pergamon Imprint, Oxford, England, from *Annals of Tourism Research*, 1993, 20: 666–84.

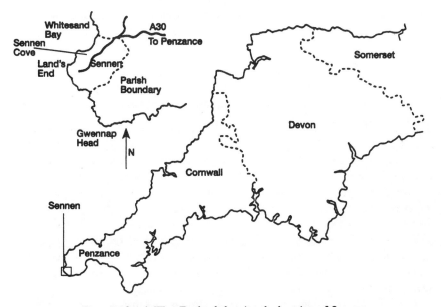

Figure 7.1 South West England showing the location of Sennen

common with the findings of Lummis (1977), the perceptions of class distinctions are recognized only as they apply to the wider society called "better class of visitor."

The paper is historical ethnography of Sennen showing that women became increasingly involved with tourism from the late 19th century, especially during economic hardship in the male-dominated fishing industry. In Sennen, the women's involvement with tourism appears to have afforded them greater economic and social significance in the parish than would have been the case if the economy had remained dominated by fishing (Kousis 1989: 318; Strathern 1985: 906). This argument is supported in part by the findings of Nadel's (1984) fieldwork in a Scottish fishing village that no longer engaged in this business. There women gained their identity through membership in kin groups (Nadel 1984: 109) and not from the restructured division of labor within the family, resulting from employment in tourism, as appears the case in Sennen. Evidence for this view is put forward later by examining three forms of relationships to which local women were party.

This study also draws on ethnographic material to demonstrate that local families, especially the women members, both identified with and differentiated themselves from their guests. The concept of the deferential dialectic (Bell and Newby 1976) is used to explain the form of relationship which existed between the tourist referred to by locals as "the better class of visitor," and by the local families whose women were involved in domestic employment in tourism. Domestic employment in tourism signifies the way in which tasks associated with household labor, food provision, housework, shopping, and laundry, characterized social relations of production and reproduction between women and visitors to Sennen (Whatmore 1991: 66). The

135

study concludes by arguing that deference to the tourist is redefined by a widening social and physical distance between the two groups, as class convergence takes place. In this respect, the role of the Ladies Lifeboat Guild in structuring relations between local women and outsiders is also examined.

A historical ethnography of Sennen

The history of Sennen in the 19th century essentially reveals the relationship between two of the occupational communities, fishing and farming, and how gradually these accommodated a third party, the tourism industry. The clearest indicator of the scale of economic and social differentiation within the parish can be found in the Census Enumerators books. Analysis of these sources confirms the relationship that has been suggested between the culture and geography of the parish (Swedland and Armelagos 1976: 5).

The working population can be broken down into three principal occupational groups that reflect the agricultural and maritime base of the economy. These were farmers, agricultural laborers, and fishermen. In addition, there was a small contingent of trades persons, coastguards, and professional people.

In the decade 1871–1881, the number of farmers rose marginally as a percentage of the working population, from 15% to 16%, while agricultural laborers fell by a similar amount, from 25% (1871) to 24% (1881). The most noticeable change in the working population was among the fishermen, who by the end of the decade represented 40% of the working population (an increase of 5% in the period between censuses).

Attachment to place has been shown to be important in creating a sense of belonging and identity among peoples (Cohen 1982). The perceived differences between the "Covers" (fishermen from the "Cove") and "Overhillers" (farmers) exemplifies this point (LaFlamme 1979: 140). To be a Cover is conditional on birthplace. The census show that fishermen were almost all born within the parish, if not in the Cove itself. In contrast, nearly half the agricultural laborers and farmers had moved to Sennen from other parishes.

The dominant position held by the local fishermen in the economy of the parish is shown in Table 7.1. In the decade between censuses the percentage number of fishermen who qualify as Covers, by birth right and livelihood, has shown a marginal increase. Residence, economic activity, and place of birth form the cornerstones of the local institution known as "being a Cover."

In sharp contrast to the solidarity of the local fishing community, the census data for agricultural laborers and farmers, collectively known as "Overhillers," show that considerable mobility took place between 1871 and 1881. While Table 7.1 shows the proportion of agricultural laborers born in the parish declined by 7.5%, the most dramatic change took place among the farmers. In the 10 years, the proportion of farmers who occupied land and had been born in Sennen fell from 42% to 25%. The landscape was increasingly occupied and farmed by men who had moved to Sennen from neighboring parishes, while the seascape and its resources became

Table 7.1 Distinction between fisherman, agricultural laborer, and farmer by birthplace

	Principal male occupational groups for the census years 1871 and 1881 by birthplace			
			Birthplace	
	1871		1881	
Occupation	Sennen	Another parish	Sennen	Another parish
Fishermen	95.4	4.6	96.8	3.2
Agricultural laborers	60.0	40.0	52.5	47.5
Farmers	42.2	57.8	25.0	75.0

almost exclusively the preserve of the Covers. These changes, looked at through the concepts of landscape and seascape, together with the evidence from the census, demonstrate that in the last quarter of the 19th century a revolution in tenure had occurred.

This examination of the labor force composition is thus far one-sided, as if women made little or no contribution to the economy (Dex 1988). But the emphasis on the male-dominated occupational groups provides historical evidence of the existence of a resource that the modern tourism industry is able to utilize: fishermen and the fishing industry. This approach leads to a paradox. In the 19th century, men were prominent in the social and economic institutions in the parish, while in the present century the economic position is reversed, with the decline in the fishing industry and the rise in tourism. However, as a consequence of the social significance of the fishermen and the attraction of the term Cover, women today, although the primary labor force in tourism, like their sisters in the 19th century, are still subordinate to the men (Collins 1984: 5). Observations in the field showed women displayed deference to men in social situations outside the home, in the public sphere. But in the private sphere, women controlled the household economy.

One reason put forward for the neglect of women has been the dominance of structural functionalist theories within sociology and anthropology that separated the family from the analysis of the relations of production in capitalist society (Beechy 1978; Collins 1984: 5; Crompton 1989: 582; Neff 1929: 17). In this context, according to Young and Willmott (1973), anthropology should seek an "alliance with history ... to study the development of the family in a wider social context" (Frankenberg 1976: 26).

The underrepresentation of women, in part, stems from the lack of available contemporary evidence. For example, it was not until the 1851 Census that marital status, by age and the pattern of migration by county, could be established. The 1891 Census provided for the first time tables showing employment patterns for married women. Only 10% of all married women in England were in employment. Caution needs to be exercised when interpreting these figures, because casual employment among women was often concealed from the enumerators or overlooked by them. This problem is particularly apparent in Sennen, as its 1881 Census returns illustrate. Common among the entries for the fishing community are those that suggest that

Table 7.2 1881 female population of Sennen by birthplace

Birthplace	Number	Percentage
Sennen	223	65.4
Land's End peninsula	68	20.0
Born outside West Cornwall	50	14.6
Totals	341	100.0

women had no official existence except through their husbands. For example, in Sennen Cove, the enumerator records John Roberts (head); occupation: fisherman; Grace Roberts (wife); occupation: "fisherman's wife"; and Richard Roberts (son) as "fisherman's son" (Census 1881: 10). The anthropologist working in Britain is in a position to overcome these problems by verifying official documentary sources against local historical research. What follows is an attempt to adopt this technique, using the 1881 Census as a base on which to build a picture of women's life and work in 19th-century Sennen.

Analysis of the census returns shows that 85% of the women in the parish at that time were born in Sennen or adjacent parishes on the Land's End peninsula. Of those women who had migrated from other parishes, 21 had come from predominantly agricultural communities (St. Buryan, Madron, Zennor, and Sancreed) to households in the townships on the plateau. Table 7.2 divides the women in the parish by the location of their birthplace. Of those women born outside West Cornwall, a number were the wives and daughters of the Coastguard and Trinity House (lighthouse) staff. In addition, a small number either gave birth or were born on Lundy Island, suggesting that wives accompanied their husbands to fishing grounds in the season. This illustrates their importance as a source of labor in processing fish, as well as bearing children, to form the next generation of fishermen and women.

Table 7.3 shows the breakdown of female employment throughout the Parish. The occupation housewife has been assigned by default to married women. The intention is to make the discussion of their role in relationship to "hidden" female employment more meaningful, housework being a form of unpaid employment. Insight into the life of these women is difficult. For the most part, one has to rely on secondary evidence for clues to their way of life.

In Sennen Cove, the contribution of women to the economy was firmly linked to fishing and in particular to the pilchard trade. This relationship was to change radically by the turn of the century. It is possible to identify accurately women referred to by informants from the census, giving a good indication of the importance of their contribution to the local fishery.

The granddaughter of a Cove woman recalled a superstition with economic logic. When the seine nets were being shot (that is encircling the fish shoal), women should not be seen outside. She recalls being told: "they could go out when the boats came

Table 7.3 The composition of the female labor force by occupation: Sennen
(1881 Census)

Occupation	Number employed	Percentage of the total number of women
Housewife	106	31.08
None recorded	103	30.20
Scholar	72	21.10
Domestic servant	30	9.00
Annuitant	7	2.05
Dressmaker	5	1.46
Charwoman	5	1.46
Barmaid	1	.30
Farmer	1	.30
Miscellaneous	11	3.05
Total	341	100.0

in." The reasoning behind this was easy to understand. The men could then be sure of a ready labor force to unload the catch. Margaret Trenary, a Cover herself, confirmed this: "The women used to help carry the fish in Cowls [fish basket carried on the back]."

This brief example suggests that the division of labor between men and women is a result of gender roles determined by culture and not gender difference. Support for the view that the difference and interaction are based on male and female economic relations outside the family can be found in the works of Rogers (1980: 17), Bell and Newby (1976), and Murrgatroyd (1982), which focus on the family unit. They argue that the fate of the women is determined by the male in the family unit and that sex typing of occupations is a reflection of social relations. Women provided an indispensable source of labor to the Parish.

The woman's contribution did not stop simply with fishing. When they were not helping men directly with the catch, they were the mainstay of what appears to have been a partial subsistence economy.

The local economy was slowly changing throughout the 19th century, partly through the new use of available capital and the application of entrepreneurial skills, and less obviously, through a change in the migrating habits and ecology of fish stocks. With the decline of the local fishery came the growing realization of an alternative use for capital, which was the land and seascape. This new form of capital was to provide the basis of the emerging tourism industry and as such increased the value attached to myths and legends, incorporating them as part of the fabric and romanticization of the dangerous occupation of fishing. The growth of a tourism industry in the local economy was founded on the visitors' wish to experience life in an occupational community. This led to the development of new forms of social relations between visitors and their host.

FORMS OF SOCIAL RELATIONS

Three forms of social relations have been found to exist between visitors and local people in Sennen: markers of the boundary between local and national systems of stratification; a relational boundary between locals and visitors; and a form of deferential interaction between tourists and local women employed to serve their needs. The three sets of relations change over time, as this discussion will show. Analysis of fieldwork records suggests that awareness of these relations is more acute among women. The reason for this is that they were, and remain, more actively involved with visitors to the parish, while the men's experiences with tourism are more remote. The women, with their basic domestic skills, provided the labor force; and the tourism industry encouraged them to enter into new relations of production with visitors (LaFlamme 1979; Stivens 1981: 118). This transformation has already been observed as occurring at the turn of the century in North Devon. Bouquet (1985: 65) says of Hartland:

> The emphasis placed on the "class of person" who stayed is interesting, since it indicates the process of transformation which was already beginning to affect the position and perception of the women who took in visitors.

Bouquet notes the new relations that developed as a result of the development of tourism:

> The people who came to stay in those days are today referred to retro-spectively as a "good class of person."
>
> (1985: 65)

Bouquet's reference to the visitor as a "good class of person" is not developed further in her book. However, the importance of this finding lies in the possibility of replicating these observations of early visitors in similar locations where anthropological fieldwork on tourism has been undertaken. Pi-Sunyer (1989: 195) illustrates the significance of the elite visitors for local people in a Catalan resort, making reference to a "real English gentleman" who "never failed to doff his hat to the women of the neighborhood." The impression given by informants was that "visitors in the past were all upper-class people" (Pi-Sunyer 1989: 195). The use of stereotypes by the host community to describe their guests has also been noted by Farrell (1979: 126) citing cases from the Pacific islands.

Similarly, fieldwork in Sennen, and in particular Sennen Cove, found that families who had become established visitors were referred to by the local people as "the better class of visitor." Encapsulated in this form of address are assumptions about attitudes and behavior toward visitors past and present and the nature of the relationship between host and guest families (Heal 1984: 68).

Pre-Second World War visitors

Social relations between local men and women and visitors have inevitably changed with the passage of time, each generation negotiating and redefining the form they take. Sennen women recognized the status of their guests and placed them firmly within the dominant stratification system (Crick 1989: 334). This was indicated by their constant reference to the "better class of visitor" or to the "gentry." Neither of these categories had previously figured largely in the local stratification system that recognized place of birth and residence as a method of assigning higher or lower status to a person. Being a Cover was perceived to be "lower" than someone born into the agricultural community "up the hill" (Overhiller).

The realization that women, who acted as intermediaries between local people and visitors, were standing metaphorically on the boundary of two different systems of stratification is shown in their comments about the better class of visitor. Margaret, a Cove woman who had served visitors, described them as ". . . a better class of person. We treated them really special . . . early visitors were the type of person you could look up to."

It is not difficult to detect elements of respect and deference to visitors in her relations with them. Margaret continued giving a detailed account of her family's relationship with the better class of visitor.

> Well, you see the women. They took the visitors in during the summer time to help keep them all in the winter; 'Cause the fishermen didn't earn any money then, you see. It was the women that looked after the visitors, the men had nothing to do with them; I lived up there with that headmaster of Cheltenham College (he had a house built overlooking the Cove). We went to live with him, mother went there as a housekeeper. She looked after the visitors in his house, that were masters from different Colleges, all summer.

Interviews conducted with surviving visitors of the pre-First World War (1905–1914) and inter-war years period (1918–1939) provide some evidence to support the view that the Cheltenham College master epitomized the type of visitor and the form of social relations described by this Cove woman.

Mr. Laseby, who could be described as one of the "better class of visitors," first came to Cornwall on his honeymoon in 1909. He stayed at the Land's End Hotel. The family spent every holiday in Sennen from 1920 to 1927. He recalled that many of the visitors would be University people. His impression at the time was that there "were never more than ten families." "If there was another party on the beach we thought it was crowded." These early visitors, according to Mr. Laseby, "lived on the fat of the land." Their diet would include Cornish cream and lobsters.

Mr. Johnson, a retired headmaster from Windsor, confirmed this account:

> As far as visitors went in those early days, I would say we were the pioneers. We were the sort of people that couldn't afford hotels and wouldn't want a hotel even if we could have. I can remember my father writing out a cheque

for £60 which covered return fare [train] to Windsor and five weeks board and lodging down here as well.

Mr. Johnson recalled some of his father's friends who came to Sennen in this period.

The first visitors were mostly professional. A lot of them were school masters or clergymen. I remember once in Sennen Church the local parson had to preach to, I think it was, something like four headmasters and three Bishops.

I can remember there were six of us down here staying with a Cove family. I think they enjoyed having us to be quite honest. Because apart from anything else we were a source of income. We used to pay by the week. My mother made suggestions about what we might have. Father being a headmaster, used to spend his morning, after breakfast until twelve o'clock writing letters.

Host and guests in the post-war period

Sennen women's perceptions of post-war bed and breakfast (and later self-catering) visitors can be drawn from the accounts of local women: Mrs. Pender, a prominent member of a Cove family, Mrs. Nicholas, a retired farmer's wife, Mrs. Beachfield, a widow, Mrs. James, a young housewife, and Mrs. Grant, who ran a self-catering accommodation business.

Mrs. Pender gave her impression of the changes in the form of hospitality being offered to the visitor in the post-war period. She argues that:

In the post-war era, Bed and Breakfast trade began – the older people who had let out apartments had nearly all retired and those who are left carried on the old tradition of "them and us" and provided meals for the new affluent working class finding it hard to change old habits.

Then in the fifties, the self-catering trade started and became very popular and is now the main trade. The Bed and Breakfast trade got a bad name when people moved into the district to make money fast – they would not provide evening meals and the visitors had to leave the house in the morning and not go back until evening [very few cafés open then].

It has not quite gone full circle but [speaking locally] the Bed and Breakfast trade is of a much higher standard, and, as in the pre-war days the men of the house had nothing to do with the visitors, today husband and wife work together (but not my husband and son!) in many cases.

The importance of bed and breakfast accommodations to the domestic economy has been recognized elsewhere in the Celtic periphery at this time. Ennew notes that "Many islanders provide bed and breakfast accommodation in their own homes, which provides extra income in the tourist season" (1980: 56). Armstrong describes a similar form of domestic tourism in the Highlands of Scotland. She says, "Twenty-two women participate in what is known as the 'bed and breakfast' tourist industry. These women convert their houses into guest houses for the season . . ." (1977: 139). Illustrations of the changes indicated by Mrs. Pender are evident from the comments of some of the women who provided hospitality, and their guests. Mrs. Nicholas had been taking in bed and breakfast guests since about 1940 at their farm situated on the parish boundary of Sennen–St. Just. She thought the war-time and post-war visitors were mostly professional types. Her remarks indicate a strong sense of mutual trust between the visitors and their host. She recalled that if someone had not come in, she would "leave the doors open" for the last visitor to return. The domestic routine of the farm kitchen formed another attraction of the farmhouse holiday. "They [her guests] would love to come and see me working, even if I was making pastries. They would help. I didn't mind them in the kitchen." Her attitude to visitors in the kitchen contrasts markedly with that of the Covers who did their best to keep them out, and in doing so drew a boundary around the domestic sphere of women's life (Ennew 1980: 79). Urry notes in this context that "people also appear to find interesting the backbreaking but unheroic household task undertaken by women." This interest in people's everyday chores is bound up with, ". . . the postmodern breaking down of boundaries, particularly between the front and back-stage of peoples lives" (1990: 107).

Mrs. Beachfield, a widow who had kept a guest-house in Sennen Cove from 1954 to 1981, was very clear about the change in her clientele:

> Well really then there was a better class of visitor, than what we get now; I used to have the Director of Rothmans tobacco people; but now (1981) you see its the ordinary, well, like ourselves, working class people that come; well its been a gradual change; it's gradually changed over the years.

Mrs. James, a Cornish woman in her early thirties, originally had come from St. Ives, had been living in Sennen for 12 years, and still did bed and breakfast. She explained that her mother was also in the holiday trade and, therefore, she had been brought up with visitors all her life (Gilligan 1984: 102–103). When she was first married, she used to take external employment during the winter and then "take the summer months off to do bed and breakfast." Her customers were mainly families.

Mr. James was very clear about how he saw the differences between the working class visitors and what he described as the "upper crust" families: "the ones that think they are well to do. The children just run wild; don't have any manners. They (the Parents) creep out of the house." When the visitors sat in the lounge, he complained that they would, ". . . pretend to read a book." He compared their behavior with that of the working class visitor: "You could tell they are working class." The implication in his next comment seemed to be that they had better manners. They would ask: "Could we have the T.V. on?"

Mrs. Grant, who owned self-catering flats, exemplifies one of the dominant forms of accommodation in Sennen in the early 1980s. Mrs. Grant believed, "You have to see holiday rents as business." The Grants indicated they were in direct competition with the old style bed and breakfast establishments. "From a business point of view, you have to think what it is going to cost for bed and breakfast." Turning to the type of visitor, she continued, "On the whole, the working class, the factory people, are grateful to you for letting the accommodation. They have saved up the whole year. I think it's up to us to ensure they are not disappointed."

Accounts drawn from the visitors in the post-war period give an insight into the way in which they experienced the modes of accommodation their host offered. Mr. and Mrs. Summerton had taken bed and breakfast accommodation with Mrs. Pender's husband's mother in Sennen Cove. Mr. Summerton described the standard of food and accommodation: "They would come for fourteen days. The cost of board would be £3.10 a week, half for children. For this they would receive a cooked breakfast and mid-day meal." He continued to give an insight into the type of accommodation; remarking that "Each family would have a sitting room and two bedrooms. All the local people did board. They used to disappear into nothing."

Over a decade later, in 1967, bed and breakfast provided by Cove families formed the main mode of budget holiday accommodation, with few other facilities. This point is illustrated by Mr. and Mrs. Banks, who first came to Sennen on holiday from the West Midlands in 1967, taking bed and breakfast accommodation with a Cove family, the Georges. The Banks' recalled:

> At that time, there weren't many people coming to the Cove. One of the few facilities for the visitor was a shop that supplied anything for the visitor, even small quantities of food for the beach.

Jim and his family from one of the commercial ports of Southern England typify the visitor associated with the growth of moderately priced self-catering accommodation in Sennen in the decade preceding the fieldwork (1970–1980). The family rented a self-catering chalet from a local family at a cost of £85.00 per week (for August 1981). Jim estimated the total cost of the holiday for his family of five to be £1,000 for the month. It was evident that they replicated their affluent working class lifestyle while on holiday. When asked what was their major interest when on holiday in Sennen, Jim replied "pubs." In contrast to the varied diet previously prepared by local women for their guests, Jim and his family appeared to consume large quantities of chips and canned beer.

As Jim prepared to leave Sennen Cove at the end of the family holiday that summer, the way this type of visitor was regarded by some local men became clear. Robert Pender, whose wife provided accommodation in the Cove, said on Jim's departure, "I couldn't care if I didn't see him again."

These accounts, by some local people and their guests, illustrate the gradual but irreversible change in the type of visitors who came to Sennen and the mode of accommodation offered.

FAMILY RELATIONS AND POST-WAR TOURISM

The involvement of women with the visitors can be seen as important in identifying the emergence of new relational boundaries between visitors and local people. By entering into a host–guest relationship with the visitors, women had established their own rules of hospitality. This relationship allowed men to maintain a social distance and thus preserve their local identity and personal space. The outlined restructuring of gender and class relations requires some attempt at theoretical explanation. The use of Bell and Newby's (1976: 154) concept of the *deferential dialectic* can be extended to account for the extra "familial personal relations" of Sennen women with their guests. Bell and Newby leave the field open for development: "we shall tend to ignore the precise nature of the articulation between productive relationships of society and the internal relationships of the family" (1976: 154). Before extending the use of this concept, it is necessary to better understand it.

The deferential dialectic concerns the forms of interaction that exist between husband and wife. Its twin elements are *identification* and *differentiation*. What Bell and Newby propose is that the relationship of marriage is by definition hierarchical, with husbands exercising traditional authority over wives. The opposing elements of the dialectic are the identification between husband and wife, which they argue to preserve stability must be positive and affective. If a careful balance between the opposing elements is not maintained, the relationship is at risk of destruction.

The "better class of visitor," it can be argued, cuts across traditional gender relations. Women come to defer not only to their husbands, but to women and men of a perceived higher class. The resentment stress within the marriage was minimized because men also recognized the visitors (reluctantly, as case material has demonstrated) as "gentry" and usually preferred to maintain a relationship of mutual respect rather than conflict.

This deference and the management of stress created by accommodating the "better class of visitor" are illustrated by the granddaughter of one of the Cove women. She said of her grandparents:

> [They] never used to call them [the visitors] Mr. and Mrs. "So and so." Always used to be Madam. I think they thought they was a bit better than we were.
>
> My Grampa, this is John White now, he used to catch crabs and things. Mother was just married. And my Granny was waiting for him to come in with some lobsters and crabs to give these people for their dinner. And he comes, here you have these crabs, the "ol lodgers shan't have em."
>
> I remember another funny thing. Once, she sent him up the butchers shop you see, to bring home meat. It was freezing cold weather ... and he came home with this great lump of meat in his hand. And my Gran say, "Your come John, I was waiting for that to cook for the lodgers." He said, "If I knowed that was for the lodgers, I'd throw it in the first bush I come to." That tells you what the men thought of the visitors here. I think they all benefited from it in the long run.

Fieldwork has demonstrated that families who took visitors were positively identified and adopted by them. This relationship between the better class of visitor and their host family was one of implicit patronage.

A local man, Jack, recalled a story often told in Sennen Cove, about the visitors. "People would ask what size shoes do you take? The old gentleman would leave a pair of shoes. This he explained, was done to humor the 'old men' who didn't really like visitors."

The impression given by locals who recalled this form of patronage was that the "better class of visitor" would leave a good quality pair of his own shoes. During the period of fieldwork, the practice of leaving new working boots was acknowledged with some embarrassment by one local man. Less subtle demonstrations of patronage occurred at Christmas time. One Cove family, the Georges, remembered being sent a food parcel, when, as Mr. George put it: "We hadn't got anything."

However, for this deferential dialectic across gender lines and between host and guest to continue, stability in relations is required. This careful balance was disrupted by the coming of the post-war visitors to the Parish. The nature of this disruption to family relations can be appreciated through the comments of a young Cove woman about the experience of her family in the post-war era. Susan began by describing her mother's role in providing bed and breakfast for visitors. She said,

> It was all basic. We never had no bathroom or nothing. Just a bowl of water in your bedroom. Then they [visitors] used to have their breakfast and go out on the beach all day. Never used to see them.
>
> The family, more or less, used to live here in the kitchen. Well you tried to keep it out of bounds. Because if they [the visitors] come in the kitchen, you'll never get rid of them.
>
> Daddy used to come home from work and they [the visitors] was in the kitchen. Couldn't sit down eat your dinner or speak or anything. You couldn't be rude or anything, but you try not to encourage it. Some people would be more interested in standing up in your kitchen than going on the beach.
>
> Perhaps they found us interesting, or else they just wanted to know. Good job they don't do it now [1982] that's all I can say. The men put up with it.
>
> But you ask anybody, I don't think anybody liked it. I think you like to keep them at what I call a friendly distance.

The form of interaction in the public sphere requires further detail.

FORMS OF INTERACTION BETWEEN HOST AND GUEST

The social structure of Sennen gradually became more complex as a result of internal and external economic and social change (Ryder and Silver 1977). The "specific interactions" documented here are "particular manifestations of larger state, class, and national politico-economic structures" (Crick 1989: 330).

Women in particular became identified as mediators between the major categories in a way that had not been possible between Covers and Overhillers, despite marriage. The provision of hospitality brought women into contact through a formalized set of relationships with their guests (Crick 1989: 331). Some of these guests eventually become *local residents* within the parish.

While on holiday in Sennen, visitors were assigned a place in the local social structure as guests, although the rules of behavior toward these guests changed over time, as local perceptions of the visitors altered. For example, an inverse relationship has been observed between the visitors' perceived social distance from their hosts and the mode of holiday accommodations. This took the following form: taking in "lodgers," who were usually seen to be "the better class of visitor," was replaced in the post-war period with accommodations for visitors in self-catering units, thus socially and physically distancing the visitors from their hosts.

Richard, a Cover whose wife and sister both took in visitors, saw the transition from bed and breakfast to self-catering accommodations as desirable. He said of contemporary visitors, "They are out there and they are out of the way." Referring back to the period when his wife took in visitors as bed and breakfast guests and he worked as a carpenter, he said disapprovingly: "You would come home and they would be in the kitchen."

From conversations with Richard it seemed that he preferred the relationship of mutual respect that existed between the better class of visitors and the Covers. Territorial boundaries had been unwritten and were rarely transgressed, the result being that good relationships have endured.

However, from the remarks made with regard to post-war visitors, it is clear that they did not enjoy the same degree of deference as was afforded the better class of visitor by the locals. This can be shown by examining the new set of relations between locals and *former visitors*, now local residents, who had retired to Sennen.

The problem faced by this now heterogeneous community is how to conduct relationships in a non-holiday setting for two of the three groups (locals and local residents). As Dogan points out, "touristic development may result in conflicts and hostilities between groups whose interests are differentially affected by tourism . . ." (1989: 226). The way in which relations between host, guests, and local residents was managed in Sennen can best be understood by looking briefly at the Lifeboat Service as a social institution acknowledged by all parties.

The Lifeboat is a complex institution of which the boat is actually only a small part. The Lifeboat brings into contact, but does not bring together, three groups: the local men who crew the boat; the women of the Ladies Lifeboat Guild who are almost exclusively local residents; and the management committee, which is an alliance of retired naval personnel and local councillors. To the visitor, the Lifeboat is the center of the community, but to Sennen people and local residents it does not symbolize village unity. Rather it represents the differentiation and conflict that exists between social groups in the parish. A discussion of the boundaries in the village that the Lifeboat symbolizes must remain the subject of a future paper. Important here is the way in which the Ladies Lifeboat Guild illustrates the complex structuring of social relations between local women and those who have moved into the Parish.

THE LADIES LIFEBOAT GUILD AND SOCIAL RELATIONS

The Ladies Lifeboat Guild in particular provides a niche for women who have moved into the parish (Armstrong 1977). The local Ladies Lifeboat Guild chairperson

147

confirmed that most of their helpers were from "up country." However, they did have one woman whose husband was on the Lifeboat crew. The well-documented practice of "newcomers" wanting to join the Guild on moving to Sennen Cove can be illustrated by the Eastfields of Cape View, Sennen Cove. Mrs. Eastfield had promised the Chairperson that she would help on the stall. This initial willingness to help is not always forthcoming, as evident from the following account. The organization of the Guild membership had recently been changed. The Chairperson recalled that originally it was based on a yearly subscription. However, the disadvantage with this system was that people only turned up on special occasions. Now the subscription is voluntary, but people have to contract to give so much service on the stall. The stall opens Monday to Friday from 10:30 to 4:30, but not Saturday. The reason it is not open Saturday is linked to the "changeover" in holiday accommodation that occurs on that day. At present, the Guild stall is run by 20 members but, as the Chairperson pointed out, "we really need six helpers for five days."

Phil, a member of the crew, summed up Sennen people's view of the Guild: "Not a local woman in it." Hugh, another member of the Lifeboat crew, crystalized the divisions within this institution as a whole, from the local point of view:

> The Lifeboat women have a stall which sells souvenirs. When there is a dinner [to which the crew, committee and ladies guild are invited] its them [local residents] who are first and foremost, not the actual crew. All they get is two pounds eighty pence a call out.

The lack of involvement of local women is not just an expression of native xenophobia, but competition between themselves and Guild members for scarce resources, monies spent by the visitor (LaFlamme 1979: 140).

Observations of some of the events the Guild organized tended to confirm this view. In the year preceding the fieldwork (1980), the Guild took in £10,000, one third from the sales of souvenirs. The Chairperson of the Guild remarked: "On a good day they sometimes took £400."

Annually there are three major events at which the Guild raise funds in addition to the regular stall. The season begins with Sennen Cove Lifeboat day, which has been celebrated in July each year since 1935. August is particularly active with a Cornish Fayre held in the first week, the male voice choir concert in the Lifeboat house at the end of the second week, and the Royal National Lifeboat Institute flag day, which is held in late August each year.

CONCLUSIONS

This paper has traced the restructuring of gender and class relations in a community that has become dependent on tourism as a means of income and employment for its inhabitants.

The restructuring of social relations in Sennen between women and men, both within the household and external to it, began with the decline of the traditional economy that was dominated by the fishery. This was accompanied by a shift in focus of production from outside the household to within, as families shared their homes with the "better class of visitor."

The embryonic development of tourism in Sennen, signified by its growth in popularity among the upper classes, was fostered by the occurrence of economic resources that could be combined: the land and seascape, the homes of local people to accommodate the visitor, and most importantly a female labor force with the skills required. Men were to become a "tourist attraction" in themselves as markers of a fishing industry, which the visitor found harsh yet romantic. In reality, men were and continue to be the victims of structural unemployment, whose value is as cultural artifacts, rather than a source of labor.

The transformation of relations of production in West Country coastal villages has been noted elsewhere (Bouquet 1985). This paper sought to address the problem using anthropological research to clarify historical documents with regard to the role of women's employment. Ethnographic research has enabled the identification of families whose women members had begun to take on the role of provider of hospitality for the better class of visitor to supplement the family income.

Three forms of social relations have been shown to exist between local women and the visitors. Prior to the Second World War, relations with the better class of visitors were deferential, indicated by the way in which informants talked about them as "gentry" and "people you could respect." The immediate post-war period brought about a broadening of the socioeconomic backgrounds of the visitors to Sennen, a change that was quickly understood and interpreted by their host. Local women were less deferential in their relations, referring to visitors as being "just like ourselves." "Over identification (otherwise referred to as the familiarity breeds contempt syndrome) may result in a denial of authority" (Bell and Newby 1976: 157) and lead to the destruction of the deferential relationship between host and guest.

The gradual convergence of perceived social distance between host and guest resulted in new relational boundaries emerging. An inverse relationship between perceived social distance and actual physical distance was signified by the demise of live-in family holidays and the growth of self-catering accommodations separate from the host family. One consequence of these changes was that it was no longer necessary for local women and men to share the same social space with their visitors. Intra-gender relations of apparent equality occurred between women perceived to be of similar social classes. The relations arose from employment in the holiday setting through the provision of hospitality as a means of income, and occurred less in other spheres of local social life. With perceived class convergence between visitor and local, deference was no longer an appropriate strategy. Instead, increased social and physical differentiation resulted.

REFERENCES

Armstrong, Karen 1977 Women, Tourism, Politics. *Anthropological Quarterly* 50(3): 135–145.

Beechy, Veronica 1978 Women and Production: A Critical Analysis of Some Sociological

Theories of Women's Work. In *Feminism and Materialism: Women and Modes of Production*, A. Kuhn and A. M. Wolpe, eds., pp. 155–197. London: Routledge.

Bell, Colin, and Howard Newby 1976 Husbands and Wives: The Dynamics of the Deferential Dialectic. In *Dependence and Exploitation in Work and Marriage*, D. L. Barker, and S. Allen, eds., pp. 152–168. London: Longman.

Bouquet, Mary 1985 *Family Servants and Visitors: The Farm Household in Nineteenth and Twentieth Century Devon*. Norwich: Geo Books.

Caplan, Pat 1989 Perceptions of Gender Stratification. *Africa* 59(2): 196–208.

Census 1871, 1881 *Civil Parish of Sennen*. London: Public Records Office.

Cohen, Anthony 1982 *Belonging: Identity and Social Organisation in British Rural Cultures*. Manchester: Manchester University Press.

—— 1985 Introduction: Symbolism and Boundary. In *The Symbolic Construction of Community*, pp. 11–15. London: Tavistock.

Collins, Irene 1984 Hardly any Women at All. Presidential lecture to the Historical Association at their Spring Conference and Annual General Meeting held at the University of East Anglia (1983). *Historian* No. 5 (Winter): pp. 3–8.

Crick, Malcolm 1989 Representations of International Tourism in the Social Sciences: Sun, Sex, Sights, Savings and Servility. *Annual Review of Anthropology* 18: 307–344.

Crompton, Rosemary 1989 Class Theory and Gender. *British Journal of Sociology* 40: 567–587.

Dex, Shirley 1988 Issues of Gender and Employment. *Social History* 13(2): 141–150.

Dogan, Hasan Zafer 1989 Forms of Adjustment: Social-cultural Impacts of Tourism. *Annals of Tourism Research* 16: 216–236.

Ennew, Judith 1980 *Western Isles Today*. Cambridge: Cambridge University Press.

Farrell, Bryan 1979 Tourism's Human Conflicts: Cases from the Pacific. *Annals of Tourism Research* 6: 122–136.

Frankenberg, Ronald 1976 In the production of their lives, Men (?) . . . Sex and Gender in British Community Studies. In *Sexual Divisions and Society: Process and Change*, Diana Leonard Barker and Sheila Allen, eds., pp. 25–51. London: Tavistock.

Gilligan, Herman J. 1984 The Rural Labor Process: A Case Study of a Cornish Town. In *Locality and Rurality*, T. Bradley and P. Lowe, eds., Norwich: Geo Books.

Heal, Felicity 1984 The Idea of Hospitality in Early Modern England. *Past and Present* 102: 66–93.

Kousis, Maria 1989 Tourism and the Family in a Rural Cretan Community. *Annals of Tourism Research* 16: 318–332.

LaFlamme, Alan G. 1979 The Impact of Tourism: A Case from the Bahama Islands. *Annals of Tourism Research* 6: 137–148.

Lummis, Trevor 1977 The Occupational Community of East Anglian Fishermen: A Historical Dimension Through Oral Evidence. *British Journal of Sociology* 28(1): 51–74.

Moore, Henrietta 1989 When God Created Women. *New Statesman and Society* March: 37–40.

Murrgatroyd, Linda 1982 Gender and Occupational Stratification. *Sociological Review* 30(4): 576–602.

Nadel, Jane Hurwitz 1984 Stigma and Separation: Pariah Status and Community Persistence in a Scottish Fishing Village. *Ethnology* 23(2): 101–115.

Neff, Wanda F. 1929 *Victorian Women: An Historical and Literary Study of Women in British Industries and Professions 1832–1850*. London: George, Allen and Unwin.

Pi-Sunyer, Oriel 1989 Changing Perceptions of Tourism and Tourist in a Catalan Resort Town. In *Host and Guests: The Anthropology of Tourism* (2nd edn), Valene L. Smith, ed., pp. 187–199. Philadelphia: University of Pennsylvania Press.

Rogers, Barbara 1980 Women and Men: The Division of Labor. In *The Domestication of Women: Discrimination in Developing Societies*. London: Kogan Page.

Ryder, Judith, and Harold Silver 1977 *Modern English Society*. London: Methuen.

Samuel, Raphael, ed. 1975 *Village Life and Labor.* London: Routledge.

Stivens, Maila 1981 Women, Kinship and Capitalist Development. In *Of Marriage and the Market: Women's Subordination in International Perspective*, pp. 112–126. London: CSE Books.

Strathern, Marilyn 1985 Women's Studies in Social Anthropology. In *The Social Science Encyclopedia*, Adam Kuper and Jessica Kuper, eds., pp. 905–907. London: Routledge.

Swedland, Alan C., and George L. Armelagos 1976 *Demographic Anthropology*. Dubuque, IA: Wm. C. Brown.

Urry, John 1990 *The Tourist Gaze: Leisure and Travel in Contemporary Societies.* London: Sage.

Van Velsen, J. 1964 Note on the Situational Analysis. In *The Politics of Kinship*, pp. xxxiii–xxix. Manchester: Manchester University Press.

Whatmore, Sarah 1991 *Farming Women: Gender, Work and Family Enterprise.* London: Macmillan.

Young, Michael, and Peter Willmott 1973 *The Symmetrical Family.* Harmondsworth: Penguin Books.

Part IV

TOURISM, UNDERDEVELOPMENT AND DEPENDENCY

8

TOURISM, DEPENDENCY AND DEVELOPMENT*

A mode of analysis

Stephen Britton

A great deal of the literature on tourism's contribution to development is devoted to either narrowly defined cost-benefit analyses, imprecise comments on the socio-cultural effects of tourism, or more technical issues such as tourist flow predictions, factors determining hotel location, and the regional impact of tourist expenditures. Clearly such studies are important. However, most of this research has a common deficiency. Discussion over the impact of tourism is typically divorced from the historical and political reality of processes that have led to the condition of under-development (de Kadt, 1979). Debate on the advantages and disadvantages of tour-ism is conducted without regard to those theories of political economy concerned with widespread, persistent poverty, and the causes of increasing inequality between and within nations.

The intention of this paper is to place the study of tourism firmly within the dia-logue on underdevelopment. It is important to investigate why it is that tourism, while bringing undoubted benefits to many poor countries, frequently also perpetu-ates already existing inequalities, economic problems and social tensions. This task requires an understanding of the underlying mechanisms, inherent in the tourist in-dustry and Third World economies, that make the promotion of tourism a highly ambiguous development strategy. To construct a framework for investigating the ar-ticulation of international tourism with Third World tourist destinations, two sets of factors need to be considered. The first consideration is an analysis of the organisa-tion and commercial structure of the tourist industry, particularly the power and dominance of certain activity components and ownership groups. The second con-sideration is that to understand how the industry manifests itself, and who benefits from tourism development, account should be taken of the organisation of eco-nomic and political power within Third World countries. This requires an appreci-ation of the historical forces that are responsible for the common characteristics of these economies.

* Reproduced with permission of Peter Lang Publishing, Inc, from Tej Vir Singh, H. Leo Theuns, and Frank M. Go (eds) (1989) *Towards Appropriate Tourism: The Case of Developing Countries*, Frankfurt am Main, Germany: Verlag Peter Lang GmbH, pp. 93–116.

THE ORGANISATION OF INTERNATIONAL TOURISM

When Third World countries participate in international tourism, they have to accept various commercial practices that typically accompany any activity that has its historical origins in the developed, or metropolitan, countries. The reason for this is simply that being first in the field gives considerable commercial advantages to pioneering firms since they define, create and supply the new industry. As the tourist industry is designed to meet, and arose out of, the recreational needs of affluent middle class citizens in the world's rich countries, metropolitan companies have become predominant in the control of international tourist movements, the definition and promotion of the tourist experience, and the organisational form by which overseas holidays are undertaken. A poor country seeking to promote tourism as a means of generating foreign exchange, increasing employment opportunities, enhancing economic independence or promoting the commercial involvement of poor sections of the community is likely to find the attainment of such goals impeded by this organisation of the tourist industry.

Since metropolitan enterprises are actually located within the principal tourist markets, they have direct contact with tourists. They therefore control the key link in the tourist flow chain (IUOTO, 1976: 41–46). Given the nature of tourist travel, particularly in packaged form, tourists may not know the type of holiday they want, nor may they be definite about which destination(s) they wish to visit (Burkart and Medlik, 1974: 213–214). A tropical island holiday, for example, may be enjoyed in any one of a dozen Pacific Island destinations. This situation puts industry intermediaries between the tourist client and the destination countries in a pivotal position. It allows metropolitan tourism corporations to influence the volume of tourist flows to any one market (where subsidiary tourist companies may be owned). It also encourages foreign interests to become directly involved in the destination country since their capital resources, expertise, market connections and control over tourist flows give them overwhelming competitive advantages over local tourism operatives.

Equally important is the fact that the great majority of tourists travelling the Pacific and elsewhere have little option but to purchase the services of metropolitan airline and cruise-ship companies. In 1978, for example, two foreign airlines were responsible for approximately 80 per cent of airline seats on the Fiji sector (IUOTO, 1976: 43–45). No island nation owns one of the six companies which have cruise-ships plying the Pacific; and few governments can afford a national airline.

Those regional carriers such as Air Pacific and Polynesian Airways which do compete for international tourists have their operations strictly confined by high operational costs, limited equipment capacity, competition from metropolitan airlines and interference by foreign management and shareholding interests (Aidney, 1977). Direct pressure by metropolitan governments also ensures protection for their national airlines (e.g. Qantas and Air New Zealand). The restricted landing and traffic rights granted to regional carriers by foreign governments prevent them gaining access to the most important tourist markets. Similarly, forcing regional carriers to fly circuitous routes to major tourist markets necessitates them charging uncompetitive fares to compensate for greater travel distances (Britton, 1979: 176–210; Kissling, 1980). Foreign transport companies, therefore, heavily influence the volume and direction of tourist flows. Airlines are particularly effective in this regard through their ability to: offer discount concessions to tour companies for any one destination; vary the allocation of seating on scheduled flights; discriminate against stop-over tourists; overfly intermediate destinations; and promote or demote a destination by changes to market advertising.

This latter factor, tourism promotion, is a vital element in the dynamics of the tourist industry. Along with transport linkages, accommodation provision and fare costs, the image or perceived attractions of a destination are the critical factors in directing tourist flows. Effective international tourist promotion, however, is undertaken by few components of the industry. Tourism as an industry is not a single, functional entity in the conventional sense. Rather it is a variety of activities or components which taken together form the tourism product group. The most important of these components are metropolitan airline, tour wholesaling and hotel chain companies. It is these sectors which most influence tourist movements and, as a consequence, undertake the most extensive advertising campaigns.

Within a tourist destination, advertising is most successfully undertaken by the larger companies which are able to absorb the high costs, and those which have direct sales and marketing links in the tourist source countries. Foreign owned and very large local companies meet these requirements. Invariably, however, these same conditions militate against small scale local operators who have limited resources, limited experience and few connections within the industry.

This situation has several important repercussions. Foreign companies, through advertising, greatly influence the image of a destination country. This typically leads tourists to perceive the host country in terms of this image and the nature of the hotel accommodation, tourist shopping, select cultural attractions and other tourist services which are publicised. All too often this results in tourists having unjustified expectations of a destination which are unable to be fulfilled (Britton, R., 1979; IUOTO, 1976: 103–105).

Marketing strategies, along with tourists' partially manipulated expectations, and the transportation modes used by airline and wholesale tour operators, all help determine the type of tourist product that is created in underdeveloped countries. More importantly, the greater the extent to which metropolitan firms promote a destination, the more incentive there is for these firms to ensure the stability and viability of their operations through direct commercial participation within the destination. In addition, because of the inability of agricultural and manufacturing producers in the underdeveloped economies to guarantee the quality and continuity of supply of inputs appropriate for "international luxury standard" tourist facilities, there is a strong reliance on imported supplies for both the construction and operation of destination tourist facilities. In Fiji, for example, 53 per cent of hotel food purchases, 68 per cent of standard hotel construction and outfitting requirements, and over 95 per cent of tourist shop wares were supplied from imports (Varley, 1977: 3; 1977, pers. comm.; Britton, 1979: 306).

Foreign tourism companies also monopolise industry managerial expertise

157

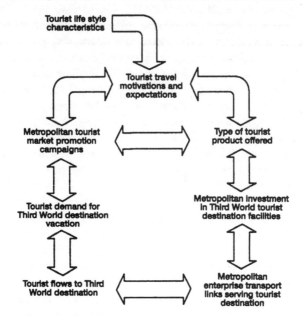

Figure 8.1 Industry inter-relationships in the generation of tourist flows

marketing skills, financial resources, and intra-industry contacts (IUOTO, 1976: 36–64, 70–71, 77–78). The apparent market competence of these metropolitan companies renders them "natural" recipients of destination government aid, co-operation and subsidisation (Hiller, 1977: 116). From the perspective of a nascent Third World industry, therefore, integration with foreign tourism capital appears both beneficial and necessary.

A series of feedback loops is thus evident within the international tourism industry (Figure 8.1).

The promotion and advertising strategies of metropolitan tourism corporations play a significant role in shaping tourist expectations. These expectations tend to conform to a type of tourist product and travel experience that best suits the priorities of these tourism firms. These priorities in turn are the key determinants of the type of tourist facilities developed in Third World tourist destinations. These take the form of luxuriously appointed, capital and energy intensive hotel resorts, the type that poor countries can least afford to build and operate. Not surprisingly, therefore, since such facilities are best planned, constructed and managed by those tourism firms with international experience, there is every incentive for metropolitan corporations to invest in, and of course profit from, Third World tourism. This is particularly evident in the hotel sector (IUOTO, 1976: 42). In the Pacific, one firm, South Pacific Properties Ltd, operates over 60 hotels throughout the region. In the Cook Islands, one hotel with substantial foreign ownership controls 50 per cent of hotel beds, and in Fiji, foreign companies are responsible for 58 per cent of tourist hotel rooms and 65 per cent of hotel sector turnover (Britton, 1981; Universal

Business Directories, 1979: 7). Underdeveloped countries participating in the international tourist trade are thus obliged to accept a high degree of foreign owner-ship, retention of tourist receipts in the metropolitan countries and leakage of foreign exchange earnings (IUOTO, 1976: 64–100). At the very least, there is likely to be a high level of expatriate management and control of key sectors of periphery tourist industries. In Fiji, more than 70 per cent of tourist expenditures are lost to the country in the form of import payments and profit repatriation; and 57 per cent of foreign hotel management positions were filled by expatriates in 1976 (Central Planning Office, 1975: 169; Britton, 1979: 239). This situation is exacerbated by the fact that tourist travel is most commonly undertaken as a package tour, since it is this organisational form that best allows foreign companies to benefit from tourist expenditures (IUOTO, 1976: 105–115).

Apart from substantially lowering travel costs, the package tour concept brings together both the conflicting psychological needs of tourists for novelty and security in a strange environment (Cohen, 1972: 166), and the commercial imperative of constituting individual touring into a standardised, repeatable and marketable prod-uct (Britton, 1981). This means of organising tourist travel has three important consequences.

As indicated above, where most of the tourists travel to a destination in group or package tours, there will be a high retention of tourist expenditures by foreign com-panies. It has been found (ESCAP, 1978: 40), for instance, that where tour packages consist of a foreign air carrier, but include local hotel and other group services, des-tination countries receive on average only 40–45 per cent of the inclusive tour retail price paid by the tourists in their home country. If both the airline and hotels are owned by foreign companies, a mere 22–25 per cent of the retail tour price will be forwarded to the destination country.

This confinement of tourists (and their expenditures) to a formalised travel ex-perience has its parallel in the spatial organisation of tourist services within the des-tination country. Many tourists are unable to fully enjoy their holiday except from a base of familiarity provided by western type hotels, transport modes, food stuffs or shops. They are unlikely to venture far from the environment provided by the packaged travel experience. In consequence, tourism tends to manifest itself as an enclave industry. On the one level tourists confine their social interaction to other tourists or to locals who behave most like westerners. Interaction with other locals and hotel staff is conducted in a manner consistent with stereotypes held by both groups (Cohen, 1972; Nettekoven, 1974). On the other hand, the luxurious, langorous surroundings of tourist accommodation and transport, as well as the prodigious expenditures of tourists, creates physical and economic barriers between visitors and hosts. This appearance of tourist enclaves is greatly accentuated if serv-ices catering to tourists are clustered in space, as is evident in Tahiti (Robineau, 1975: 67–74). This is also the case in Fiji where 95 per cent of all tourist expenditures occur in the Lautoka–Suva tourist belt on the main island of Viti Levu (Britton, 1980a).

The reduction of the travel experience to a relatively standardised product can also have the effect of putting a destination into a marginal market position. A country's tourist product may be easily substituted by foreign companies and tourists for that of an alternative destination (IUOTO, 1976: 101–103). One of the characteristics of tourism in the South Pacific is that a multitude of destinations sell an essentially undifferentiated product. With foreign companies largely responsible for advertising, international transport, global wholesaling of tours, and the provision of international hotel chains, no one tourist destination is able to gain satisfactory control over the viability of its own tourist facilities.

At the risk of oversimplification, it can be said that such an industry structure ensures that Third World destinations have a largely passive and dependent role in the international system. Foreign multinational companies directly serve, and partially create, the demand and the means by which tourists consume Third World tourist products.

Destination countries on the other hand are the recipients of tourists. They provide the novelty and superficial rationale for an overseas holiday. The central problem, then, for Third World destinations, is the essentially inequitable relationship inherent in this international system. Immobile tourism facilities in tourist destinations rely on foreign corporations to supply both the tourist environment and the tourists to ensure viability. The flow of tourists to these destinations is achieved by either gaining the co-operation of foreign interests or by national bargaining power over factors affecting the profitability of these foreign interests. For, in the final analysis, metropolitan tourism corporations could direct tourists to alternative destinations. The influence and negotiating power of destination countries is further weakened by the fact that control of a very high proportion of tourist flows is in the hands of only a few corporations in each metropolitan market (IUOTO, 1976: 102–103).

If the tourist trade is so organised that metropolitan interests rather than tourist destination interests predominate, it can also be shown that a similar inequitable distribution of commercial power and attendant benefits occurs within destination countries. This is primarily due to the form of economic and political arrangements that predominate in Third World countries.

THE STRUCTURE OF PERIPHERAL CAPITALIST ECONOMIES

Focusing on the enterprise organisation of international tourism, the industry's structure can be conceptualised as a three-tiered hierarchy. At the apex are metropolitan market countries in which are located the headquarters of those transport, tour, hotel and tourism supplying companies which dominate the lower levels of the industry hierarchy. At the intermediate level, in the tourist destinations of the underdeveloped countries, are the branch offices and associate commercial interests of metropolitan firms operating in conjunction with their local tourism counterparts. At the base of the pyramid lie those small scale tourism enterprises of the destination country which are marginal to, but dependent upon, the tourist companies of the intermediate level.

It is no coincidence that this hierarchy is a microcosm of a more universal organ-isational form: that of the multinational company (Hymer, 1972: 49–50). By its very nature, international tourism encourages the participation and creation of multi-national companies (Britton, 1981). However, the economic and political implica-tions of promoting development strategies that involve such foreign participation are but part of a still wider problem. That is the set of inequitable linkages between developed and underdeveloped countries, the core and the periphery, referred to as relations of dependency between dominant and subordinate nations and classes (Dos Santos, 1973; Obregon, 1974).

The dependent status of Third World, or peripheral, countries has its origins in a common historical experience of colonialism. Colonialism articulated peripheral countries with the global capitalist economy in such a way that the former had im-posed upon them forms of production, social organisation and trading patterns de-signed to meet the economic and political requirements of the colonial powers (Amin, 1974; Beckford, 1972; Frank, 1978). Because of the pervasive impact that co-lonial and imperialist institutions had on the periphery, and the extensive economic and strategic gains which accrued to western powers, the process of industrialisation for a few countries and the underdevelopment of the remaining majority can be re-garded as part of the same set of processes.

As a corollary of their integration with the core capitalist economies, underdeve-loped countries are now characterised by specific forms of distorted development. The economic organisation (mode of production) and associated political institu-tions of underdeveloped countries can be identified as a distinct type of social formation – that of a peripheral capitalist economy (Amin, 1976). This designation applies where a capitalist ("modern") enclave sector, comprised typically of mon-opolistic firms, is deeply integrated into an economy and linked to a variety of non- or petty capitalist ("traditional") subsistence sectors. The resulting linkages between the two sectors are primarily a response to the needs and pace of expansion of the capitalist sector. In other words, peripheral capitalist economies are characterised and defined by a *combination* of capitalist and non-capitalist forms of production under the domination of the former.

This structure can also be envisaged as one where capitalist sector firms have few linkages among themselves, but are, in fact, an integral part of the developed economies where their company head offices are located (Amin, 1974: 259). These firms profit from the commodities they manufacture, mine or grow in the periphery, as well as from cheap and ample labour. There are many examples where large foreign concerns used bases in the Pacific to supply industrial commodities to their processing plants in developed countries. In the process, Pacific Island economies were profoundly restructured by such companies as: the Colonial Sugar Refining Company's (Australia) sugar plantations in Fiji, Lever's Pacific Plantations Ltd (U.K.) copra plantations in the Solomons, Société de Nickel's (France) mining operations in New Caledonia, and the Deutsche Handels und Plantogen Gesellschaft der Sudsee Inseln zu Hamburg's (Germany) copra plantations in Papua New Guinea (Brookfield, 1972: 34–44).

161

Those countries which make up the core and periphery of the world cannot be seen as comprising simply of separate independent economies. They can also be regarded as specialised "sectors", performing different functions, in one global system dominated by the developed countries (Wallerstein, 1972: 2). That is, there exists an inequitable international division of labour in which the practices governing such factors as specialised commodity production, international trade, as well as technology, migration and capital flows, result from the actions of companies, institutions and governments in the core countries. Furthermore, these pressure groups, through their close association with dominant local political and commercial classes, are able to encourage political decisions over economic policy, commercial practices and labour legislation consistent with their interests. The close liaison with local political groups, one which persists despite the divergence of national as well as elite vested interests from foreign parties, is itself a legacy of the economic, political and ideological transformation of traditional societies through the "total institution" of colonialism (Frank, 1972; Witton, 1974). Equally important is the fact that it is only members of the local elite classes who have the power to bargain with either foreign industry or government representatives and to implement policies appropriate to such interests.

These characteristics of peripheral social formations have a considerable impact on the way an industry which has its origins in the core countries, such as tourism, is integrated into the local economy. The organisation of production in the form of foreign dominated export enclaves means that firms in these sectors heavily influence the economic milieu of local productive activities outside these enclaves. Since governments in the periphery are so dependent on the dominant enclave sectors for the economic viability of the state (and their own political legitimacy) the operation and expansion of these monopolistic interests occurs in a sympathetic "investment environment". The allocation of financial aid, the provision of infrastructure, the orientation of administrative services and the passing of licensing, labour and marketing regulations all proceed in accordance with the requirements of the dominant sectors.

A large differential is then created between productive resources, public support and political influence accorded to the strategic enclave industries and the remainder of the economy. The historical advantages held by foreign and local elite owned companies perpetuate a situation where other small scale or petty producers cannot evolve to a point where they effectively compete with the favoured firms or benefit from public support services (Bienefeld, 1975: 64). At the same time, the distortions in the economy, especially the selective allocation of resources and stunted accumulation of surpluses for productive purposes, result in a progressive inability of the economy to meet the food, shelter, social welfare and employment needs of an increasingly impoverished, marginalised majority of the population. One consequence of these two tendencies is that only privileged commercial and political groups in the periphery, along with foreign interests, are in any position to coordinate, construct, operate and profit from the development of a new industry – such as tourism. In the absence of any concerted government intervention, this situation manifests itself as increasing economic and social polarisation. This occurs at three scales: between metropolitan and periphery *states*; between dominant capitalist and marginal petty producer and subsistence *sectors* within the periphery; and between dominant and subordinate *enterprises* within any one sector or industry.

There is one other issue of importance. Not only does the distorted allocation of resources within a peripheral economy lead to the marginalisation of much local enterprise, it also results in a series of contradictory processes that cause very serious problems of economic and political management. Only passing reference is possible here on the scope of these problems. Mention can be made, however, of a few problems which have more or less relevance to Pacific Island economies (Shand, 1980). Colonial labour policies and the creation of, or support for, inequitable land tenure systems have rendered many rural dwellers landless and dependent on wage labour. At the same time, traditional agriculture has been ignored by governments with the shift of resources to export crops and away from domestic food production. Yet the commercial imperatives of dominant sector firms and the narrow allocation of public funds have meant that the rate of employment creation has fallen far short of employment demand. In addition, the neglect of domestic food production, the local manufacture of luxury goods, conspicuous consumption on the part of the local rich, and the westernisation of consumer preferences all create excessive demands for imports. Simultaneously, these countries suffer from the outflow of capital accompanying the presence of foreign companies, dependence on a narrow range of primary exports, adverse terms of trade for these commodities, restricted access to metropolitan markets, and the domination of international transport and commodity processing by metropolitan firms. All these factors operate to drastically reduce the capacity of peripheral economies to generate foreign exchange. The inability to achieve a favourable balance of payments is exacerbated by dependence on high cost liquid fuels and reliance on foreign sources of development capital.

Two pressing development priorities of Pacific and other underdeveloped countries are, therefore, the needs to create employment opportunities and generate foreign exchange. The former is critical if general social welfare is to be improved and social instability is to be avoided, which would constitute a threat to existing holders of political power. The latter priority is essential since without overseas funds the purchase of imported foods, consumer goods, fuels and capital equipment, all necessary for various development programmes, would be curtailed with the consequent frustration of rapidly rising local expectations for improved living standards.

However, because of the internal organisation of peripheral economies, and the disproportionate distribution of power in the international market place, the alternative policy options for achieving these development goals are highly circumscribed. The selective, historic utilisation of resources and interference in social institutions in the periphery by foreign powers have been the prime causes of these problems. The class interests of peripheral government members, limited local resources, and the deeply embedded presence of foreign companies mean that peripheral governments have little room to manoeuvre. They are forced to seek continued reliance on development strategies that revolve around a dependence on foreign public and private capital. Thus in Fiji, for example, over 80 per cent of business turnover in 1973 was attributable to foreign companies (Annear, 1973: 43). Yet the growth sectors of the economy will be forestry, tourism and copper mining – all industries heavily reliant on foreign capital.

It is in the context of these national and international development issues that the potential of the tourist industry to contribute to development should be considered. The underlying proposition of this paper is that a close examination of the tourist industry can be assumed to reflect processes within, and the characteristics of, a peripheral capitalist economy.

A CONCEPTUAL MODEL OF TOURISM IN A PERIPHERAL CAPITALIST ECONOMY

From surface observation, it would seem that the development of a tourist industry in a peripheral economy would benefit all its participants. This appears so even considering tourism's three-tiered international hierarchical structure. As the diagram in Figure 8.2 shows, the superficial two-directional flows linking each sub-component, at each level of the hierarchy, act to the apparent mutual advantage of all the system's interlocking parts.

As the preceding discussion has indicated, however, the dynamic of tourism is such that key components of the industry have a powerful influence on the way peripheral destinations are integrated with the international tourist industry. This is

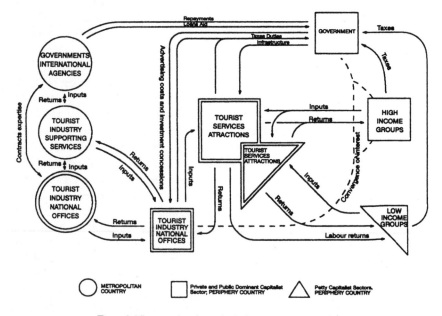

Figure 8.2 International tourist industry interaction flows

164

because of their strategic role in defining what constitutes the tourist products and their involvement in transporting tourists. Similarly, the form of structural organisation, as well as the attendant economic imperatives and inequities characteristic of peripheral countries, result in a reliance on foreign (tourism) capital and a tendency for already privileged groups to take advantage of the opportunities created by tourism development. These processes are summarised in the schematic model shown in Figure 8.3.

Metropolitan companies determine the organisation and operation of tourism,

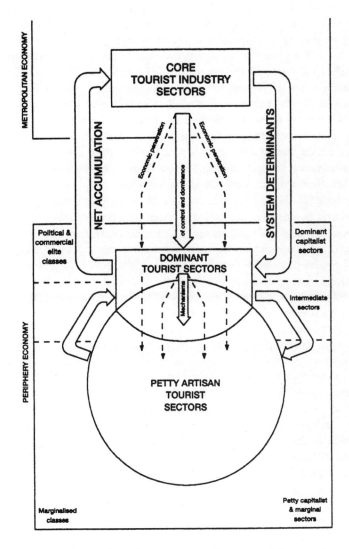

Figure 8.3 International linkages between metropolitan and periphery tourism sectors

both internationally and in the dominant sectors of the periphery, through a series of "system determinants". These include ownership of international transport, the wholesaling of package tours, overseas tourism investment, and the manipulation of tourist preferences and expectations. Various commercial practices or "mechanisms of control and dominance" ensure that the interests of foreign and, secondly, local dominant capitalist firms are protected at all levels of the hierarchy. Such practices take the form of control over tourism technology, industry expertise, product pricing and design, and commercial advantages from economies of scale and consequent bargaining power. Control of these facets of the industry allows dominant sector tourist firms to subordinate, or penetrate, the spheres of operation of petty producer and artisan enterprises. This is achieved in several ways, including control over pricing and franchise rights of tourist goods sold by petty entrepreneurs, as well as control of marketing and communications networks required to successfully retail tours, tourist attractions and handicrafts. Dominant firms also hold influence over the costing and wholesaling of many other products and services essential to petty producers (e.g. tools to make handicrafts), and the physical provision of opportunities (e.g. hotels) that create income earning opportunities.

The ultimate consequence of the imposition of these system determinants and controls, from the highest down to the lowest levels of the industry hierarchy, is that while all participants in the industry profit to a degree, the overall direction of capital accumulation is in a reverse direction *up* the hierarchy.

Incorporated in this model is the ideal of the tendencies towards social, spatial and economic polarisation and selective integration that are inherent in the peripheral capitalist mode of production. The concrete expression of peripheral tourism at the national level, of course, can have various permutations depending on the nature of pre-existing centre–periphery linkages, divergence or otherwise of metropolitan and periphery elite interests, and the nature of the tourist product sold. Given the fundamental internal organisation of the industry, however, the actual operation of tourism in a peripheral economy can be conceptualised as an enclave, as depicted by the simplified diagram in Figure 8.4.

The flow of tourists proceeds up urban and enterprise hierarchies in the metropoles to an international point of departure, usually a major or capital city, in which are located the headquarters of metropolitan tourism corporations and associated non-tourism companies. Tourist arrival points in the periphery are typically the primate urban centres of ex-colonies, now functioning as political and economic centres of independent countries. Within these urban centres are located the national headquarters of foreign and local tourism companies and retail outlets of travel, tour, accommodation, airline, bank and shopping enterprises. If on package tours, tourists will be transported from the international terminals to hotels and resort enclaves. The transport, tour organisation and accommodation phases will be within the confines of the dominant sector tourism companies. Tourists will then travel between other resort enclaves and return to the primary urban areas for departure back to the metropole or on to other tourist destinations. While resident in the resort enclaves, tourists will make brief excursions into the urban and rural petty

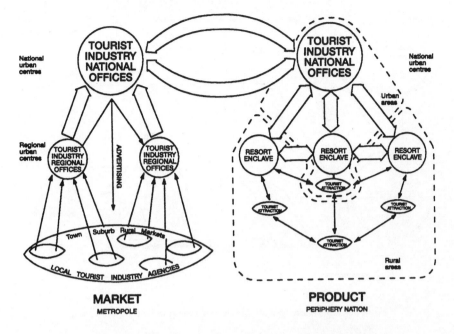

Figure 8.4 An enclave model of tourism in a periphery economy

producer and subsistence sectors of the economy for the purchase of shopping items, entertainment, sightseeing and other services.

This enclave model of the tourist industry emphasises the capacity of dominant tourism sectors to control tourist expenditures through the control of tourist movements, to the relative exclusion of the petty producer sectors. This exclusion of lower order tourism enterprises from the mainstream of the industry's operations is diagrammatically presented in Figure 8.5. The hypothetical distribution of tourists' expenditures is seen to parallel the differential control of tourist movements and the capacity to cater to tourist demands by dominant and petty producer sectors. For any one destination country, the size of the sectoral flows will vary considerably, depending among other things on the distance (cost) of travel between the metropole and periphery, the type and extent of tourism ground plant, and the type of tourist attractions available. Not all tourist destinations have duty-free shopping, for example, whereas other destinations may have extensive handicraft industries.

Using this model, a series of important generalisations can be made concerning the likely performance and role of tourism within a peripheral economy. These generalisations can be summarised in the form of eight propositions that may be used to guide investigation and research.

1 The establishment of a tourist industry in a peripheral economy will not occur from evolutionary, organic processes within that economy. Development will

167

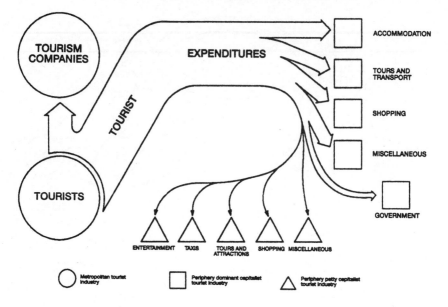

Figure 8.5 The generalised distribution of tourist industry expenditure

occur either by demand from overseas tourists and new foreign company investment, or from the extension of foreign interests already present in the destination country.

2 While the peripheral economy provides the setting, metropolitan enterprises largely dictate the form and characteristics of the tourist experience, or product, offered in a new destination. Hence the incorporation of tropical islands, and the intangible images they are imbued with, into the international tourism system as one alternative holiday package.

3 Since foreign companies are most important in defining what constitutes a tourist product, tourist services in a peripheral destination are likely to be owned and provided by these firms. Where local entrepreneurs collaborate in the provision of these services, they will be members of a privileged commercial elite. This is because only these groups command the financial and other resources necessary to provide such specialised, capital intensive facilities. The likelihood of co-operative enterprise between foreign and local elite groups is encouraged by the fact that, apart from expatriates, members of the local elite are often the only persons in the host society that have any appreciation of the recreational and lifestyle requirements of usually white, affluent, overseas tourists.

4 The direction of capital accumulation in the industry will be up the three-tiered hierarchy: from petty producers, to local dominant sector enterprises, and ultimately overseas to foreign tourism corporations. This pattern derives from the capacity of foreign and national dominant sector firms to provide and control

essential industry inputs. Such control allows these firms, and key activity components of the industry, to determine the economic milieu within which other enterprises have to function.

5 The hierarchical pattern of commercial power based on enterprises can be redefined in terms of a specialisation of functions within the industry. Metropolitan companies, with their command over resources and location within tourist markets, provide the most vital services such as package tours, international transport, marketing, communications facilities (ticketing and reservations), luxury hotel chains, and financial services (travellers cheques, insurance). For tourist services located at the destination, foreign firms are essential suppliers of duty-free goods, food, drink, management services, hotel construction materials, hotel furniture, kitchen and refrigeration equipment and a myriad of other products. Local counterparts to foreign firms, where they exist, will also predominate in the accommodation, tour, travel agency and wholesaling sectors.

Enterprises of the third, lower, tier of the industry are thus likely to be involved in tourism in one of three ways. They may provide services which lie outside the commercial interests of dominant sector firms. Options here would include handicraft production and petty transport services (taxis, rickshaws). Alternatively they may attempt to provide services similar to those offered by dominant sector firms. But because of their small scale, limited access to key inputs, and entrepreneurial inexperience, only services of poor quality, limited appeal and low cost are provided. Such activities would involve budget accommodation, localised tours, village organised tourist attractions, and the retailing of cheap souvenirs. Enterprises involved in either of these types of activities are limited in their capital accumulation potential.

The third possibility is to provide services which complement tourist services and attractions controlled by dominant sector and other enterprises. Possibilities here would include acting as shopping guides, or the retailing of duty-free consumer goods supplied by large wholesaling and importing companies. Invariably, however, the pricing and franchise rights accompanying these products will restrict profits. Another avenue would be the "professional native" function of locals as guides and dance troupes for tours, hotels and cultural display centres. Again, the remuneration from such activities will be limited, in this case by the contract conditions imposed by the companies owning the performing environment.

6 Following from 4 and 5 above, the structure of the tourist industry at the international and destination levels is tending towards monopolistic organisation. Control of the key processes within the industry is being concentrated into the hands of metropolitan tourism capital in general, and large tourism corporations in each industry sector in particular. This is most evident in the spheres of airline, cruise ship, travel, hotel and tour operations (Britton, 1981; IUOTO, 1976: 48–55).

7 Most peripheral tourist destinations were previously integrated into the international economy to provide supplies of raw material commodities to colonial powers. The encouragement of tourism by governing elites in the periphery is

likely to be in response to contradictions caused by this selective development of their export enclave dependent economies. Tourism, particularly as a foreign exchange generator, may help alleviate highly restricted income generating capacities in these economies while, at the same time, continue to serve the economic interests of foreign capital and national elites.

8 Finally, the historical expansion of metropolitan capital into the periphery was closely associated with the development of transport networks (Harvey, 1975). The inclusion of Third World destinations into the international tourist industry is likely, therefore, to be closely associated with the spatial patterns of these transport infrastructures. Since technological revolutions in transport have been the basis of mass tourism, those peripheral economies with appropriate colonial infrastructure networks, or those located on new transport route networks established by metropolitan enterprises, are most likely to be incorporated into the international tourism trade.

CONCLUSION

If research on the relationship between tourism and development is to maximise its contribution to both understanding the nature of the subject and providing the basis of policy formulation, then the issues raised in this paper are of relevance. Tourism does provide many benefits to poor countries (Britton, 1979: 11–38). The emphasis here, however, is on not whether tourism is economically advantageous in aggregate terms, but to whom these advantages accrue. If in "development" one includes the goal of reducing inequalities and redistributing social goods according to priorities of basic needs, then the distributive aspect of tourism is of central importance.

As it stands, the tourist industry, because of the predominance of foreign ownership, imposes on peripheral destinations a development mode which reinforces the characteristics of structural dependency on, and vulnerability to, developed countries. This is particularly true for small Pacific Island economies (Britton, 1980b). This situation results in the flow of tourists in a destination being channelled within the commercial apparatus controlled by large scale foreign and national enterprises which dominate the industry. The greatest commercial gains, therefore, go to foreign and local elite interests. By far the greatest majority of locals can only participate in tourism through wage labour employment or small, petty enterprises. Though highly differentiated in their activities, these enterprises have one common attribute: their income generating potential is severely limited. If destination governments wish to ensure greater and more widely spread benefits from tourism, then the organisation of the tourist industry, and the distribution of power within their own countries, requires consideration.

REFERENCES

Aidney, D. 1977 Address by the Chairman, Board of Directors, Air Pacific; Paper presented to the 16th Annual Fiji Tourism Convention. Suva: Fiji Visitors Bureau.

Amin, S. 1974 *Accumulation on a World Scale: A Critique of the Theory of Under-development.* New York: Monthly Review Press.

—— 1976 *Unequal Development: An Essay on the Social Formations of Peripheral Capitalism.* Hassocks: Harvester Press.

Annear, P. 1973 Foreign Private Investment in Fiji. In *Fiji – A Developing Australian Colony.* A. Rokotuivuna *et al,* eds. North Fitzroy: International Development Action, 39–49.

Beckford, G. 1972 *Persistent Poverty: Underdevelopment in Plantation Economies of the Third World.* Oxford, London.

Bienefeld, M. 1975 The Informal Sector and Peripheral Capitalism: the case of Tanzania. *Institute of Development Studies Bulletin,* 6(3): 53–73.

Britton, R. 1979 The Image of the Third World in Tourism Marketing. *Annals of Tourism Research,* 6(3): 318–329.

Britton, S. 1979 Tourism in a Peripheral Capitalist Economy: The Case of Fiji. Unpublished Ph.D. thesis. Canberra: Australian National University.

—— 1980a The Spatial Organisation of Tourism in a Neo-Colonial Economy: A Fiji Case Study. *Pacific Viewpoint,* 21(2): 114–165.

—— 1980b Tourism and Economic Vulnerability in Small Pacific Island States. In *The Island States of the Pacific and Indian Oceans: Anatomy of Development.* R. Shand, ed. Canberra: Australian National University, 239–263.

—— 1981 International Tourism and Multinational Corporations in the Pacific. In *The Geography of Multinational Corporations.* M. Taylor and N. Thrift, eds. London: Croom Helm, 252–274.

Brookfield, H. C. 1972 *Colonialism, Development and Independence: The Case of the Melanesian Islands in the South Pacific.* Cambridge, London.

Burkart, A. J. and Medlik, S. 1974 *Tourism: Past, Present and Future.* London: Heinemann.

Central Planning Office 1975 *Fiji's Seventh Development Plan.* Suva: Government Printer.

Cohen, E. 1972 Towards a Sociology of International Tourism. *Social Research,* 39(1): 164–182.

De Kadt, E. 1979 Social Planning for Tourism in the Developing Countries. *Annals of Tourism Research,* 6(1): 36–48.

Dos Santos, T. 1973 The Crisis of Development Theory and the Problem of Dependence in Latin America. In *Underdevelopment and Development.* H. Bernstein, ed. Harmondsworth: Penguin, 57–80.

Economic and Social Commission for Asia and the Pacific (ESCAP) 1978 The Formulation of Basic Concepts and Guidelines for Preparation of Tourism Sub-Regional Master Plans in the ESCAP Region. *Transport and Communications Bulletin,* 52: 33–40.

Frank, A. G. 1972 *Lumpenbourgeoisie: Lumpendevelopment.* New York: Monthly Review Press.

—— 1978 *Dependent Accumulation and Underdevelopment.* London: Macmillan.

Harvey, D. 1975 The Geography of Capitalist Accumulation. *Antipode,* 7(2): 9–21.

Hiller, H. O. 1977 Industrialism, Tourism and Island Nations, and Changing Values. In *Social and Economic Consequences of Tourism in the South Pacific.* B. H. Farrell, ed. Santa Cruz: Centre for South Pacific Studies, University of California.

Hymer, S. 1972 The Multinational Corporation and the Law of Uneven Development. In *International Firms and Modern Imperialism.* H. Radice, ed. Harmondsworth: Penguin, 37–62.

International Union of Official Travel Organisations (IUOTO) 1976 *The Impact of International Tourism on the Economic Development of the Developing Countries.* Geneva: World Tourism Organisation.

Kissling, C. 1980 *International Civil Aviation in the South Pacific: A Perspective.* Canberra: Development Studies Centre, Australian National University.

Nettekoven, L. 1974 Mass Tourism from the Industrial Societies to the Developing Countries. *Economics,* 10: 121–137.

Obregon, A. 1974 The Marginal Pole of the Economy and the Marginalised Labour Force. *Economy and Society,* 3(4): 393–428.

Robineau, C. 1975 The Tahitian Economy and Tourism. In *A New Kind of Sugar: Tourism in the Pacific.* B. R. Finney and K. A. Watson, eds. Honolulu: East–West Centre, 61–78.

Shand, R. ed. 1980 *The Island States of the Pacific and Indian Oceans: Anatomy of Development.* Canberra: Australian National University (Development Studies Centre Monograph).

Universal Business Directories 1979 *Cook Islands Visitors Guide.* Auckland: UBD Centre.

Varley, R. M. 1977 The Impact of Tourism on the Fiji Economy. *Research Scheme R3079.* London: Overseas Development Ministry.

Wallerstein, I. 1972 Dependence in an Interdependent World: The Limited Possibilities of Transformation within the Capitalist World Economy. *African Studies Review,* 17(1):1–26.

Witton, R. A. 1974 Dualism or Development: Colonial Society as a Total Institution. *International Journal of Comparative Sociology,* 15(1):107–113.

9

CLOSE ENCOUNTERS OF THE THIRD WORLD*

Cecilia A. Karch and G. H. S. Dann[1]

INTRODUCTION

If we accept the notion of encounter as "any joint act between two or more persons where the interactants are maintaining symbolic, visual, auditory and at times, tactile contact" (Goffman, 1961), then clearly social encounters are always problematic since "we literally do not know what to do with respect to another person until we have established his meaning for us and our meanings for him" (McCall and Simmons, 1966). The arrival at such consensus is a complex process, one which, analytically at least, continually oscillates between the statics of cognition and the dynamics of expression. A working agreement whereby interaction can take place occurs between the parties if there is sufficient congruence between the two processes. Central to such theory is the concept of "role." Role is envisaged as the prime unit of analysis, adding an intermediary dimension to the philosophical problem of one and many and an alternative to the alleged dichotomy of sociological holism and atomism (Dawe, 1970). Theoretically, role is introduced to overcome the divergence of society and the individual (Dahrendorf, 1968) and to void the impasse of analyses found to be incapable of transcending the group or personality (Cooley, 1929; Mead, 1934). Nevertheless "role" itself is not without difficulty: the abundance of introduced distinctions [e.g. "ascribed-achieved" (Linton, 1936), role-set (Merton, 1968), "latent-manifest" (Gouldner, 1970), etc.] demonstrates its multiple interpretation among those holding differing, and even similar, theoretical positions.

The following seeks to highlight some often overlooked considerations of role negotiation by focusing on an area where many traditional assumptions may be called into question. The topic to be examined is that of the "beachboy" phenomenon on the Caribbean island of Barbados. The data for the study were recently collected over an 18-month period, and are based on semiparticipant observation[2] and a series of unstructured interviews with beachboys, tourists, hotel managers, security guards, police officials, and medical practitioners.

Before embarking on the analysis, however, it is first necessary to contextualize the phenomenon under investigation. Barbados in many respects is typical of

* Reproduced with permission of Plenum Publishing Corporation, from *Human Relations*, 1981, 34: 249–68.

developing small island societies which have moved from a traditional monocrop plantation economy to one of heavy reliance on mass tourism. Ironically, recent (1966) political independence has witnessed an even greater dependency, predicated on the arrivals of North American and European visitors. At the level of human interaction, encounters between guest and host can be seen to imply superiority and subservience with respect to class, culture, and race (cf. Linton, 1971; Matthews, 1977). Whether such relationships be viewed as customer–clerk, master–slave, demander–supplier, etc., they form part and parcel of an overriding syndrome of dependence. They should therefore be understood within the framework of the wider international relations existing between the metropolitan and developing worlds. The tourist is needed for his dollars. He comes. He pays for a service, and more often than not the host compromises himself socially by rendering it to him. This type of situation is well illustrated by the analogous "hustling" of the black beachboy who offers his sexual services to white visitors in return for cash or kind. As the name suggests, a beachboy is a young member of the host community, generally between the ages of 16 and 30, who initiates his trade on those beaches frequented by tourists (in practice the south and west coasts of the island). Significantly, it is for this age group that unemployment reaches its highest level (over 25%). Following the occupational pattern in Barbados, beachboy activities are also of a temporary and seasonal nature. While few assume the "career" of beachboy on a full-time basis, there are many who combine it with such ancillary occupations as deck-chair attendant, watersports operator, or peddler of black coral. However, the multiple manifestation of external guises under which the phenomenon operates does not present so much of a problem for the host community as its lack of institutionalization. Unlike female prostitutes, beachboys are not territorially assigned. Nor for that matter are they given a position (status) with attendant cultural obligations (role). Confusion over their identity and bisexuality together with their ambiguous legal status only adds to the dilemma of beachboys vis-à-vis their own society. Operating in such a limbo, the situation becomes even more complex when the role of beachboy is interpreted by first-time visitors. Many of them are simply unaware of the phenomenon and are confused when confronted with it. They are even more perplexed by the suggestion of payment for services rendered. Those more acquainted with the situation may still be puzzled at what strikes them as a situation of role reversal. The phenomenon thus raises several questions over the identities of those involved in tourist–beachboy encounters. In this paper we focus on role imputation and the closely associated areas of altercasting, role improvisation, and presentation of self.

ROLE IMPUTATION

By the beachboy

The successful imputation of a role to alter is based on the ability to infer accurately from appearance and those clues furnished by external behavior. If we assume for the sake of simplicity here as elsewhere that encounters between white tourists and

black beachboys[3] are on a one-to-one basis,[4] then clearly the direction of the imputation is important. From the analogous case of female prostitution we would expect
the beachboy to make the initial approach to a potential client. In sizing up his mark
several inferences are drawn in relation to her availability and disposition. Dress, although obviously limited in a beach setting, can provide clues in relation to degree
of scantiness and provocation. The beach towel, if it carries the insignia of a hotel,
will be viewed as less promising than if it bears the name of an apartment hotel, a
less security-conscious establishment providing greater access to the room of the
guest. Sunglasses, title of magazine or book, tan lotion, etc., are all weighed up in the
balance of favorable or unfavorable impression. Language too offers scope for the
beachboy's assessment of the situation. Past experience has taught him that there is a
hierarchy of willingness among female tourists ranging from Quebecoise (the most
cooperative) to American (the least).[5] In this connection one of our beachboy
informants, when asked which female tourists were found to be the most readily disposed, answered:

> The French Canadian, man. Dem girls does come down here specially for fun
> and games. Dem girls does boast to their friends back home of the sweet
> foopin' we guys does give dem. Then you find their friends does want to come
> down too to taste some of de action.

Interestingly enough, marital status of the potential client does not loom large in the
calculations of the beachboy. This is probably related to supply and demand. Only 1
in 8 peak-season tourists are single and female. Of these less than a quarter are in the
same age group as beachboys. If beachboy activities were solely limited to this group,
then it would mean that on average each beachboy would enjoy the company of just
two female tourists per year. That there are contrary indications suggests that beachboys overstep the traditional barriers of compatibility associated with age and marital status. More positively and blatantly we have overheard conversations such as:

Beachboy: "Excuse me but you look lonely."
Tourist: "No I'm not."
Beachboy: "Who is with you then?"
Tourist: "My husband."
Beachboy: "What about your sex life; is he satisfying you?"

By the tourist

So far we have considered the beachboy's assessment of the situation. Now let us
look at the same encounter from the position of the tourist. Over the last twenty
years Barbados has built up a thriving tourist industry largely due to the reputed
friendliness of its people. Many visitors to the island remark on this quality and
compare the place favorably with other destinations in the Caribbean which they
perceive to be characterized by more overt racial tension. Tourism planners have
capitalized on this asset and marketed Barbados accordingly. With Barbados' annual

175

tourist arrivals outstripping the local population of approximately 270,000, we may say that their attempts have been successful. Sociologically, we can also highlight the expectational value associated with the host's claim to friendship. Tourists, due to the anticipatory socialization efforts of advertisers at the pretrip stage, are disposed to the greetings and smiles of locals. The remarks of their friends returning from a vacation in the island tend to reinforce the claim at the word-of-mouth level. Consequently, when a beachboy approaches a tourist, it is quite natural for the latter to interpret the gesture as typically amicable. The first-time visitor makes little attempt to establish the identity of the beachboy. The role imputed to him is simply one of "friendly native" with whom greetings can be exchanged and light conversation entertained. Racial diversity at this stage of the encounter may present few problems to the tourist, who has been informed that 95% of the Barbadian population is black. Nor does she reflect on her own typical minority status in such a situation.

DIFFICULTIES OF ROLE IMPUTATION

Role imputation by the beachboy tends to hinge on his ability to infer the dispositions of the tourist in order to cast her successfully in the role of "client." To bring about this end the beachboy wishes to give the impression of assigning the role of "Juliet" in the hope that he will be recognized as "Romeo." By contrast, when the tourist imputes a role to the beachboy, wittingly or otherwise, she tends to do so in terms of his perceived lower status. The incongruency of the role imputation process raises a number of difficulties, all of which focus on individual abilities to assume the role of alter.

Successful role taking is determined by several factors (McCall and Simmons, 1966). We will look at just a few. In the first place it depends on the accurate *imputation of motive or purpose*. If the beachboy incorrectly attributes sexual desire to the other party, then the perceived insult normally terminates the encounter. Rarely does he attempt renegotiation of identities if he has genuinely missed his mark. From our observations, such failure on the part of the beachboy is usually accepted with a shrug of the shoulders and a farewell pleasantry. Sometimes, however, the beachboy underestimates the willingness of the tourist; on these occasions he finds that the tempo of her "loving" response[6] is unconducive to what he has come to appreciate as the commercial benefits of a gradual relationship, one in which encounters are cumulative. Somehow he must discover a happy medium between the above extremes. The imputation of motive to the beachboy by the tourist is also problematic. If he has been cast in the role of "friend," then his subsequent advances may simply be attributed to cultural differences.

Even though tactile behavior is often regarded as taboo among strangers in the tourist's country of origin, close contact may be reinterpreted as permissible in a setting which has not been institutionalized by the tourist. For example, back home she may avoid the glance of her fellow male passengers in the office elevator, and instead join them in looking upwards at the registration of the floor count. Similarly, in selecting her seat in the library she often chooses a corner suitably distanced from the next reader. In such settings social spacing is normatively established. By contrast, open beaches may be seen as containing no proscribed areas; particularly where, as in Barbados, they are legally considered to be of unlimited access to both members of the host and guest communities.

If, however, the beachboy's advances are overestimated by the tourist, then he is likely to be thought of as "fresh." Here the tempo of the potential relationship is deemed inappropriate this time by the tourist. In such an event the beachboy is likely to be perceived as a gigolo. Once more the striking of a balance somewhere between the above two positions requires successful negotiation of identities. However, it is one matter to impute motives but quite another to have the appropriate discourse for so doing. *Vocabularies of motive* vary according to the situation and the definer. In this regard the open locale of the encounter can both aid and hinder the approaches of the beachboy. To attribute sexual motives to a sunbather is possibly easier in a "natural" setting, even though the intimacy demanded is usually denied. As a compromise strategy the beachboy organizes his terrain as a place for initial contact, prefering to establish future appointments elsewhere.

More importantly, perhaps, vocabularies of motive may also be nonverbal. Here the mutual task revolves around cue giving, whether facially or bodily, or both. The following we have observed is a typical beach situation. The female tourist has been walking back and forth along the shore. She lies down to sunbathe, untying her top in the process. She begins to read a magazine. The beachboy, who has been eyeing her for some time from further afield, passes by whistling. He smiles. She looks up and returns the smile. He comes over and squats in front of her, while gazing into her eyes the whole time.

Beachboy: "Sorry to disturb you, but do you think you could let me have a cigarette?"

Tourist: "Sure, go ahead."

She offers him a cigarette and he feigns no matches. She lights both and places one in his mouth. All the time he is looking at her and she at him. By now their faces are only a few inches apart. As she removes her hand she lightly brushes it against him. Here the tourist displays body signals to which the beachboy responds. Mutual eye and body contact are used by both parties to the encounter. The familiar cigarette ploy used by prostitutes is exploited to the full and represents an interesting case of role reversal. In this instance it is the male who asks for the cigarette and the woman who lights it for him.

Role taking also varies with the ability to *empathize*. Gordon (1969) reminds us that the appreciation of the feelings of alter ranges from that based on universal experience (e.g. grief over the loss of a loved one) to that found at cultural and subcultural levels. Effective communication becomes even more difficult at the levels of role and individual personality. In the tourist–beachboy encounter empathy can be quite problematic. A fundamental reason for the difficulty lies in the polarity of basic roles including those of sex and race. Generally where these roles are different between the parties to an encounter the only basis for understanding is previous counterrole performance or vicarious experience. The former derives from interaction with those of the opposite sex or of different racial origin, and the various lessons learned on those occasions.

However, in the tourist–beachboy situation the learning process itself tends to place the beachboy at a greater advantage. He bases his counterrole performance on past encounters in the same or similar locale. She, on the other hand, particularly as this is a first-time experience, must rely on analogous situations in her country of origin. In other words, the environment factor differs in her case. Although she may have learned a great deal from her limited dating experience with males of a different ethnic origin back home, various cultural norms and sanctions obtaining in *that* milieu place limits on any such relationship. Moreover, where it is frowned upon it often represents an inhibitory factor to deeper communication. Sudden transference to a world of greater permissiveness can thus represent a greater cultural shock to her. By contrast the beachboy is more familiar with the situation and its possible outcomes.

The deficiency in the beachboy's counterrole performance is more likely to depend on his ability to relate to Barbadian women. *De facto* rejection in this sphere (usually on economic grounds and associated lack of future prospects) may induce him to overcompensate in his relationship with the tourist as an alternative. Negotiation of racial barriers is often quite skillfully manipulated by the beachboy. Take the situation at a hotel bar, where the beachboy sitting in the corner has been selecting his target for some time.

The woman approaches the bar and orders a drink.
Beachboy: "Hi there."
Tourist: "Hi."
Beachboy: "I called to you earlier, but when you didn't answer I thought maybe it was because you don't like us black boys."
Tourist: "What me? I'm no racist. Here why don't you have a drink with me?"
Beachboy: "No let me buy you one."

Here the beachboy capitalizes on the liberalism he flatteringly attributes to the tourist. Of course he didn't call her earlier, but it is a useful introduction to the remark which follows. He completes the coup by offering to buy the drink himself, thereby reducing any possible guilt feeling remaining from faulty imputation. It is also a way of minimizing his perceived relative deprivation. Rapport is thus established. By the next round of drinks the beachboy permits the insisting woman to place the order. After a few rounds of rum punch the tourist is far less inhibited and is swayed by the further interspersed suggestions of the beachboy.

Successful role taking also tends to vary with the actual or perceived social distance between the parties constituting a relationship. Although social distance is conceptually multifaceted it is the racial factor which largely predominates in tourist–beachboy encounters. The black beachboy comes from a neocolonial society where color is still regarded in terms of the interassociated variables of class and socioeconomic status. Upward social mobility becomes increasingly possible as he ascends the pyramidal color stratification structure. In this sense the current situation differs little from the possibilities offered in the past by planter–slave miscegenation and its consequent opportunities for freedom. In the tourism setting, access to the white tourist represents a similar breakthrough to the world with which she is associated. When she can be accompanied to her private room in a hotel this act in itself is replete with sociohistorical significance. Formerly hotels in Barbados were clubs which practiced racial discrimination. Even today in some quarters a "de facto" color bar is perpetuated and there is little mixing between races or intermarriage among members of the local population. The mere presence of the beachboy on tourist premises thus represents a personal triumph in having overstepped such racial boundaries. The enhanced feeling of freedom is also reinterpreted by the beachboy in terms of the benefits that association with whites can bring. It is no mere coincidence that he often accepts payment for his services in the form of luxury items or the promise of an immigration visa and job in the metropolis. The white tourist stands as a symbol of escape from the drudgery of the Third World and access to a better life. Finally, conquest of the white female is perceived in terms of acceptance and integration into the society she represents. Hence, the beachboy, in imputing a role to the tourist, must ensure that this includes a minimization of social distance based on race and that any racial tolerance demonstrated by the tourist is consequently maximized. The tourist, on the other hand, may be completely unaware of the beachboy's aspirations. Instead she may regard him simply as an unemployed youth seeking to "make a fast buck." Should she decide to go along with this temporary diversion, it is quite likely that she does so in terms of an amusement on which she can spend her discretionary income. If she is unaware that payment for services is required she may simply look on the affair as one of harmless flirtation. Whatever her attitude, then, clearly much needs to be negotiated between herself and the beachboy in order that a working consensus be achieved.

Labeling is another area which can hinder successful role taking. Where this is conventionally applied it often represents an uncritical response to stereotypes. In the tourist–beachboy situation the use of labels can often mitigate a common understanding of the relationship. One reason for this hinges on the association of labeling with deviance. Understandably the tourist is more prone to labeling than the beachboy. Her ignorance of the phenomenon provides a partial explanation for her resorting to labels. The beachboy, of course, is not exempt from labeling tourists. However, in coming to an assessment of a potential client, he tends to do so from a position of greater experience. From past encounters he has come to appreciate the differences between easier and more-difficult customers. His tendency to label comes to the fore whenever he identifies the facility of his approach with national characteristics of the tourist.

Summing up those from the United States, a beachboy remarked:

Man dem chicks brother is a problem sometimes. Yuh carry dem out, yuh spen' yuh money 'pon dem, and when yuh get back to de apartment, day jus' tell yuh "thank for the evenin." Dey gon' along inside and don' even look back. Man I don' dig dose Yankees.

It is also interesting to note that often the beachboy accepts the tourist's labeling of himself at the initial stages of the encounter. Later he tries to renegotiate his mistaken identity. He only attempts to do so at an early point in a potential relationship he normally finds to be overabrasive and unconducive to its continuance.

DYNAMICS OF ROLE IMPUTATION

So far we have only considered the difficulties of role imputation in tourist–beachboy encounters from a static and cognitive viewpoint. The picture, however, remains incomplete without the inclusion of the expressive and dynamic factors involved. Essentially the beachboy is the party who takes the initiative in successfully managing his role-set repertoire of impressions in such a way that his envisaged rewards are maximized. He is constantly modifying his own behavior in the light of the responses he receives from his initial imputation of role to the tourist. Here we consider the subsequent dynamics of the situation under four headings: (1) altercasting; (2) modification of his own imputed role; (3) presentation of self; and (4) fantasy.

Altercasting

Earlier we noted that if the beachboy is to be successful in his endeavors then he must ensure that the tourist plays the role of "willing client." At the same time he has to manage the impression of casting her in the role of "friend" or "lover." In other words, alter's role of "client" is intended to be latent and that of "lover" manifest. For the beachboy it is important that the tourist does not discover that the latter is but a means to the former. His objective is to delay the revelation that payment is required for his services until they are rendered. Premature disclosure affects the whole basis of working consensus and the dynamics of role imputation. If the tourist becomes aware that she is merely a "client" it is unlikely that she will continue to impute the role of "friend" to the beachboy. Instead he will be thought of in terms of "male prostitute," and, in the majority of cases, the relationship will be either terminated or suspended for renegotiation of identities. Maintaining alter in the role of "friend" or "lover" presents a number of difficulties for the beachboy. Essentially they derive from the tourist's unfamiliarity with the situation. Back home she probably experiences little difficulty in accepting the role of "companion" from her boyfriend. As the evening progresses through dinner, drinks, and dancing she is prepared for the role of "friend" and "lover."

However, the smooth transition from companion to lover is more problematic in the Barbadian context. The tourist may never have dated a black man before. Despite her protestation to the contrary her attitudinal background may be quite racist. She is also unsure about the feeling of the host country to mixed relationships. If these factors weigh too heavily on her mind and the associated phobias cannot be easily dispelled by the beachboy, then he will not succeed in this altercasting. He must therefore try and reassure her that Barbados is a model of racial harmony where blacks and whites freely integrate. If he has anticipated the tourist's problems he may already have taken her to one of the island's night clubs frequented by other beach-boys. There he will point out a number of mixed couples dancing (without of course revealing the identity of the black party).

Modification of imputed role

It will be remembered that the beachboy has been cast in the role of "friend" by the tourist and that this is insufficient for his purposes. It barely distinguishes him from a lifeguard or any other smiling member of the host community. If he is to obtain a "lover" response (i.e. a client response) from the woman then he has to shift his own role toward that of "Romeo" once more. To this end he must isolate his mark and relocate the activity. Naturally the first procedure is only required when the woman is accompanied (usually by another female friend). The one-to-one basis of the relationship becomes established as soon as the beachboy has made satisfactory arrangements for the other party (often by bring-ing in another beachboy for the occasion). In other words, the reduction of a triad to a dyad (Simmel, 1950) is a condition for the subsequent assumption of a more amor-ous role. Relocation of the activity is necessary for several reasons. First, the beach is too public for prostitute–client behavior. Second, it does not provide the tourist with sufficient means to distinguish the identity of the beachboy from other friendly locals. Third, it is a daytime location, and temporarily inappropriate for more roman-tic activity. Instead the beachboy usually manages to meet the tourist on her "home" ground. Then he will decide whether it is necessary to take her out for the evening or whether the operation can proceed more swiftly.

The dialogue proceeds along such lines as the following:

Stage One

Beachboy:	"How are you enjoying your holiday?" (establishes tourist identity)
Tourist:	"Fine so far." (friendly response)
Beachboy:	"Where are you from then?" (request for classificatory information)
Tourist:	"London." (information given to imputed friend)
Beachboy:	"London. I thought so. Oh I know London well." (relates most landmarks and one or two lesser known areas from his repertoire of learned facts, thus establishing common ground, even though he may not actually have been to London)

Stage Two

Beachboy:	"Is this your first visit to Barbados?" (request for *vital* information)
Tourist:	"Yes, and I just love beaches." (indicates lack of experience)
Beachboy:	"This one is nothing. I'll show you another one which is much

	better. Say, how would you like to see the rest of the island?" (attempt to relocate activity)
Tourist:	"Sounds great. So far I've only been on this beach near the hotel." (accepts and points out alternative location)
Beachboy:	"You mean that one over there? That's not far. Tell you what. I'll come by this evening after dinner and meet you in the bar about eight thirty." (relocates activity)

Presentation of self (cf. Goffman, 1959)

Once the beachboy manages to have his imputed role modified to that of "Romeo" it is essential that his actions match his new identity. Any performance blunder or failure to act out his line convincingly will only raise doubts in the tourist's mind and create further insecurity about the developing relationship. Additionally, through his behavior the beachboy has to maintain the "lover" image which he has assigned to the tourist and avoid any premature lapse into the role of "client." Beachboy performance blunders tend to be inversely related to the publicity of locale and degree of isolation attained. For instance, if he takes the tourist to a night club he increases the risk of meeting fellow beachboys and their clients. If the former are less adroit in their performances than himself, they may be detected for what they really are by the woman accompanying him. Consequently, if he is to maintain his impression of "Romeo," he must either ignore their comradely greetings, convincingly wave them aside or provide an adequate explanation for them. Similarly it is imperative that he prevent his own client from overhearing the proposals or deciphering the techniques being used by his colleagues on their marks. The beachboy should also be prepared to deal with the more embarrassing situation of meeting a former client.

Confronted by such a variety of obstacles, why then does the beachboy frequent such a public place with his client? One reason, and probably the main one, is that the beachboy must cover himself economically. If he discovers during the appointment encounter that he has erred in his assessment of his client's willingness to respond to his overtures or that she will require too many engagements to bring the relationship to a satisfactory conclusion, then the night club provides the ideal location for discarding her and selecting another more amenable customer. Time invested without payment by the first tourist can thus be recouped with the second. From our observations at night clubs in Barbados it would seem that client-switching is quite a common occurrence. It is usually settled among beachboys using the neutral territory of the bathroom as the center for their negotiations. If crossing of pitches is involved (i.e. taking a client from a nonassigned table or seat) sometimes arbitration from a senior member of the fraternity is required.

Performance blunders are often connected with the profit motive. Where the beachboy acts overhastily in seeking to maximize his gains, he runs the risk of unmasking the prostitute in himself and the client in the tourist. To counteract this tendency he should, therefore, be prepared to pay a few entrance fees and to buy one or two drinks as investment towards his future remuneration. His reluctance to do so

182

may well diminish his role of "Romeo" in the eyes of the tourist, normally accustomed to being treated by her date. A blunder here may of course be redeemed by assuming the role of "poor little black boy." Nevertheless, its associated diminished status can be as disadvantageous to the objectives of the beachboy as the discovery by a female skier that outside the winter season her favorite instructor is no more than a farm hand.

Fantasy

In addition to the efforts of the beachboy the successful outcome of a tourist–beachboy relationship is also heavily dependent on the fantasy of the tourist. Without this ingredient the beachboy will surely fail to impute the latent role of "client" or to cast her in the role of "lover." The remodification of his own role and its subsequent performance is only convincing to the extent that it is desired on the part of the tourist. Although tourist fantasy is too complicated a topic to treat here, and indeed has been dealt with by one of us elsewhere (Dann, 1976), its connection with predisposition can briefly be commented upon.

In many ways a vacation in the pretrip stage is viewed as total project (cf. Schutz, 1972) by the tourist. Values and motives are weighed and matched and a destination is finally selected. Part of the decision process relates to sexual fantasy, together with the envisaged permissiveness of the resort and the most appropriate means to its realization. For example, if a woman travels on her own or with a female friend, the beachboy is more likely to impute sexual motives to her than if she is accompanied by her family and children. His estimate is based on the probability that the tourist has freed herself from customary ties as a deliberate choice and that this act of liberation is connected with curiosity. The fantasy itself relates to the anticipatory experience of enacting culturally proscribed behavior. The myth of black sexual prowess heightens the desire to overcome culture taboo. The possibility of fulfillment is offered in terms of a destination where the fantasizer is unknown and in the context of the temporary and inconsequential nature of encounters which tourism provides. Nor should we forget the fantasy of the beachboy. His quest for the white woman is also laden with fantasy. In this context the hypotheses of Fanon (1967) come to mind:

> Out of the blackest part of my soul, across the zebra striping of my mind, surges the desire to be suddenly *white*. I wish to be acknowledged not as a *black* but as a *white*. Who but a white woman can do this for me? By loving me she proves that I am worthy of white love. I am loved like a white man. I am a white man.

DEPENDENCY OF TOURIST–BEACHBOY ENCOUNTERS

So far we have examined the difficulties experienced with regard to the cognitive and expressive aspects of tourist–beachboy encounters. In this final section we argue that the inherent problems may be viewed from the overall perspective of dependency characteristic of such relationships. Furthermore, we suggest that many other types of role encounter share this same property, albeit to a more limited degree. We do so under three headings: (1) asymmetrical encounters; (2) deference; and (3) exchange.

Asymmetrical encounters

The relationship is said to be asymmetrical where the attendant obligations and expectations are structured so as to favor differentially one of the two parties (Goffman, 1961). In other words, the resulting balance of power yields a pattern of superordination and subordination (Simmel, 1950). Examples are not difficult to find. That of doctor–nurse, for instance, illustrates a skewness in rights and duties *vis-à-vis* status and role-sets. Tourist–beachboy encounters are similarly weighted. The difficulties experienced in role imputation are indicative of the respective attempts to bring sufficient equilibrium to the relationship in order that a working consensus be negotiable.

Nevertheless, the two parties negotiate from very different positions. The tourist has few obligations or commitments to members of the host community, who, directly or otherwise, have been paid to ensure that she enjoys her vacation. She may be politely asked to comply with various cultural norms (e.g. dressing for shopping expeditions) but their actual binding force is negligible. Besides, she always has the final option of selecting an alternative resort if she feels overly pressured by such constraints. This leverage itself is perceived as a position of power by the inhabitants of those destinations who compete for her presence. It is unlikely, therefore, that the beachboy will be successful in his initial attempt to reduce the tourist to the diminished status of client. For him it will be more than a satisfactory beginning to be treated symmetrically as friend. Remember she is white and he is black. She, if married, has assumed the class of her husband, and if not then that associated with her job. However lowly her occupation in the country of origin, at least it represents a job and a relatively high standard of living when compared to destinations in the Third World. Moreover, it is likely to be subjectively viewed by a Barbadian as placing her above the confines of the working class, which he objectively identifies with his own background and that of 70% of his compatriots.

The asymmetry of tourist–beachboy encounters is further reinforced by its underlying relationship of dependence. The beachboy, dependent on the visitor's cooperation, is a microcosm of his country's reliance on the capricious choices of potential tourists. There are no viable alternatives to the friendly smile as a symbol's gesture of dependence.

Essentially the beachboy attempts to realize an increase in his own status and a parallel diminution in that of the tourist. The difficulty of such a task perhaps explains why the majority of tourist–beachboy encounters are multiple and dangling (Denzin, 1974). Here in common with most relationships the fewer the encounters the higher will be the role negotiation involved. A possible exception is that of a light-hearted occurrence where humor itself tends to derive from the lowering of status. The teacher who loses face in front of the class as victim of a practical joke acknowledges the situation by his own laughter. The skillful beachboy can capitalize on an analogous quality of tourism if he exploits the happy-go-lucky attitude of the tourist. The "laugh–some fun–no recriminations" ploy can and does work in his favor and achieves a far more serious goal.

Alternatively, he can bring the tourist to his own level by providing her with an opportunity for ego enhancement. The tenured university professor does not *need* to help the junior faculty member by taking an interest in his research. Nevertheless, occasional (and often genuine) concern can boost his own image among his colleagues. Similarly (and again provided the relationship is as temporary as that offered by tourism) the visitor may feel that she has become more than a mere tourist by "fraternizing with the natives."

Deference

Here deference is understood as the symbolic means whereby appreciation or regard is displayed by one individual for another (Goffman, 1956). As no person gives deference to himself, he must of necessity seek it from others in his day to day encounters. Where it forms part of a regular pattern, we are in the presence of ritual (Denzin, 1974). Deference is not limited to asymmetrical relationships but can take place between equals, and, on occasion, even be shown to a subordinate. Often too, deference is displayed to the *office* or *category* rather than, or in spite of, the present incumbent (e.g. a private saluting a corporal). Whether or not it is given in good faith, initial esteem or evaluation of alter is indicative of the acceptance of his identity and signals that interaction can proceed further.

Typical deference is manifested in the rituals of avoidance and presentation. The former refers to the maintenance of an appropriate distance from alter's personal and ideal sphere of privacy (Simmel, 1950). Generally the holder of a superior position is permitted to be more familiar than his subordinate counterpart (e.g. psychiatrist–patient). Familiarity also includes touching behavior. While avoidance ritual focuses on prescription, presentation ritual involves prescription regarding the deference shown to alter.

Tourist–beachboy encounters are particularly interesting with respect to deference behavior. The comparatively short duration of the vacation and its originality for the first timer preclude the tourist's daily familiarity with such encounters. Additionally, if her anticipatory role or counterrole performance is minimal (cf. Cavan, 1972), the activity will assume an unritualistic quality. Denzin (1974) refers to this as "routine," a term which we have avoided for its ambiguous condition of regularity.

By contrast the beachboy will tend to view a given relationship as forming part of a regular and cumulative series of experiences and look upon an encounter as a ritual occasion. We therefore have a situation which alternates between unfamiliarity and ritual dependent on its definer, a possibility overlooked in the literature. Yet its presence here should alert us to analogous encounters where the same "novice–old hand" distinction operates (e.g. the new boy on his first day at school, a passenger flying for the first time, etc.).

It is also worthwhile noting the significance of deference in the development of tourist–beachboy encounters. Clearly the tourist must be encouraged to proceed with the business at hand, and for this she must be adequately motivated. Deference can supply such motivation, provided that she does not detect its being given in bad faith. Once she realizes that it is being shown to her as a white body or client, rather than as a person, there is little likelihood of a furtherance of the relationship. Hence another reason for the beachboy's postponement of a discussion of payment for service rendered.

The beachboy may also capitalize on the deference shown to him. If he is acknowledged as "friend" he can take this as a signal for the continuation of his efforts. The fact that deference is being displayed in an asymmetrical relationship is sufficient encouragement for him that the relationship is entering the phase of equilibrium necessary for the achievement of his ends.

The dynamics of tourist–beachboy encounters also illustrate the possibility of progressing from a position of avoidance to one of presentation. Although the above distinction rather assumes a static dichotomy, here we have more of a continuum model of behavior. The prescriptions surrounding tactile taboo eventually give way to a situation where they are replaced by prescriptions of normative compliance in which touching of the superordinate by the subordinate is permitted. The barriers surrounding the previously envisaged sphere of personal privacy are dismantled by intimacy. The significance of the event becomes even more marked with the realization that such barriers are constituted by class, culture, and race. For analogous occurrences we usually have to turn to the world of deviance (cases of sacrilege, incest, etc.). Yet, interestingly enough, such is the nature of tourist–beachboy encounters that they are generally not viewed as deviant by the participants. Travel to an area where one is not known permits the escape from a vetoing environment to one of *perceived* relaxation of rules.

Exchange

Social exchange has been described as referring to voluntary actions that are motivated by the return they are expected to bring, and typically do in fact bring, from others (Blau, 1964). It may be distinguished from economic exchange in that the envisaged rewards are often unspecified, there is no exact price, and the project itself is often nonrational (Pareto, 1935). The returns anticipated from social exchange are themselves social in nature. These range from the establishment of individual trust and friendship to more universal social approval and esteem. The fascination of tourist–beachboy encounters stems from the appreciation that they constitute a relationship of *mixed* exchange, one generally not envisaged by the tourists, but one which nevertheless they should be prepared to entertain conceptually. Let us briefly elaborate.

The beachboy views this relationship with a tourist as conditional to the obtaining of an extrinsic goal. If the visitor is particularly attractive that is an added bonus.

However, it does not represent or detract from his principal objective of remuneration. In order to prepare the tourist for eventual financial compensation, he attempts to create in her mind the anticipatory image of a friend who provides a range of services for her. These include buying drinks, paying admission charges, and acting as a general guide and unofficial courier. In this way he hopes that a norm of reciprocity will become established and that the tourist will begin to feel obliged towards him. However, if he gives too much or bestows too many favors upon her, she will tend to view herself in the position of subordinate. He therefore tries to maneuver a course of action whereby she will actually wish to remunerate him as one friend bestowing a gift upon another – a situation of greater equality. For this he needs to create and manage the impression of trustworthiness, a quality which can be tested, for example, in her keeping of an appointment. Hopefully, also, he can instill a sense of commitment in her, one which in most relationships of this nature is deeper in one party than the other. He commences with simple riskable acts of friendship whose outcome is clearly cordiality or hostility (Lévi-Strauss, 1957). If the former, he gradually builds up his approaches to maximize the situation from his role repertoire.

The first-time tourist, on the other hand, imputes no such monetary motive when forming a relationship with a beachboy. For her the encounters represent acts of *social exchange*. Her rewards, though less obvious, nevertheless form part of her definition of the situation, and to that extent are real. Here, multiple motives range from curiosity and fantasy fulfillment to the winning of social approval. The last mentioned is particularly interesting, for the presence of an audience once a one-to-one relationship has become established is quite inappropriate. This is not to say that the audience has to become purely fictitious. Even during an encounter a tourist can contemplate her future action of "trip-dropping" to a real group of peers back home. In this regard she views her current project as future perfect and acts in accordance with a range of probable outcomes. Schutz (1972) would maintain that here the linkage of her motivation and action is no different from any other type of social behavior. For example, a man who intends going next door to borrow a newspaper regards his ensuing action as a project and already imagines himself talking to that person. If asked (after the event) why he went to his neighbor's house he can supply a sociological *valid* reason *because* he has projected his future action before it has yet taken place. Moreover, it is by understanding this and similar processes that we come to grips with true causality. If reflection on the tourist–beachboy encounter teaches us something more about "verstehen" then the exercise should have been worthwhile.

CONCLUSION

We have examined tourist–beachboy encounters from an interactionist standpoint as we believe this approach highlights their inherent problematic. Furthermore, the perspective underlines the dynamic process of role imputation in a relationship which moves to a situation of greater equality before returning to its original qualities. We have argued also that in many ways such encounters typify an overriding dependency that has come to be associated with tourism in the Third World, a dependency which is typified in the beachboy's philosophy that the end justifies the means. While the means taken strives to reduce dependence in order that the relationship may progress, the outcome of the projected course of action only further emphasizes the reliance of the host upon the guest. The alleged increase in international understanding which tourism encounters are reputed to yield may in fact be as unrealistic as the suspension of the distinction between the developing and developed worlds. Essentially, a mutual equality between those who have discretionary income to spend upon beachboys and the objects of their pleasure only last as long as the encounters. The beginning and the end underline the permanence of dependency. The means are perhaps as unreal and transitory as tourism itself.

NOTES

1 Requests for reprints should be sent to Graham Dann, Department of Sociology, University of the West Indies, Cave Hill, Barbados.
2 Full participant techniques on some occasions were debarred on ethical grounds.
3 Here we exclude the relationship of black tourist–black beachboy for the declared lack of mutual interest by the parties concerned. The attachments of black or white visitors to white beachboys are not considered here for their comparatively low incidence.
4 Rarely do two or more beachboys entertain a tourist. Similarly infrequent are contacts between one beachboy and a number of female visitors.
5 Approximate percentages of visitor arrivals are as follows: Canada 45%, U.S.A. 25%, U.K.–Europe 20%, other 10%. French-speaking Canadians constitute about half of that country's tourists.
6 By contrast the intolerance shown by prostitutes towards "lovers" or "dawdlers" occurs within the framework of a single encounter (cf. Winick and Kinsie, 1971).

REFERENCES

Blau, P. *Exchange and power in social life.* New York: Wiley, 1964.
Cavan, R. Self role adjustment during old age. In A. Rose (Ed.), *Human behaviour and social processes.* London: Routledge, 1972, pp. 526–536.
Cooley, C. *Social organization.* New York: Scribner's, 1929.
Coulson, M. *Approaching sociology.* London: Routledge & Kegan Paul, 1970.
Dahrendorf, R. *Essays in the theory of society.* Stanford: Stanford University Press, 1968.
Dann, G. The holiday was simply fantastic. *Revue de Tourisme,* 1976, **3**, 19–23.
Dawe, A. The two sociologies. *British Journal of Sociology,* 1970, **21**, 207–218.
Denzin, N. The methodological implication of symbolic interactionism for the study of deviance. *British Journal of Sociology,* 1974, **25**, 269–282.
Fanon, F. *Black skin, white masks.* New York: Grove Press, 1967.
Goffman, E. The nature of deference and demeanour. *American Anthropologist,* 1956, **58**, 473–502.
Goffman, E. *The presentation of self in everyday life.* New York: Doubleday, 1959.
Goffman, E. *Encounters.* Indianapolis: Bobbs-Merrill, 1961.
Gordon, R. *Interviewing: Strategy, techniques and tactics.* Homewood, IL: Dorsey Press, 1969.

Gouldner, A. Cosmopolitans and locals. In G. Oscar and G. Miller (Eds.), *Sociology of organizations*. New York: Collier-Macmillan, 1970, pp. 477–482.

Lévi-Strauss, C. The principle of reciprocity. In L. Coser and B. Rosenberg (Eds.), *Sociological theory*. New York: Macmillan, 1957.

Linton, N. Tourism and race relations: A third world perspective. Paper presented to the Caribbean Ecumenical Consultation on Development, 1971.

Linton, R. *The study of man*. New York: Appleton-Century-Crofts, 1936.

McCall, G., and Simmons, J. *Identities and interactions*. New York: Collier-Macmillan, 1966.

Matthews, M. Radicals and third world tourism, a Caribbean case. *Annals of Tourism Research*, 1977, *V*, special number, 20–29.

Mead, G. *Mind, self and society*. C. Morris (Ed.). Chicago: University of Chicago Press, 1934.

Merton, R. *Social theory and social structure*. Glencoe: Free Press, 1968.

Pareto, V. *Mind and society*. (A. Buongiorno and A. Livingston, Trans.) New York: Harcourt, Brace, 1935.

Schutz, A. *The phenomenology of the social world*. (G. Walsh and F. Lehnert, Eds. and Trans.) London: Heinemann, 1972.

Simmel, G. *The sociology of Georg Simmel*. K. Wolff (Ed.). Glencoe: Free Press, 1950.

Winick, C., and Kinsie, P. *The lively commerce: Prostitution in the United States*. Chicago: Quandrangle Books, 1971.

Part V

TOURISM AND SOCIAL INSTITUTIONS

10

THE CHANGING ECONOMICS OF THE TOURIST INDUSTRY*

John Urry

INTRODUCTION

The relationship between the tourist gaze and those industries which have been developed to meet that gaze is extremely problematic.

Initially, it should be noted that almost all the services provided to tourists have to be delivered at the time and place at which they are produced (see Urry, 1987). As a consequence the quality of the social interaction between the provider of the service, such as the waiter, flight attendant or hotel receptionist, and the consumers is part of the 'product' being purchased by tourists. If aspects of that social interaction are unsatisfactory (the offhand waiter, the unsmiling flight attendant, or the rude receptionist), then what is purchased is in effect a different service product. The problem results from the fact that the production of such consumer services cannot be entirely carried out backstage, away from the gaze of tourists. They cannot help seeing some aspects of the industry which is attempting to serve them. But furthermore, tourists tend to have high expectations of what they should receive since 'going away' is an event endowed with particular significance. People are looking for the *extra*ordinary and hence will be exceptionally critical of services provided that appear to undermine such a quality.

Other features of tourist industries are likely to produce difficulties for the producers of such services. Such services cannot be provided anywhere: they have to be produced *and consumed* in very particular places. Part of what is consumed is in effect the place in which the service producer is located. If the particular place does not convey appropriate cultural meanings, the quality of the specific service may well be tarnished. There is therefore a crucial 'spatial fixity' about tourist services (see Bagguley, 1987). In recent years there has been enormously heightened competition to attract tourists. In relationship to Britain there has been a 'Europeanisation' of the tourist market and increasingly a 'globalisation'. So while the producers are to a significant extent spatially fixed, in that they have to provide particular services in particular places, consumers are increasingly mobile, able to consume tourist services on a global basis. The industry is inevitably competitive since almost every place in the

* Reprinted with permission of Sage Publications Ltd from John Urry (1991) *The Tourist Gaze: Leisure and Travel in Contemporary Societies*, Newbury Park, CA: Sage, pp. 40–65.

world could well act as an object of the tourist gaze. Such services are inherently labour-intensive and hence employers will seek to minimise labour costs. A variety of strategies is employed to bring this about but some at least will result in tarnishing or wholly undermining the extraordinary character of the consumers' tourist gaze.

The emphasis on the quality of the social interaction between producers and consumers of tourist services means that developments in the industry are not explicable simply in terms of 'economic' determinants. As will be shown later in the book it is also necessary to examine a range of cultural changes which transform people's expectations about what they wish to gaze upon, what significance should be attached to that gaze, and what effects this will have upon the providers of the relevant tourist services. This is an industry which has always necessitated considerable levels of public involvement and investment and in recent years this has increased as all sorts of places attempt to construct or reinforce their position as favoured objects of the tourist gaze. The economics of tourism cannot be understood separately from the analysis of cultural and policy developments, just as work in tourist industries cannot be understood separately from the cultural expectations that surround the complex delivery of such services. Work relationships in tourist industries are significantly culturally defined.

In this present chapter attention will be directed to some of the more obvious recent developments in what can loosely be termed the changing political economy of the tourist industry. The next section gives a brief account of the concept of positional goods, the main economic concept used to account for the economics of tourism, before moving on to the changing UK tourist industry, noting particularly its tendencies to globalisation; and to some of the main changes in the political economy of overseas tourism.

THE SOCIAL LIMITS OF TOURISM

The economist Mishan presents one of the clearest accounts of the thesis that there are fundamental limits to the scale of contemporary tourism (1969; and see Urry, 1990, on the following). These limits derive from the immense costs of congestion and overcrowding. In the 1960s Mishan perceptively wrote of: 'the conflict of interest . . . between, on the one hand, the tourists, tourist agencies, traffic industries and ancillary services, to say nothing of governments anxious to augment their reserves of foreign currencies, and all those who care about preserving natural beauty on the other' (1969: 140). He quoted the example of Lake Tahoe, whose plant and animal life had been destroyed by sewage generated by the hotels built along its banks. A 1980s example would be the way in which the coral around tourist islands like Barbados is dying because of the pumping of raw sewage into the sea from the beachside hotels and because locals remove both plants and fish from the coral to sell to tourists.

Mishan also notes that here is a conflict of interests between present and future generations which stems from the way in which travel and tourism are priced. The cost of the marginal tourist takes no account of the additional congestion costs imposed by the extra tourist. These congestion costs include the generally undesirable effects of overcrowded beaches, a lack of peace and quiet, and the destruction of the scenery. Moreover, the environmentally sensitive tourist knows that there is nothing to be gained from delaying a visit to the place in question: if anything, it is the opposite. There is a strong incentive to go as soon as possible – to enjoy the unspoilt view before the crowds get there! Mishan's perspective as someone horrified by the consequences of mass tourism can be seen from the following: 'the tourist trade, in a competitive scramble to uncover all places of once quiet repose, of wonder, beauty and historic interest to the money-flushed multitude, is in effect literally and irrevocably destroying them' (1969: 141). His middle-class elitism is never far from the surface. For example, he claims that it is the 'young and gullible' who are taken in by the fantasies dreamt up by the tourist industry.

His main criticism is that the spread of mass tourism does not produce a democratisation of travel. It is an illusion which destroys the very places which are being visited. This is because geographical space is a strictly limited resource. Mishan says: 'what a few may enjoy in freedom the crowd necessarily destroys for itself' (1969: 142). Unless international agreement is reached (he suggests the immensely radical banning of all international air travel!), the next generation will inherit a world almost bereft of places of 'undisturbed natural beauty' (1969: 142). So allowing the market to develop without regulation has the effect of destroying the very places which are the objects of the tourist gaze. Increasing numbers of such places are suffering from the same pattern of self-destruction. One resort that has recently been thought to be so damaged is St Tropez, the place initially made famous by Brigitte Bardot. She now claims that it is being swept by a 'black tide of human filth'; that tourists 'are mediocre, dirty, ill-mannered and rude'; and that she intends 'leaving it to the invaders' (see Rocca, 1989).

This pessimistic kind of argument is criticised by Beckerman, who makes two useful points (1974: 50–2). First, concern for the effects of mass tourism is basically a 'middle-class' anxiety (like much other environmental concern). This is because the really rich: 'are quite safe from the masses in the very expensive resorts, or on their private yachts or private islands or secluded estates' (Beckerman, 1974: 50–1). Second, most groups affected by mass tourism do in fact benefit from it, including even some of the pioneer visitors who find available services that were previously unobtainable when the number of visitors was rather small. Hence Beckerman talks of the 'narrow selfishness of the Mishan kind of complaint' (1974: 51).

This disagreement over the effects of mass tourism is given more theoretical weight in Hirsch's thesis on the social limits to growth (1978; also see the collection Ellis and Kumar, 1983). His starting point is similar to Mishan's: he notes that individual liberation through the exercise of consumer choice does not make those choices liberating for all individuals together (1978: 26). In particular he is concerned with the positional economy. This term refers to all aspects of goods, services, work, positions and other social relationships which are either scarce or subject to congestion or crowding. Competition is therefore zero-sum: as any one person consumes more of the good in question, so someone else is forced to consume less. Supply cannot be increased, unlike the case of material goods where the processes of economic growth can easily produce more. People's consumption of positional goods is inherently relational. The satisfaction derived by each individual is not infinitely expandable but depends upon the position of one's own consumption to that of others. This can be termed coerced competition. Ellis and Heath define this as competition in which the status quo is not an option (1983: 16–19). It is normally assumed in economics that market exchanges are voluntary so that people freely choose whether or not to enter into the exchange relationship. However, in the case of coerced consumption people do not really have such a choice. One has to participate even though at the end of the consumption process one is not necessarily better off. This can be summarised in the phrase: 'one has to run faster in order to stay still'. Hirsch cites the example of suburbanisation. People move to the suburbs to escape from the congestion of the city and to be nearer the quietness of the countryside. But as economic growth continues, the suburbs get more congested, they expand, and so the original suburbanites are as far away from the countryside as they were originally. Hence they will seek new suburban housing closer to the countryside, and so on. The individually rational actions of others make one worse off and one cannot avoid participation in the leap-frogging process. No one is better off over time as a result of such coerced consumption.

Hirsch clearly believes that much consumption has similar characteristics to the case of suburbanisation, namely that the satisfaction people derive from it depends on the consumption choices of others. This can be seen most clearly in the case of certain goods which are scarce in an absolute sense. Examples cited here are 'old masters' or the 'natural landscape', where increased consumption by one leads to reduced consumption by another (although see Ellis and Heath, 1983: 6–7). Hirsch also considers the cases where there is 'direct social scarcity': luxury or perhaps snob goods which are enjoyed because they are rare or expensive and possession of them indicates social status or good taste. Examples here would include jewellery, a residence in a particular part of London, or designer clothes. A third type Hirsch considers is that of 'incidental social scarcity': goods whose consumption yields satisfaction which is influenced by the extensiveness of use. Negative examples here would include the purchase of a car and no increase of satisfaction because of increased congestion, as everyone does the same; and the obtaining of educational qualifications and no improved access to leadership positions because everyone else has been acquiring similar credentials (Ellis and Heath, 1983: 10–11).

It is fairly easy to suggest examples of tourism that fit these various forms of scarcity. On the first, the Mediterranean coastline is in the condition of absolute scarcity where one person's consumption is at the expense of someone else. On the second, there are clearly many holiday destinations which are consumed not because they are intrinsically superior but because they convey taste or superior status. For Europeans, the West Indies, West Africa and the Far East would be current examples, although these will change as mass tourist patterns themselves alter. And third, there are many tourist sites where people's satisfaction depends upon the degree of congestion. Perhaps the best example would currently be Greece. Hirsch quotes a middle-class professional who remarked that the development of cheap charter flights to such a previously 'exotic' country meant that: 'Now that I can afford to come here I know that it will be ruined' (1978: 167).

Although I have set out these different types of positional good identified by Hirsch the distinctions between them are not fully sustainable and they merge into each other. Furthermore, there are a number of major difficulties in Hirsch's argument. It is ambiguous about what is meant by consumption in the case of much tourism. Is it the ability to gaze at a particular object if necessary in the company of many others? Or is it to be able to gaze without others being present? Or is it to be able to rent accommodation for a short period with a view of the object close at hand? Or is it the ability to own property with a view of the object nearby? The problem arises because of the importance of the gaze to touristic activity. A gaze is after all visual, it can literally take a split second, and the other services provided are in a sense peripheral to the fundamental process of consumption, which is the capturing of the gaze. This means that the scarcities involved in tourism are more complex than Hirsch allows for. One strategy pursued by the tourist industry has been to initiate new developments which have permitted greatly increased numbers to gaze upon the same object. Examples include building huge hotel complexes away from the coastline; the development of off-peak holidays so that the same view can be gazed upon throughout the year; devising holidays for different segments of the market so that a wider variety of potential visitors can see the same object; and the development of time-share accommodation so that the facilities can be used all of the year.

Moreover, the notion of scarcity is problematic for other reasons. I shall begin here by noting the distinction between the physical carrying capacity of a tourist site and its perceptual capacity (see Walter, 1982). In the former sense it is clear when a mountain path literally cannot take any more walkers since it has been eroded and has effectively disappeared. Nevertheless, even in this case there are still thousands of other mountain paths that could be walked along and so the scarcity only applies to *this* path leading to this particular view, not to all paths along all mountains.

The notion of perceptual capacity changes the situation. Walter is concerned here with the subjective quality of the tourist experience (1982: 296). Although the path may still be physically passable, it no longer signifies the pristine wilderness upon which the visitor had expected to gaze. Thus its perceptual carrying capacity would have been reached, but not its physical capacity. Walter goes on to note that perceptual capacity is immensely variable and depends upon particular conceptions of nature and on the circumstances in which people expect to gaze upon it. He cites the example of an Alpine mountain. As a material good the mountain can be viewed for its grandeur, beauty and conformity to the idealised Alpine horn. There is almost no limit to this good. No matter how many people are looking at the mountain it still retains these qualities. However, the same mountain can be viewed as a positional good, as a kind of shrine to nature that individuals wish to enjoy in solitude. There is then a 'romantic' form of the tourist gaze, in which the emphasis is upon solitude, privacy and a personal, semi-spiritual relationship with the object of the gaze.

Barthes characterises this viewpoint as found in the *Guide Bleu*; he talks of 'this bourgeois promoting of the mountains, this old Alpine myth . . . only mountains, gorges, defiles and torrents . . . seem to encourage morality of effort and solitude' (1972: 74).

Walter discusses the example of Stourhead Park in Wiltshire, which illustrates:

> the romantic notion that the self is found not in society but in solitudinous contemplation of nature. Stourhead's garden is the perfect romantic land-scape, with narrow paths winding among the trees and rhododendrons, grot-toes, temples, a gothic cottage, all this around a much indented lake. . . . The garden is designed to be walked around in wonderment at Nature and the presence of other people immediately begins to impair this.
>
> (1982: 298)

When discussing Mishan I noted his emphasis that 'undisturbed natural beauty' con-stituted the typical object of the tourist gaze. However, this is in fact only one kind of gaze, what I will call the 'romantic'. There is an alternative: the 'collective' tourist gaze, with different characteristics. Here is a description of another Wiltshire house and garden, Longleat:

> a large stately home, set in a Capability Brown park; trees were deliberately thinned . . . so that you can see the park from the house, and house from the park. Indeed the house is the focal point of the park . . . the brochure lists twenty-eight activities and facilities. . . . All this activity and the resulting crowds fit sympathetically into the tradition of the stately home: essentially the life of the aristocratic was public rather than private.
>
> (1982: 198)

Such places were designed as public places: they would look strange if they were empty. It is other people that make such places. The collective gaze thus necessitates the presence of large numbers of other people, as were found for example in the seaside resorts previously discussed. Other people give atmosphere or a sense of carnival to a place. They indicate that this is *the* place to be and that one should not be elsewhere. And as we saw, one of the problems for the British seaside resort is that there are not enough people to convey these sorts of message. As Walter says: 'Brighton or Lyme Regis on a sunny summer's day with the beach to oneself would be an eerie experience' (1982: 298). It is the presence of other *tourists*, people just like oneself, that is actually necessary for the success of such places, which depend upon the collective tourist gaze. This is also the case in major cities, whose uniqueness is their cosmopolitan character. It is the presence of people from all over the world (tourists in other words) that gives capital cities their distinct excitement and glamour (see Walter, 1982: 299).

Large numbers of other tourists do not simply generate congestion, as the pos-itional good argument would suggest. The presence of other tourists provides a market for the sorts of service that most tourists are in fact desperate to purchase, such as accommodation, meals, drink, travel and entertainment. New Zealand is an interesting case here. Once one leaves the four major cities there are almost no such facilities because of the very few visitors in relation to the size of the country. The contrast with the Lake District in north-west England is most striking, especially given the scenic similarity.

Thus Hirsch's arguments about scarcity and positional competition mainly apply to those types of tourism characterised by the romantic gaze. Where the collective gaze is to be found there is less of problem of crowding and congestion. And indeed Hirsch's argument rests on the notion that there are only a limited number of objects which can be viewed by the tourist. Yet in recent years there has been an enormous increase in the objects of the tourist gaze, far beyond Mishan's 'undisturbed natural beauty'. Part of the reason for this increase results from the fact that contemporary tourists are collectors of gazes and appear to be less interested in repeat visits to the same auratic site. The initial gaze is what counts.

Those who do really value solitude and a romantic tourist gaze do not see this as merely *one* way of regarding nature. Instead they attempt to make everyone sacralise nature in the same sort of way (see Walter, 1982: 300–3). Romanticism, which as we previously noted was involved in the early emergence of mass tourism, has become widespread and generalised, spreading out from the upper middle classes, although the notion of romantic nature is a fundamentally invented and variable pleasure. And the more its adherents attempt to proselytise its virtues to others, the more the conditions of the romantic gaze are undermined: 'the romantic tourist is digging his [*sic*] own grave if he seeks to evangelize others to his own religion' (Walter, 1982: 301). The romantic gaze is an important mechanism which is helping to spread tourism on a global scale, drawing almost every country into its ambit as the romantic seeks ever-new objects of that gaze, and minimising diversity through the extension of what Turner and Ash term the 'pleasure periphery' (1975).

The contemporary tourist gaze is increasingly signposted. There are markers which identify the things and places worthy of our gaze. Such signposting identifies a relatively small number of tourist nodes. The result is that most tourists are concentrated within a very limited area. As Walter says: 'the sacred node provides a positional good that is destroyed by democratisation' (1982: 302). He by contrast favours the view that there are 'gems to be found everywhere and in everything . . . there is no limit to what you will find' (Walter, 1982: 302). We should, he says, get away from the tendency to construct the tourist gaze at a few selected sacred sites, and be much more catholic in the objects at which we may gaze. Undoubtedly this has began to occur in recent years, particularly with the development of industrial and heritage tourism. However, there is little doubt that Walter's analysis of the class character of the romantic gaze is highly persuasive, and I will end this section on the economic theory of tourism by noting his thoroughly sociological analysis of the pervasiveness of the romantic as opposed to the collective gaze and the consequential problem of the positional good of many tourist sites:

professional opinion-formers (brochure writers, teachers, Countryside Commission staff, etc.) are largely middle class and it is within the middle class

that the romantic desire for positional goods is largely based. Romantic solitude thus has influential sponsors and gets good advertising. By contrast, the
largely working class enjoyment of conviviality, sociability and being part of a
crowd is often looked down upon by those concerned to conserve the environment. This is unfortunate, because it . . . exalts an activity that is available
only to the privileged.

(Walter, 1982: 303)

GLOBALISATION AND THE ECONOMICS OF TOURISM

We have already seen that the English seaside resort went into decline in the mid-
1960s, at the moment when mass tourism, at least in Europe, became internationalised. There has continued to be massive growth of international tourist flows. By
1984 there were almost 300 million tourist arrivals worldwide and international tourism was the second-largest item in world trade. International tourist receipts increased 47.6 times over the period 1950–84. About three-quarters of this involves
industrial countries, with the USA, Japan, Germany and Canada having a large defi-
icit, and Austria, France, Italy, Spain and Switzerland large credits (IMF, 1986: Table
C-5). In the case of Britain, overseas earnings and expenditure more or less balanced
in the early 1980s, having been well in credit in the 1970s. Britain ranks after the
USA, Spain, Italy and France in terms of overseas earnings from tourism. But in the
mid- to late 1980s a strong deficit developed, resulting from increased visits being
made by UK citizens abroad, the tendency for UK holiday-makers to increase their
expenditure when abroad, and the fact that overseas visitors to Britain are now
spending relatively less, in part because fewer of them now come from North
America (Department of Employment, 1988).

This internationalisation of tourism means that we cannot explain tourist patterns
in any particular society without analysing developments taking place in other countries. The internationalisation of tourism especially in Europe means that every tourist site can be compared with those located abroad. So when people visit somewhere
in their own country they are in effect choosing not to visit a site abroad. The internationalisation of tourism means that all potential objects of the tourist gaze can be
located on a scale, and can be compared with each other.

The result of such internationalisation is that different countries, or different
places within a country, come to specialise in providing particular kinds of objects to
be gazed upon. An international division of tourist sites has emerged in the last
decade or two. Britain has come to specialise in its tourism on history and heritage
and this affects both what overseas visitors expect to gaze upon, and what attracts
UK residents to spend time holiday-making within Britain. Moreover, this internationalisation of holiday-making is more advanced and developed in Britain than in
most other countries. This is partly because of the early and innovative development
of the package or inclusive holiday in Britain, and partly because of the availability
of an exceptional number of historical sites suitable for attracting large numbers of
overseas tourists. Just as the UK economy in general is an open economy so this is

specifically true of tourism. I shall briefly consider the nature of the package holiday industry before considering the main features of the UK domestic holiday industry.

Tour operators based in Britain seem to sell their inclusive or package holidays at a considerably cheaper price than in comparable European countries. In a sample survey of 57 hotels in Spain, Portugal and Greece, it was noted that in the case of 40 it was a British tour operator that offered the lowest price. In 39 out of 57 hotels a German company charged the most (Milner, 1987: 21). British-based companies have been particularly effective at reducing unit costs and at generating a huge market in Britain. There are now about 11 million package holidays sold each year (compared with around 5 million in 1980). The main reason why the inclusive holiday has had such an impact in Britain is because of the early emergence of integrated companies, the tour operators, who made spectacularly successful use of the new technologies of jet transport and computerised booking systems (see Reynolds, 1989: 330–3, on Thompson Holidays, which is incidentally part of the Canadian media and marketing multinational).

The expansion of the package tour industry in Britain has been accompanied by a considerable degree of concentration. In 1988 the Thompson Group (which includes Portland and Skytours) accounted for 28 per cent of the market and International Leisure Group (which includes Intasun and Club 18–30) 17 per cent. The third major company with about 8 per cent of the market, Horizon, has recently been taken over by Thompson (see *The Economist*, 27 August 1988: 59–60). The strength of the main tour operators partly results from the fact that either they own their own airline (Thompson owns Brittannia, for example), or that they are themselves owned by an airline (British Airways owns Martin Rooks). When Thompson takes over Horizon, the new airline will be larger than Swissair and twice as big as SAS (*The Economist*, 1988: 59; see Goodall, 1988: 28–9, on vertical integration in much of the European tourist industry). Indeed, to a significant extent, the tour operators tend to make more money through their subsidiary airline than through the inclusive holiday operations, where it is claimed that the profit to the operator may be as little as £1–2 per £275 holiday (see Williams, 1988; *The Economist*, 27 August 1988).

Yet there are still many small companies providing inclusive holidays. In 1988 there were 679 tour operators licensed with the Civil Aviation Authority. However, the share taken by the large companies has steadily increased. It seems that this is partly because they have began to cater for more exotic destinations and for more specialised tastes – for what I will discuss later as the 'post-tourist'. Thompson, for example, has twenty-four brochures orientated to different market segments and the purchase of Horizon will probably involve further market segmentation (Williams, 1988; *The Economist*, 27 August 1988).

In a survey of their readership in 1987 *Holiday Which?* found that none of the largest three tour operators were placed into even the top twenty in terms of popularity. The people surveyed preferred smaller, more specialised operators. But in a different survey Lewis and Outram found that clients could perceive little difference between tour operators (1986: 209–10). They also found a high level of satisfaction with the operators among their respondents, who had particularly positive attitudes towards the convenience and value aspects of their package holiday (Lewis and Outram, 1986: 213). This perhaps runs counter to the image of such holidays which has been effectively summarised by Pile. He quotes an industrial psychologist who writes of such holidays that:

> they are based upon all the most stressful features of 20th century life. It is amazing that people never question why they are going on them. . . . They leave a comfortable, civilised home and basically go to live on a beach for a fortnight in acute discomfort. . . . They arrive in a strange hotel where they can't even eat when they want to.
>
> (Pile, 1987)

A number of important changes will affect this sector of the tourist industry. First, the inclusive holiday will enable much more distant destinations to become accessible to a mass market, such as the Caribbean for £300 or the Far East for £400–500.

Consumers will in future be able to put together much more flexible packages, a kind of holiday 'mix n' match' or what the industry terms 'Free and Independent Travel' or FITs (see Hart, 1988: 19). This will mean that travel agents must become more skilled than surveys suggest they are at present (see Welsh, 1989, for a damning indictment of the leading travel agencies).

At present the markets for inclusive holidays are regulated or organised by the central aviation authority within each country (see *The Economist*, 27 August 1988). However, after 1992 and the formation of a single market, the tour operators in Europe will operate in each of the major countries to a much greater extent. This will have the effect of increasing competition and reducing the level of concentration within a single country. It will be interesting to see whether this is associated with increased vertical integration, with the operators continuing to purchase travel agents, hotels and airlines (see Buck, 1988, on the complex relationship between tour operators and travel agents).

And finally, this is an industry where new technologies are particularly appropriate because of the immense informational and communication problems involved. The most recent innovation has been the adoption of interactive videotext. This was first introduced in 1979 and has now spread to about 90 per cent of all UK travel agencies, the best known being Thompson's TOPS system (see Bruce, 1987: 115). In the future one development will be systems which permit customers to 'self-serve' themselves with tickets and other standardised products. Indeed if these were available in the home then it is possible to envisage a 'paperless travel agency'. Expert systems developments would enable the prospective traveller to provide some parameters of intended travel and then allow the computer to generate a number of possible consumer products (Bruce, 1987: 117–18).

In 1989 in Britain a million fewer inclusive holidays were sold by the main operators than in the previous year. And it is predicted that the number will be below 9 million in 1990 (Barrett, 1989). Yet the number of overseas holidays taken by UK residents continues to rise (8 per cent in 1988–9). It seems that with increased leisure time people are increasingly moving away from the somewhat standardised package holiday and seeking out a wider variety of forms of leisure activity, including independent travel (see McRae, 1989). This will force tour operators to develop the more flexible kinds of travel arrangements discussed earlier. There has been a marked increase recently in seats-only flights, partly because of the demand for more flexibility and partly because of the growth in overseas property ownership (see Ryan, 1989, for a very useful survey of future developments). These factors are also likely to force the tour operators to seek greater quality control over all aspects of the holiday, something they have not been very successful in achieving so far: 'The result has been inconsistent quality. There is no Marks and Spencer in tourism. As a result the operators have been losing the up-market, high value-added segment of the trade' (McRae, 1989).

Barrett also suggests that the switch to independent travel 'is partly a reaction to the "naffness" of package holidays', that they are no longer viewed as fashionable or smart (1989). So far I have looked at some features of the industry concerned with transporting British and other north European holiday-makers mainly to certain south European countries, within which there is intense spatial concentration (see Goodall, 1988: 25–6). In the next section I shall consider the organisation of the tourist industry in societies that host large numbers of such visitors. But before this I shall examine some features of the tourist industry concerned with the provision of services *within* Britain.

I have noted that there is currently a large tourism deficit in Britain. The response from the minister with (at the time) special responsibility for tourism, John Lee, was as follows:

> Hopefully the British public will increasingly appreciate that holiday taking in the United Kingdom is not only a stimulating and enjoyable experience but will help our balance of payments as well. Any potential visitors from overseas should realise that there can be few countries in the world offering such a range of heritage, countryside, resorts, sporting and cultural opportunities and attractions (in England 30% of our existing visitor attractions have been opened since 1980, 60% since 1970).
>
> (Lee, 1988)

Not only does the minister make it clear that 'heritage' is the first feature likely to appeal to overseas visitors, but also that there has been an exceptional investment in new tourist attractions.

As far as overseas visitors are concerned there is a huge potential market. Only 7 per cent of US citizens possess a passport; this figure would merely have to reach 10 per cent for there to be very large increases in the number of potential US visitors (see Cabinet Office, 1983). Of foreign visitors coming to Britain only about 20 per cent visit the seaside, and even Blackpool attracts relatively few for all its efforts to market itself as an international leisure centre. Only 7 per cent of overseas visitors visit 'funfairs and entertainment parks' (BTA/ETB Research Services, 1988: 58).

The preferences of foreign visitors are highly localised, most being attracted to

London and to various small inland towns and cities. In fact about 80 per cent began their visit in London and this is reflected in, for example, the fact that 40 per cent of those visiting London West End theatres are foreign visitors (BTA/ETB Research Services, 1988: SWET, 1982). The build-up of overseas visitors in London in 1984–5 produced no significant deflection to provincial centres that were not already attracting significant numbers of such visitors. Indeed it may be that if there is no shift in visitor patterns and if hotel space in London remains in short supply, then some overseas visitors will divert to other destinations in Europe rather than to the rest of Britain (see Jeffrey and Hubbard, 1988).

Over 75 per cent of overseas visitors made their own travel arrangements rather than booking an inclusive tour. There has been no increase in the absolute numbers coming to Britain on inclusive tours (BTA/ETB Research Services, 1988).

There has been no increase in the number of new visitors to Britain and all of the increase has been in 'repeat' visitors. Such 'repeats' are more common amongst business travellers, West Europeans, visitors aged over 35 and people travelling outside London (BTA/ETB Research Services, 1988).

At present the six most popular leisure pursuits for overseas visitors are visiting shops or markets (82 per cent), restaurants or cafés (77 per cent), churches, cathedrals, etc. (69 per cent), historic sites/buildings (69 per cent), museums and art galleries (64 per cent), and historic cities or towns (62 per cent) (BTA/ETB Research Services, 1988, and see Key Note Report, 1987). Overseas visitors were also asked which facilities provided good value for money. All bar one facility were thought to provide good value by a higher proportion than those thought to be offering not good value. The one exception were London hotels, where three times as many visitors thought that they provided poor value compared with those who thought they provided good value (BTA/ETB Research Services, 1988). This low level of satisfaction reflects the very strong demand for accommodation in London. Average room occupancy rates are the highest in the country, 78 per cent, compared with 47 per cent in seaside towns, 51 per cent in the countryside and 57 per cent in England overall in 1984 (Bagguley, 1987: 16). A number of other features of the economic and geographical organisation of UK tourism will now be considered.

An initial point is that accommodation in Britain is still mainly provided in small units. In the early 1980s it was estimated that there were almost 10,000 hotels, 20,000 guesthouses, 18,000 bed-and-breakfast outlets, 4,800 farms providing accommodation, 300 motels and 90 holiday camps (Bagguley, 1987: 8). Only 1.5 per cent of hotels had more than 200 rooms, while almost 85 per cent had fewer than 50. In a slightly later survey it was calculated that of the 0.5 million hotel bedrooms in Britain, 30 per cent were corporately owned and 70 per cent owned by independents, most with fewer than 50 bedrooms (Slattery and Roper, 1986). Small capital is thus of enduring significance in the provision of tourist accommodation in Britain (see Goffee and Scase, 1983: my account here is taken from Bagguley, 1987). Amongst these small capitals four different sectors can be usefully distinguished.

There are the 'self-employed' with no outside labour and a reliance upon family labour and trade-based skills. They are economically marginal, formal management skills are weak, and there is no divorce of ownership from control. Many bed-and-breakfast units are of this type. There is the category of 'small employers' who are distinguished from the self-employed by the intermittent employment of outside labour. Many guesthouses and small hotels are of this sort. There are 'owner controllers' who do not use family labour and employ outside labour which may receive a considerable degree of training. Levels of capital investment are higher and there are more formal means of managerial control. These are less economically marginal, although there is no divorce of ownership and control. Many so-called country house hotels where the proprietor is directly involved are of this sort. Finally there are 'owner directors', where there is considerable capital investment, formal training, and the separation of ownership and control. This is most common amongst those city centre hotels which are not part of any chain.

All these different types of small capital in the hotel and catering industry show enormous vulnerability in the market. Between 1980 and 1986 121,700 hotel and catering establishments were established in the sense that they came on to the register for paying value-added tax (VAT); and yet a staggering 115,900 came off the register (Smith, 1988: 22). A quarter of such establishments close within two years, half within four. About one-quarter do survive while the remaining three-quarters are in a continuous state of flux. The rate of turnover is considerably higher in this industry than in most others, which means that there is great job insecurity. The Hotel and Catering Industry Training Board estimates that a quarter of the workforce will lose or leave their job within a year (Smith, 1988: 22). The category of 'owner controllers' appears best able to survive. Drew Smith, former editor of *Good Food Guide*, maintains that: 'it is the personally run restaurant [and hotel] that has bucked the trend, often by the sheer dedication of the owner. At the other end of the market this has usually been because the family has kept the freehold of the site' (1988: 22). This therefore is an industry of enormous volatility. And although there are some very large operators, as we will see, there are very few restrictions on entry (and exit!). Drew Smith estimates that 50 new restaurants open each month in London. Indeed there are almost certainly large variations in the propensity to start new restaurants and hotels in different parts of the country. Cornwall, for example, seems to have relatively weak traditions of local entrepreneurship in all sectors of the economy including tourism, although most tourist-related enterprises are in fact small. In a study of Looe it was found that there was a very high level of geographical mobility amongst entrepreneurs. For example, 90 per cent of the owners of hotels and guesthouses came from outside the south-west region, the majority coming from the south-east (Hennessy *et al.*, 1986: 16). Cornwall in general seems important in attracting entrepreneurs rather than generating its own.

One major influence on the kinds of tourist-related activities found in different areas (London compared with Looe) is the growth of business tourism and its differential impact. There was a 30 per cent increase in the number of business 'nights' spent in Britain between 1980 and 1984. This makes up for the more or less static number of hotel 'nights' demanded by domestic holiday-makers over the same period (Bagguley, 1987: 13–18). Business tourism is made up of a number of components: genuine meetings held in other parts of the country; conferences – in 1984 there were 14,000 international conferences of which the majority were held in Britain; and the provision of travel as a non-taxable perk – something provided by about one-third of the UK's top companies (see Williams and Shaw, 1988b: 19).

The effect of business tourism in particular has generated periodic hotel building booms. For example, in the late 1960s and early 1970s there was a spate of hotel building especially in London following the 1969 Hotel Development Incentive Scheme. In March 1970 there were 61 hotels planned for London, all but 7 with more than 75 bedrooms. This was three times as many as were planned for traditional 'holiday areas' (Bagguley, 1987: 17). However, in the early 1980s the pattern was quite different. Only 3 per cent of new hotels completed between 1981 and 1984 were in London and the largest proportion, 38 per cent, were located in 'small towns' (BTA/ETB, 1985). By 1988 (January–June) investment in new hotels is fairly evenly spread amongst the different tourist regions, with London getting about 10 per cent. Interestingly though, most expenditure on hotel refurbishment is to be found in London, as is the largest expenditure on the building of 'themed attractions'. Overall about one-third of all tourist investment is currently taking place in London (BTA/ETB Research Services, 1988: Table A).

It was noted earlier that much accommodation in Britain is provided by a very large number of small hotels and guesthouses. It seems that the number of these declined between 1951 and 1971 (Stallinbrass, 1980). In recent years the data are less clear although it is certain that there has been an increase in the share of accommodation provided by the very large hotel groups. The major chains in 1984 were Trusthouse Forte, Ladbroke Group, Crest and Thistle (Key Note Report, 1986, and see Shaw *et al.*, 1988).

The most significant development in the past few years has been the development of hotel consortia. In 1984 there were 170 Best Western hotels in Britain, 103 in the Inter Hotel consortium and 22 Prestige hotels (*Financial Times*, 2 January 1984; Bagguley, 1987: 19–22). Such consortia grew up in the 1960s to compensate for the decentralised structure of capital and the problems this created for capital accumulation and systematic cost cutting. Corporate growth is difficult given the low levels of concentration, small unit size and the varying combination of meals, drinks and accommodation for which demand is both seasonal and volatile (see Litteljohn, 1982; Slattery *et al.*, 1985; Bagguley, 1987). Consortia enable hotels to gain economies of scale and hence to be able to compete much more effectively with the economies obtainable by the large chains listed earlier. Consortia fall into a number of different types:

marketing consortia: to provide access to a corporate marketing department on the basis of either regional grouping (such as Thames Valley Hotels) or a specific market segment (such as Prestige);

marketing and purchasing consortia: apart from marketing economies they also negotiate reduced prices for bulk orders (Best Western, Inter Hotel);

referral consortia: to provide a national or international system of referrals particularly connected with airlines (British Airways Associate Hotels);
personnel and training consortia: to provide common training and personnel functions (Concord Hotels);
reservations systems: to provide a system of national or international reservations often linked to various other tourist offices (Expotel Hotels).

Some of the groups just considered are highly internationalised. In 1986, for example, Trusthouse Forte owned 793 hotels worldwide, as well as many other catering-related establishments (see Williams and Shaw, 1988b: 26). The second largest British group, Ladbrokes, also own 140 Hilton International hotels. The largest international groups in 1986 were Holiday Inns with 1,907 hotels and Quality Inns with 801. Since then Holiday Inns sold first in 1988 control of all their non-North-American hotels, and then in 1989 their North American hotels, to the UK company Bass (see Jamieson, 1989). Holiday Inns control 10 per cent of all hotel beds in the USA. Bass have not bought outright ownership since most of the hotels are franchised. They now have control of 1,700 hotels worldwide.

Grand Metropolitan is also a major international force but recently it has sold the whole of the huge Intercontinental Hotels group for £1.3 billion to the Japanese company Seibu Saisson (the *Guardian*, 14 November 1988). Another major company operating in Britain is Novotel which is French-owned, while the largest hotel company in West Germany is Queens Moat Houses which was also the sixth largest British group (in 1984).

It is useful to distinguish between a number of different kinds of hotel group. First, there is the relatively specialist hotel chain such as Queens Moat Houses. Second, there is the more general hotel and catering group such as Trusthouse Forte which, apart from its 700 or so hotels, also owns Little Chef and Happy Eater and operates 200 Kentucky Fried Chicken outlets (the *Guardian*, 14 November 1988). Overall it has reduced its degree of diversification since 1970 (see Gratton and Taylor, 1987: 48–50). Third, there are hotel chains such as Sheraton, which is part of the American conglomerate ITT, or in Britian Rank hotels which are part of the much wider Rank group, which includes Butlins and Top Rank bingo (Gratton and Taylor, 1987: 41–4). Fourth, there are the hotels owned by tour operators or travel agents. This is more common in Western Europe than North America – Thompson in Britain or Vingressor in Sweden would be good examples. And finally, there are the hotel chains owned by airlines: Pan-Am, for example, used to own Intercontinental Hotels (see Williams and Shaw, 1988b: 26).

Internationalisation of leisure provision also occurs in other parts of the tourist-related services. This can be seen most obviously in the case of McDonalds, as well as in the rather larger one-off investments. Currently, for example, the Walt Disney Company is building a $2 billion theme park near Paris to be called Euro Disneyland. This will be a 5,000-acre development with 500 hotel rooms, shopping centres, offices, up-market housing, camp grounds, golf courses and a rail link – a full 'destination resort' rather than a 'day trip' as at the Disneyland in California (for more detail see Shamoon, 1989).

The tourist-related industries analysed here are thus intensely competitive. Moreover, although the very large operators can move their capital around in response to changing market pressures, most of the industry cannot, certainly in the short run. This is an industry with great spatial fixity, an exceptional degree of decentralisation, and immense volatility of taste. Since many visitors to hotels and restaurants are only there because of the attractions available in the area so there may well be much support given to enhancing local facilities, to increasing the number or attraction of potential objects of the tourist gaze. Hence, tourist-related capital may well be in favour of large-scale public investment by local authorities to provide new or enhanced objects upon which visitors can gaze.

In the past decade or so companies involved in the industry have pursued a number of cost-reduction strategies.

1 The expansion of self-serviced accommodation which in Britain in 1984 accounted for 152 million nights, compared with 91 million nights in serviced accommodation (Bagguley, 1987: 18).
2 The extensive use of information technology (IT). This results from the fact that slack resources in the industry such as a hotel bed cannot be stored; the immense volatility of demand; and the fact that tourist units are necessarily geographically dispersed. IT enables various kinds of network to be established between potential consumers and the many locally specific and decentralised units.
3 Changing the labour input especially via the growth of part-time female employment. The proportion of such workers in the UK hotel and catering industry rose from 32.5 per cent in 1971 to 44.7 per cent in 1981 (Census of Employment).
4 Economising on costs through joining a group or consortium.
5 Closing down in slack periods or alternatively trying to generate extra business during the often very long off-season.
6 Generating extra income by improving the quality of the product provided, such as better meals, more trained labour or *en suite* facilities.

THE OVERSEAS IMPACT OF TOURISM

Tourist development outside Britain has had a broad economic, social and cultural impact. Drawing in part on the concept of the positional good discussed earlier, I shall consider some Mediterranean countries, and shall follow this with comments on North America and South-East Asia.

As we saw previously, there are complex relationship between tourists and the indigenous populations of the places at which those tourists gaze; the resulting artificiality of many tourist attractions arises from the particular character of the social relations that come to be established between 'hosts' and 'guests' in such places (see Smith, 1978, for much of the following). There are a number of determinants of the particular social relations that are established between 'hosts' and 'guests'.

1 The *number* of tourists visiting a place in relationship to the size of the host population and to the scale of the objects being gazed upon. For example, the geographical size of New Zealand would permit many more tourists to visit without either environmental damage or an undesirable social effect. By contrast the physical smallness of Singapore means that extra tourists cannot easily be accommodated except by even more hotel building which would only be possible by demolishing the remaining few Chinese shophouses which in the past have been one of the main objects of the tourist gaze. Similarly the magnificently preserved medieval city of Dubrovnik has an absolute physical limit determined both by the city walls and by the fact that over 4,000 people live within those walls. The resident population cannot but be numerically overwhelmed during the height of the summer season.

2 The predominant *object* of the tourist gaze, whether it is a landscape (the Lake District), a townscape (Chester), an ethnic group (Maoris in Rotorua, New Zealand), a life-style (the 'wild West'), historical artefacts (Canterbury Cathedral or Wigan Pier), bases of recreation (golf courses at St Andrews), or 'sand, sun and sea' (Majorca). Those tourist activities which involve observation of physical objects are clearly less intrusive than those which involve observing individuals and groups. Moreover, within the latter category, the observation of the private lives of host groups will produce the greatest social stress. Examples here include the Eskimo, or the Masai who have responded to the gaze by charging a '£ for car' for visits to their mud huts. By contrast, where what is observed is more of a public ritual then social stress will be less pronounced and indeed wider participation may be positively favoured, as in various Balinese rituals (see Smith, 1978: 7).

3 The *character* of the gaze involved and the resulting spatial and temporal 'packing' of visitors. For example, the gaze may be something that can take place more or less instantaneously (seeing/photographing New Zealand's highest mountain, Mount Cook), or it may require prolonged exposure (seeing/experiencing the 'romance' of Paris). In the case of the former, Japanese tourists can be flown in for a visit lasting just a few hours, while the experience of the romance of Paris will necessitate a longer and 'deeper' immersion.

4 The *organisation* of the industry that develops to service the mass gaze: whether it is private or publicly owned and financed; whether it is locally owned or involves significant overseas interests; whether the capital involved is predominantly small or large scale; and whether there are conflicts between the local population and the emergent tourist industry. Such conflicts can occur around many issues: conservation as opposed to commercial development, the wages to be paid to locally recruited employees, the effects of development on local customs and family life, what one might call the 'trinketisation' of local crafts, and how to compensate for the essential seasonality of labour (see Smith, 1978: 5–7).

5 The effects of tourism upon the *pre-existing agricultural and industrial activities*. These may range from the destruction of those activities (much agriculture in Corfu); to their gradual undermining as labour and capital are drawn into tourism (parts of

Spain); to their preservation as efforts are made to save pre-existing activities as further objects to be gazed upon (cattle farming and hence grazing in the Norfolk Broads).

6 The economic and social *differences* between the visitors and the majority of the hosts. In northern Europe and North America tourism creates fewer strains since the mass of 'hosts' will themselves be 'guests' on other occasions. It may be that tourism can in a rather inchoate way develop 'international understanding'. The shift in public attitudes in Britain towards a pro-Europeanism in the 1980s is difficult to explain without recognising that some role is played by the European tourism industry and the way in which huge flows of visitors have made Europe familiar and unthreatening. Elsewhere, however, there are usually enormous inequalities between the visitors and the indigenous population, the vast majority of whom could never envisage having either the income or the leisure time to be tourists themselves. These differences are reinforced in many developing countries by the nature of the tourist development, which appears to be exceptionally opulent and highly capitalised, as for example in many hotels in India, China, Singapore, Hong Kong and North Africa, partly because there are so few service facilities otherwise available to either visitors or the host population.

7 The degree to which the mass of visitors demand *particular standards of accommodation and service*, that they should be enclosed in an environmental bubble to provide protection from many of the features of the host society. This demand is most marked amongst inclusive tour visitors, who not only expect western standards of accommodation and food but also bilingual staff and well-orchestrated arrangements. Such tourists rarely leave the security of the western tourist bubble and to some degree are treated as dependent 'children' by the tourist professionals (see Smith, 1978: 10–11). In some cases the indigenous culture actually is dangerous, as in Sicily, in parts of New York, and recently in Florence. This demand is less pronounced amongst individual exploring 'travellers', poorer tourists such as students, and those visitors for whom 'roughing it' is part of their expected experience as tourists.

8 The degree to which the *state* in a given country actively seeks to promote tourist developments or alternatively endeavours to prevent them. Good examples of the former are Spain, Tunisia and Hawaii which are all actively developing a fully fledged tourist culture where large numbers of tourists have become part of the 'regional scenery' (Smith, 1978: 12). By contrast many of the oil states have for moral/social reasons explicitly decided to restrict tourism by refusing visas (Saudi Arabia is a good example). Likewise during the Cultural Revolution in China the state actively sought to prevent the growth of tourism. When this changed in the early to mid-1970s western visitors were so unusual that they were often applauded as though they were royalty.

9 The extent to which *tourists can be identified and blamed* for supposedly undesirable economic and social developments. This is obviously more common when such visitors are economically and/or culturally and/or ethnically distinct from the host population. It is also more common when the host population is experi-

encing rapid economic and social change. However, such change is not necessarily the outcome of 'tourism'. In the case of Tonga, for example, it is not the annual influx of visitors but rather gross overpopulation which accounts for the high inflation rate. And yet of course it is much easier to blame the 'nameless, faceless foreigner' for indigenous problems of economic and social inequality (see Smith, 1978: 13). Moreover, some local objections to tourism are in fact objections to 'modernity' or to modern society itself: to mobility and change, to new kinds of personal relationships, to a reduced role of family and tradition, and to different cultural configurations (see discussion in Welsh, 1988, of the 'Ecumenical Coalition of Third World Tourism').

The social impact of tourism will thus depend on the intersection of a wide range of factors. For example, great concern has been expressed about the likely consequences of tourism in various Mediterranean countries. The growth of tourism in the Mediterranean is one of the most significant economic and social developments in the postwar period. It is a particularly striking symbol of postwar reconstruction in Europe, or at least in Western Europe.

There is a high income elasticity of demand for tourist services and as incomes have grown in West Germany, France, Scandinavia, the Low Countries and the British Isles, so there has been a more than corresponding increase in demand for overseas travel. Western Europe in fact accounted for 68 per cent of all international tourists in 1984 (Williams and Shaw, 1988a: 1). In response to such demand the countries of southern Europe have developed enormous tourist industries. And those industries have been particularly cost effective, which in turn has lowered the real cost of overseas travel and hence led to further expansion of demand. Spain was the first and has remained the largest of the Mediterranean destinations. The number of foreign visitors increased from 6 million in 1960 to over 40 million by the early 1980s (Williams *et al*, 1986: 5). Other major destination countries are France, Italy, Greece, Portugal, Yugoslavia and, most recently, Turkey. In 1984 tourism receipts accounted for over 4 per cent of the national income of Spain and Portugal, over 3 per cent of that of Greece, and 2.6 per cent of that of Italy (Williams *et al*, 1986: 13; it should also be noted that the highest proportion in Europe is in fact Austria, with 7.8 per cent). Overall, the Commission of the European Community argues, tourism generates a net distribution of wealth from northern to southern Europe, and especially to Spain and Italy. It should, however, be noted that developments in such countries tend to be highly geographically concentrated. In the case of Spain for example 26 per cent of all high-season employment in the Balearic Islands is to be found in hotels although attempts are now being made to encourage travel into inland Spain (Valenzuela, 1988). Likewise in Yugoslavia 86 per cent of all tourists stay in coastal resorts (Williams and Shaw, 1988b: 36).

The problematic effects of such tourist developments in at least some of the countries concerned are fairly well known. They result from the huge number of tourists and their seasonal demand for services, the deleterious social effects particularly resulting from the gendered work available, the geographical concentration of visitors, the lack of concerted policy response, the cultural differences between hosts and guests, and the demand by many visitors to be enclosed in expensive 'environmental bubbles'. One place that has been overrun by tourists is Florence, where the resident population of 500,000 accommodates 1.7 million visitors each year. This has led to the plan to remove the city's academic, commercial and industrial functions from the centre and to turn it over entirely to tourism. It would mean the 'Disneyfication of Florence' (Vulliamy, 1988: 25).

Robert Graves has written of the similar tourist transformation of Majorca:

> the old Palma has long ceased to exist; its centre eaten away by restaurants, bars, souvenir shops, travel agencies and the like . . . Huge new conurbations have sprung up along the neighbouring coast. . . . The main use of olive trees seems to be their conversion into . . . salad bowls and boxes for sale to the tourists. But, as a Majorcan wag remarked, once they are all cut down we will have to erect plastic ones for the tourists to admire from their bus windows.
>
> (1965: 51)

A recent UN report has moreover suggested that there is a very serious threat to the whole Mediterranean coastline. At present, with 100 million visitors (in 1985), it is the world's most popular tourist destination. The UN report suggests that this number could increase to 760 million in 2025, thereby placing a huge strain upon food, water and human resources. The growth of existing coastal cities needs to be dramatically slowed down (the *Guardian*, 2 November 1988). But the opposite is occurring with Turkey, the latest country to develop as a major tourist destination. The immediate attraction for local investors in Turkey is that most revenue comes in the form of foreign exchange which protects them against a domestic inflation rate of around 80 per cent. Private sector tourist investment in 1988 was more than double that for 1987 although interestingly the source of much of that investment was the state-owned Turizm Bankasi (Bodgener, 1988). Turkish tourism has so far involved the proliferation of some exceptionally ugly unplanned developments, such as those in Bodrum and Marmara, which may have to be demolished fairly soon. One specialist operator, Simply Turkey, withdrew from selling holidays in Gumbet because it was 'No longer small and pretty, it is a sprawling building site, noisy and dusty, with a beach not large enough to cater for its rapid development' (quoted in Whitaker, 1988: 15). In other words, it is no longer suitable for the up-market holidays that company had been selling. So far the demand has grown very rapidly, with 400,000 inclusive holidays to Turkey sold in 1988. The impact of such rapid growth is felt particularly keenly because this area of south-west Turkey has always attracted considerable numbers of individual travellers due to the exceptional quality of its antiquities. Turkey is hence poised between the conflicting interests of mass tourism and a more socially select tourism.

The second most important area of tourist activity worldwide is North America. Developments here are, interestingly, different from Europe. Central to North American tourism has been the car, the highway, the view through the windscreen and the commercial strip. Jakle talks of how, in the postwar period, cities, towns and rural areas were all remade in what he calls 'universal highway order' (1985: ch. 9 for the following). In 1950 80 per cent of all long-distance trips were made by automobile and by 1963 43 per cent of American families took long vacation trips each year, averaging 600 miles.

There was a rapid improvement of the quality of the road system, to cope with faster travel and higher traffic volume. Unfortunately there was little to see from the new roads except the monotony of the road itself. John Steinbeck wrote that 'it will be possible to drive from New York to California without seeing a single thing' (quoted in Jakle, 1985: 190). Roads came to be built for the convenience of driving, not for the patterns of human life that might be engendered. The ubiquitousness of the radio and to some extent of air conditioning in American cars insulates the passengers from almost all aspects of the environment except the view through the windscreen (see Wilson, 1988).

And this view reveals almost nothing because even townscapes consist of commercial strips, the casual eradication of distinctive places and the generation of a standardised landscape. Jakle terms this the production of 'commonplaceness'. The commercial strips are common places lacking the ambiguities and complexities that generally make places interesting. They were 'unifunctional landscapes' which became even more uniform in appearance as large corporations developed which operated chains of look-alike and standardised establishments (such as McDonalds, Howard Johnson, Col. Saunders, Holiday Inn and so on). The automobile journey has become one of the icons of postwar America, reflected in novels like Kerouac's *On the Road* or the film *Easy Rider*. In *Lolita* Humbert Humbert concludes 'We have been everywhere. We have seen nothing' (quoted in Jakle, 1985: 198).

Probably the most famous tourist site in North America is Niagara Falls. Reaction to it has always involved superlatives (see Shields, 1990, for the following). Observers reported themselves lost for words. It was an exotic wonder; it had an immense natural aura. However, a series of transformations that have taken place have rendered Niagara as a series of different objects of the tourist gaze. First, in the late nineteenth century the Falls became the most favoured of places for honeymoons and for courtship more generally. Shields links this to the way in which the Falls constituted an admirable liminal zone where strict social conventions of the bourgeoisie were relaxed under the exigencies of travel and relative anonymity. The historic association of waterfalls with passion, whether of love or death, further enhanced the salience of such a zone. Travellers expected the Falls to be exceptional, a place where the limits of ordinary experience were transcended. The trip was analogous to a pilgrimage. Nathaniel Hawthorne wrote of going 'haunted with a vision of foam and fury, and dizzy cliffs, and an ocean tumbling down out of the sky' (quoted in Shields, 1990). More recently, however, the honeymoon has been emptied of its symbolic liminal status. It has become a meaningless nuptial cliché, referring to nothing but itself. All the emphasis at the Falls is placed on the props, on honeymoon suites and heart-shaped 'luv tubs'. The Falls now stand for kitsch, sex and commercial spectacle. It is as though the Falls are no longer there as such: they can only be seen through their images.

Thus the same object in a physical sense has been transformed by a variety of commercial and public interests. The nature of the gaze has undergone immense changes. In the eighteenth century the Falls were an object of intense natural aura; in the nineteenth century they functioned as a liminal zone gazed upon and deeply experienced by courting couples; and in the later twentieth century they have become another 'place' to be collected by the immensely mobile visitor for whom the gaze at the Falls stands for spectacle, sex and commercial development.

A related kind of development has been the growth of so-called 'sex-tourism' in South-East Asia. In South Korea this has been specifically encouraged by the state. Its main form consists of the kisaeng tour specifically geared to Japanese businessmen (see Mitter, 1986: 64–7). Many Japanese companies reward their outstanding male staff with all-expenses tours of kisaeng brothels and parties. South Korean ministers have congratulated the 'girls' for their contribution to their country's economic development. Other countries with a similarly thriving sex industry are the Philippines and Thailand. In the case of the former the state encourages the use of 'hospitality girls' in tourism, and various brothels are recommended by the Ministry of Tourism (Mitter, 1986: 65). Package tours organised in conjunction with a Manila agent include preselected 'hospitality girls'. Of the money earned only about 7–8 per cent will be retained by the women themselves. In Thailand it is calculated that there are 500,000 women working in the sex industry, with perhaps 200,000 in Bangkok alone (see Lea, 1988: 66–9). Particular processes which have helped to generate such a pattern are: the exceptionally strong set of patriarchal practices which cast women as either 'madonna/virgin' or 'whore'; the belief amongst people from affluent countries that women of colour are more available and submissive; the high rate of incest and domestic violence by fathers/husbands in some such societies; and rural depopulation which draws people into the cities looking for any possible work (see Enloe, 1989: ch. 2, for more detail, especially on attempts by women to organise to protect the prostitutes).

Similar factors apply elsewhere but such patterns are less obvious. Singapore provides an interesting contrast. In the advertising material provided for tourists there are no references to sex-tourism. The only clubs listed are various discos and Asian-style shows. Singapore is nevertheless an extremely successful object of the tourist gaze but this has been achieved by playing down its exotic character. Much of the emphasis in the publicity material is on Singapore's attractions as a modern shopping centre, and there is indeed an extraordinary complex of shopping centres along the now wholly misnamed Orchard Road. Singapore has also transformed many of the old areas of Chinese shophouses into modern hotel complexes, including what is claimed to the the tallest hotel in the world. This has been built next to the world-famous Raffles Hotel, which has an impressive range of historical and literary connections as well as superb colonial atmosphere. However, it was announced in 1989 that the Raffles is going to be shut for a year or so while it is 'modernised'. Singapore is 'in the east' but not really any more 'of the east'. It is almost the ultimate modern city and does not construct itself as 'exotic/erotic' for visitors.

An interesting development in Singapore is the leisure island of Sentosa, which is best entered by cable car. A wholly artificial environment is being created although much of it has a (vaguely) colonial appearance including especially the new ferry terminal. It is only when one gets close to it that one realises that it is built almost entirely of plastic. The capacity of the island is very large and visitors are transported from site to site by a purpose-built monorail. This curious combination of the ultramodern and the historical pastiche will no doubt be seen as a model tourist development for other parts of Asia and Africa.

CONCLUSION

It is clear that the effects of tourism are highly complex and contradictory, depending on the range of considerations outlined earlier. Not surprisingly there has been much discussion about the desirability of tourism as a strategy for economic development in so-called developing societies. This raises many difficult issues, which I am not going to enter into here. Instead I shall make some points of clarification.

It is important to appreciate that the growth of tourism in developing countries, such as 'game tourism' in Kenya, 'ethnic tourism' in Mexico, 'sports tourism' in the Gambia and so on, does not simply derive from processes internal to those societies. Such a development possibility results from a number of external conditions: technological changes such as cheap air travel and computerised booking systems; developments in capital including the growth of worldwide hotel groups (Ramada), travel agencies (Thomas Cook), and personal finance organisations (American Express); the widespread pervasion of the 'romantic' gaze so that more and more people wish to isolate themselves from the existing patterns of mass tourism; the increased fascination of the developed world with the cultural practices of less developed societies; the development of the tourist as essentially a 'collector' of places often gazed upon and experienced on the surface; and the emergence of a powerful metropolitan lobby concerned to promote the view that tourism has a major development potential (see Crick, 1988: 47–8).

The economic benefits from tourism are often less than anticipated. Much tourist investment in the developing world has in fact been undertaken by large-scale companies based in North America or Western Europe, and the bulk of such tourist expenditure is retained by the transnational companies involved; only 22–25 per cent of the retail price remains in the host country (Lea, 1988: 13; and see de Kadt, 1979). In Mauritius, for example, 90 per cent of foreign exchange earned from tourism is repatriated to companies based elsewhere. This repatriation is particularly likely to happen with the presently high level of vertical integration in the industry – in 1978, for instance, just sixteen hotel groups owned one-third of all hotels in the developing countries (see Crick, 1988: 45). An interesting exception to this pattern of repatriation (in addition to Singapore and the central role of Singapore Airlines) is Yugoslavia and the near monopoly on inclusive tours at least from Britain exercised by the state-owned Yugotours.

A further problem, again avoided in Singapore, occurs where tourism accounts for a really high proportion of the national income of the country. Some of the islands in the West Indies experience this difficulty. It means that if anything serves to undermine tourist demand, an enormous loss of national income will result. This is also what happened, for example, in Fiji in 1987 following military coups (see Lea, 1988: 32–6, particularly on the scale of advertising needed to restore consumer confidence, especially in Australia). It must also be asked: development *for whom?* Many of the facilities that result from tourism (airports, golf courses, luxury hotels and so on) will be of little benefit to the mass of the indigenous population. Likewise much indigenous wealth that is generated will be highly unequally distributed and so most of the population of developing countries will gain little benefit. This does of course depend on patterns of local ownership. Finally, much employment generated in tourist-related services is relatively low-skilled and may well reproduce the servile character of the previous colonial regime, what one critic has termed 'flunkey training' (quoted in Crick, 1988: 46).

However, it has to be asked whether many developing countries have much alternative to tourism as a development strategy. Although there are serious economic costs, as well as social costs which I have not even considered here, it is very difficult in the absence of alternatives to see that developing societies have much choice but to develop their attractiveness as objects of the tourist gaze, particularly for visitors from North America, Western Europe and increasingly from Japan (see Dogan, 1989, on an assessment of costs and benefits).

REFERENCES

Bagguley, P. (1987) *Flexibility, Restructuring, and Gender: Changing Employment in Britain's Hotels*, Lancaster Regionalism Group Working Paper No. 24.

Barrett, F. (1989) 'Why the tour operators may face their last supper', *The Independent* 7 November.

Barthes, R. (1972) *Mythologies*, London: Jonathan Cape.

Beckerman, W. (1974) *In Defense of Economic Growth*, London: Jonathan Cape.

Bodgener, J. (1988) 'Bright spot on the landscape', *Financial Times* 8 December.

Bruce, M. (1987) 'New technology and the future of tourism', *Tourism Management* 8: 115–20.

BTA/ETB (1985) *Hotel Development in the UK*, Mimeo.

BTA/ETB Research Services (1988) *Overseas Visitor Survey*, London.

Buck, M. (1988) 'The role of travel agent and tour operator', in B. Goodall, and G. Ashworth (eds) *Marketing in the Tourism Industry*, pp. 67–74, London: Croom Helm.

Cabinet Office (Enterprise Unit) (1983) *Pleasure, Leisure, and Jobs: The Business of Tourism*, London: HMSO.

Crick, M. (1988) 'Sun, sex, sights, savings, and servility', *Criticism, Heresy and Interpretation* 1: 37–76.

Department of Employment (1988) *Overseas Travel and Tourism – September 1988*, Employment Department press notice.

Dogan, H. (1989) 'Forms of adjustment: sociocultural impacts of tourism', *Annals of Tourism Research* 16: 216–36.

Ellis, A. and Heath, A. (1983) 'Positional competition, or an offer you can't refuse', in A. Ellis and K. Kumar (eds) *Dilemmas of Liberal Democracies*, pp. 1–22, London: Tavistock.

Ellis, A. and Kumar, K. (eds) (1983) *Dilemmas of Liberal Democracies*, London: Tavistock.

Enloe, C. (1989) *Bananas, Beaches and Bases*, London: Pandora.

Goffee, R. and Scase, R. (1983) 'Class entrepreneurship and the service sector: towards a conceptual clarification', *Service Industries Journal* 3: 146–60.

Goodall, B. (1988) 'Changing patterns and structures of European tourism', in B. Goodall and G. Ashworth (eds) *Marketing in the Tourism Industry*, pp. 18–38, London: Croom Helm.

Gratton, C. and Taylor, P. (1987) *Leisure Industries: An Overview*, London: Comedia.

Graves, P. (1965) *Majorca Observed*, London: Cassell.

Hart, J. (1988) 'A package for Christmas', *Signature* November/December: 18–19.

Hennessy, S., Greenwood, J., Shaw, G. and Williams, A. (1986) *The Role of Tourism in Local Economies: A Pilot Study of Looe, Cornwall*, Tourism in Cornwall, Department of Geography, University of Exeter.

Hirsch, F. (1978) *Social Limits to Growth*, London: Routledge & Kegan Paul.

IMF (1986) *Balance of Payments Statistics*, 37, Part 2, Washington, DC: IMF.

Jakle, J. (1985) *The Tourist*, Lincoln, NE: University of Nebraska Press.

Jamieson, B. (1989) 'Bass checks into the penthouse suite', *The Sunday Telegraph* 27 August.

Jeffrey, D. and Hubbard, N. (1988) 'Foreign tourism, the hotel industry, and regional economic performance', *Regional Studies* 22: 319–30.

de Kadt, E. (1979) *Tourism: Passport to Development?*, Oxford: Oxford University Press.

Key Note Report (1986) *Tourism in the UK*, London: Keynote Publications.

—— (1987) *Tourism in the UK*, London: Keynote Publications.

Lea, J. (1988) *Tourism and Development in the Third World*, London: Routledge.

Lee, J. (1988) Press Notice. Employment Department, 30 November.

Lewis, B. and Outram, M. (1986) 'Customer satisfaction with package holidays', in B. Moores (ed.) *Are They Being Served?*, pp. 201–13, Oxford: Philip Allan.

Litteljohn, D. (1982) 'The role of hotel consortia in Great Britain', *Service Industries Journal* 2: 79–91.

McKay, I. (1988) 'Twilight at Peggy's Cove: towards a genealogy of "Maritimicity" in Nova Scotia', *Borderlines* Summer: 29–37.

McRae, H. (1989) 'The clouds gather over the sea and sun package', *The Guardian* 12 August.

Milner, M. (1987) 'Where the squeeze is not only holidaymakers', *The Guardian* 20 August.

Mishan, E. (1969) *The Costs of Economic Growth*, Harmondsworth: Penguin.

Mitter, S. (1986) *Common Fate, Common Road*, London: Pluto.

Pile, S. (1987) 'You'll have no fun rushing to the sun', *The Observer* 16 May.

Reynolds, H. (1989) *The 100 Best Companies to Work for in the UK*, London: Fontana/Collins.

Rocca, T. (1989) 'Bardot, scorns "Black tide of filth" in St. Tropez', *The Guardian* 10 August.

Ryan, C. (1989) 'Trends, past and present in the holiday industry', *Service Industries Journal* 9: 61–78.

Shamoon, S. (1989) 'Shares for sale in Euro Disneyland', *The Observer* 5 March.

Shaw, G., Greenwood, J. and Williams, A. (1988) 'The United Kingdom: market responses and public policy' in A. Williams and G. Shaw (eds) *Tourism and Economic Development*, pp. 162–79, London: Belhaven Press.

Shields, R. (1990) *Place on the Margin*, London: Routledge.

Slattery, P. and Roper, A. (1986) *The UK Hotel Groups Directory: 1986–7*, London: Cassell.

Slattery, P., Roper, A. and Boer, A. (1985) 'Hotel consortia: their activities, structure, and growth', *Service Industries Journal* 2: 192–9.

Smith, D. (1988) 'A fine old stew in the kitchens', *The Guardian* 12 November.

Smith, V. (1978) 'Introduction', in V. Smith (ed) *Hosts and Guests*, pp. 1–14, Oxford: Blackwell.

Stallinbrass, C. (1980) 'Seaside resorts and the hotel accommodation industry', *Progress in Planning* 13: 103–174.

SWET (Society of West End Theatres) (1982) *Britain at its Best: Overseas Tourism and the West End Theatre*, London.

The Economist (1988) 'Travel companies: a holiday for two', 27 August: 59–60.

Turner, L. and Ash, J. (1975) *The Golden Hordes*, London: Constable.

Urry, J. (1987) 'Some social and spatial aspects of services', *Environment and Planning D: Society and Space* 5: 35–55.

—— (1990) 'The consumption of tourism', *Sociology* 24: 23–35.

Valenzuela, M. (1988) 'Spain: the phenomenon of mass tourism', in A. Williams and G. Shaw (eds) *Tourism and Economic Development*, pp. 39–57, London: Belhaven Press.

Vulliamy, E. (1988) 'Squalid renaissance', *The Guardian* 16 April.

Walter, J. (1982) 'Social limits to tourism', *Leisure Studies* 1: 295–304.

Welsh, E. (1988) 'Are locals selling out for a bowl of gruel?', *The Sunday Times* 11 December.

Welsh, E. (1989) 'Unmasking the special agents', *The Sunday Times* 26 February.

Whitaker, R. (1988) 'Welcome to the Costa del Kebab', *The Independent* 27 February.

Williams, A. and Shaw, G. (1988a) 'Tourism and development: introduction', in A. Williams and G. Shaw (eds) *Tourism and Economic Development*, pp. 1–11, London: Belhaven Press.

—— (1988b) 'Western European tourism in perspective', in A. Williams and G. Shaw (eds) *Tourism and Economic Development*, pp. 12–38, London: Belhaven Press.

Williams, A., Shaw, G., Greenwood, J. and Hennessy, S. (1986) *Tourism and Economic Development: A Review of Experiences in Western Europe*, Tourism in Cornwall Project, Department of Geography, University of Exeter.

Williams, I. (1988) 'Profits take a holiday', *The Sunday Times* 30 June.

Wilson, A. (1988) 'The view from the road: nature tourism in the postwar years', *Borderlines* 12: 10–14.

11

TOURISM AND THE FAMILY IN A RURAL CRETAN COMMUNITY*[1]

Maria Kousis

INTRODUCTION

Since much of the literature on the family tends to emphasize industrialization as an agent of change, family change under tourism remains a relatively unexplored topic. In this highly debated area, the arguments parallel those of the larger theoretical arena for issues of social change. The discourse revolves around the respective merits of ideological versus material factors.

Two main paradigms encompass social change and the family (Ismael 1979). The first perspective is functionalist and is epitomized by the works of Ogburn (1955) and Goode (1963). The second is Marxist which did not become active until the late 1970s. The former model emphasizes the waning of the extended family under industrialization. However, it is largely ahistorical; it only traces vague links between internal family processes and the external world of experience. The latter model has only recently begun to explore the systematic linkages between the mode of production and labor as categories of family analysis. Schneider and Schneider (1984) illustrate its approach in their study of a Sicilian rural town.

Goode (1963, 1982) cautiously stresses ideological factors. Shorter (1975) and Stone (1981) also identify with these factors, emphasizing the impact of changing values and political ideas on the family. One of Goode's (1982: 175–176) major conclusions is that "in all parts of the world and for the first time in world history, all social systems are moving fast or slowly toward some form of the conjugal family and also toward industrialization." Hence, he argues that what was once functional for traditional (i.e. peasant) life is now dysfunctional under industrialization. His search thus focuses on how the contemporary family serves industrialized society.

At the other end of the spectrum are theorists such as Tilly and Scott (1978), Hareven (1982), Spagnoli (1983), and Levine (1984) who stress economic factors. They argue that industrialization created structural changes which led to the appearance of new family patterns.

Recently, Thornton and Fricke (1987) have examined the impact of social and economic change upon family structure in the West, China, and South Asia. They

* Reprinted with kind permission of Elsevier Science Ltd, Pergamon Imprint, Oxford, England, from *Annals of Tourism Research*, 1989, 16: 318–32.

attempt to show how factors such as industrialization, urbanization, demographic change, advances in education, and the growth of income change the family. They conclude that while all of the foregoing lead to changes in the structure of family relationships, industrialization is still the cause behind most of the other factors. Reinforcing Goode's observation that explanations of family change should not rely on concepts like industrialization and urbanization, the study points out the need for research on how ideological and religious trends affect family change.

Although up to now most studies of family change have concentrated on the experiences of the industrialized regions of the West, in nonindustrialized societies activities such as tourism or new agricultural methods constitute alternative agents of growth which most likely have profound effects on family life.

Regarding tourism's specific effects on the family, there has been little empirical analysis. Nevertheless, the issue has received some attention in the literature dealing with the sociology of tourism. In her work on tourism in the Swiss Alps, Frey (1976), for example, discusses the increase in exogamy and the weakening of lateral kin ties. Greenwood (1972) and Loukissas (1977) briefly examine the decrease in lateral kin ties in Spanish and Greek islands, respectively. Greenwood's study of Fuenterrabia shows that

> ... older ties of co-operation and mutual aid between families began to lapse because most families had the economic strength to fend for themselves without incurring obligations to neighbors. In addition, families entered into stiff competition in patterns of consumption. ...
>
> (1972: 86)

In a study of the Mykonian family, Stott (1973) provides evidence from a Greek peasant community that changed in response to tourism. The presentation, while not subscribing to any particular theory of family change, nevertheless reaches similar conclusions to those of Goode. The Mykonian family has experienced a decline in the influence of family control, and parental power over children's courtship and marriage has weakened. Furthermore, the dowry system has lost the prevalence it once held under subsistence farming, and with tourism more women have independent jobs. Finally, young men and women in the community now enjoy more sexual freedom. In a later work on the same community, Stott (1978) finds that Mykonian women emulate Athenian modes and look down on those local males who emulate foreign behavior and date tourist women. Similarly, Boissevain's (1978) studies of Malta and Costa del Sol show that many local young men emulate tourist behavior and engage in sexual activities with North European women.

Cohen (1984), in a review paper on the sociology of tourism, addresses the family in the context of the division of labor between the sexes, thus indicating the lack of systematic research on family change under tourism. He makes specific mention of the Boissevain and Serracino-Inglott (1979) study which demonstrates the way in which tourism increases employment opportunities for young women, and how this in turn affects the division of labor within the household, the status of women in the family, and parental control over children.

SCOPE OF PAPER AND RESEARCH APPROACH

The purpose of this paper is to explore the changes of the family in the rural community of Drethia on the Greek island of Crete, in light of its transformation from a peasant to a contemporary community via tourism development. Data collection covered a one-year period, most of which was spent in the community, with occasional visits to Iraklion (the largest city in Crete) and Athens in order to gather additional relevant information.

Primary data were collected from the mayor's office, the District Bureau of Deeds and Mortgages, and the local church archives. The mayor's office manages the community's list of residents, the registers for births, marriages and deaths, as well as low class tourist accommodation ownership. Church records chronicle baptisms, marriages, and deaths, while the archives in the Bureau of Deeds contain information on property transfer contracts.

Formal interviews were abandoned when, after repeated attempts, the locals felt very uneasy. Participant observation became thus a laborious yet invaluable portion of the research. All waking hours were spent with different families in their homes, stores, or fields. Conversations and informal interviews were carried out in an orderly fashion, without inconveniencing the locals. Twenty life histories of people in different age and occupational groups were successfully added to the data set. Secondary data, collected from the Hellenic Chamber of Hotels, the Regional Development Bureau of Crete, and the National Statistical Service of Greece, supplemented the material used for this paper.

The period of the past thirty years up to the very early 1980s was segmented into three time phases for which family issues were examined: the pre-tourism or subsistence farming phase (1950–1964); the early tourism phase (1965–1972) that witnessed the appearance of the first hotel in the area; and the full-scale tourism phase (1973–1983), when the dominant economic activity in the community was the tourist service industry. The community's family profiles during its peasant and contemporary periods are sharply contrasted to illuminate the changes.

After a brief background on tourism development in Greece, the paper discusses eight aspects of family life: economic changes and the family, family size, family control, marital arrangements and the dowry system, marital ages, endogamy versus exogamy, the position of women, and sexual freedom.

TOURISM AND FAMILY CHANGE

The social and economic effects of international tourism in Greece are phenomena of the past two decades and they are likely to have had their greatest impacts in subsistence farming communities such as Drethia.

In the 1950s, approximately 70% of Cretans were subsistence farmers living in rural areas. By the early 1980s, however, tourism in Crete had grown at a much faster pace than in most other Greek regions, to the extent that the service sector now generates more than half of the island's gross regional product. This transformation inevitably brought about dramatic social changes.

Since the early 1960s, the Greek government had carefully studied Crete, in an attempt to explore its potential for the systematic development of tourism. This initiative was heavily expanded by the military government of 1967 which drafted tourism development plans, assigned incentives (loans, grants, subsidies) to relevant investment projects, and declared regions of development priority (KEPE 1981).

Several parts of Crete, Drethia included, were among the "top priority" regions to take advantage of the various investment opportunities, comprising such operational incentives as tax reliefs and discounted insurance premiums. The government undertook the infrastructural projects itself. The necessary supply organizations for package tourism hotel groups, and transport and tour operators, originated from developed tourist-sending countries.

As a result, Drethia has become one of the most attractive tourist resorts in Crete (Kousis 1984). Since this community has undergone significant socioeconomic changes between its subsistence farming and its contemporary phases, it constitutes an appropriate site for the study of family change.

Economic changes and the family

To highlight part of the socioeconomic impact associated with the advancement of tourism in Drethia, Table 11.1 presents the major occupational changes of heads of households across the three phases. In the pre-tourism period, Drethia was a relatively small community of subsistence farmers, some small shopkeepers, and a few civil servants. Wage earners were almost nonexistent as shown in Table 11.1. During this period, the village consisted of 430 families, the great majority of whose members worked in their own fields. Although land holding data show that some fifteen families were better off, property was not concentrated in the hands of a few families, and there were no rural bosses.

Data from the Bureau of Deeds show clearly how land ownership has changed

Table 11.1 Occupational distribution of active male heads of households in Drethia (1950–1982)[a]

| Occupations | Percent active male heads by phase | | |
	Pre-tourism (1950–64)	Early tourism (1965–73)	Full-scale tourism (1974–82)
Farmer	82.3	47.9	24.4
Government employee	5.4	2.4	3.9
Self–Family employed	10.6	28.9	27.1
Wage earner	1.7	20.8	44.6
Total percent (*N*)	100 (350)	100 (534)	100 (639)

[a] On the basis of data courtesy of Drethia Community Office, the District's Bureau of Deeds and personal interviews with locals. Figures are calculated by adding the relative three active male birth cohorts for each phase. The actual numbers of families are higher. Households headed by women are not included; they were rare and housewife was the main occupation cited.

for the last three decades. Between 1965 and 1972, land ownership changed significantly, given that a large part of the coastal area was purchased by a handful of outsiders from other parts of Crete, Athens, and abroad. The vendors, local families, were usually obliged to give up their fields, either through expropriation laws or because of their impoverished status. Overall, changes in land ownership during the early-tourism period resulted in the loss of part of the subsistence base for a large number of local families. At this time, 422 locals sold 389 fields totaling 340 stremmata (1 stremma = 0.25 acres) to 11 buyers. Once the coastal land purchases were made, almost all of the buyers started building large tourist accommodation complexes. This had further implications for the locals. Many of them began to sell their labor by doing construction-related work. With tourism growth, the number of farmers decreased drastically, while those for small shopkeepers and wage earners increased considerably (Table 11.1).

The expansion of tourist-service businesses occurred during the full-scale tourism phase. Consequently, the unskilled and semiskilled labor supply increased even more in those years, most of it coming out of subsistence farming. At this phase, a new pattern of capital ownership significantly altered the community's traditional dynamics. Increases in local land sales took place, when 124 sellers gave up 103 coastal zone fields (totaling 52 str.) to eight buyers. More importantly, however, the major source of income in Drethia became tourist bed ownership, 80% of which accrued to outside investors who owned the largest and most expensive hotels.

Toward the end of the 1970s (during the full-scale tourism period) five very different socioeconomic groups emerged from those of the pre-tourism phase. They were based on wealth holdings. The highest bracket group comprised the six largest land buyers and/or owners of "A" class hotels. The remaining 730 local families were separated into four groups. The first, a group of six families, were buyers of coastal zone land, owners of "B" and "C" class hotels and large tourist-item and food stores. The second, a group of 13 families, included owners of "D" and "E" class hotels, as well as medium-size tourist shops. About 165 families owning lower forms of tourist accommodations and other businesses made up the third group. The remaining 546 families were active mainly as tourist business employees (maintenance personnel, gardeners, waiters, cooks, etc.) and subsistence farmers. Thus, the occupational distribution of Drethian heads of households today reflects the new pattern of capital ownership during the 1970s and early 1980s.

Proletarianization (Kousis 1985; Tilly 1984), induced by the loss of the local base of subsistence for part of the inhabitants, the introduction of new economic activities reshaping employment opportunities, and the establishment of a changed socioeconomic structure, safely characterizes the transformation which took place in Drethia during the postwar period, and which affected the lives of local families.

Family size

The size of the nuclear family household in Drethia did not undergo significant changes in the postwar period. Figure 11.1 shows the number of children born by

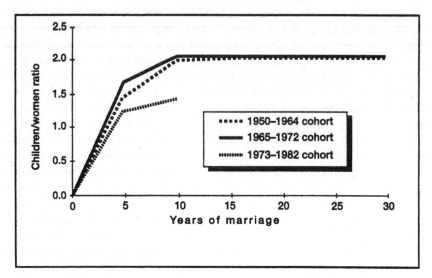

Figure 11.1 Children born in Drethia by years of marriage for three cohorts

years of marriage during each phase. The appropriate comparison between the first and third cohorts cannot be made, since the last cohort have not completed their childbearing. However, judging from childbearing in the early years of marriage, the more recently married women seem destined to bear fewer children.

The work environment introduced by tourism has contributed to the decrease in family size. On the one hand, wage earners cannot afford to have more than two children when confronted with the problems of future employment opportunities and a rising standard of living. On the other hand, since family businesses will usually be inherited by the owners' children, parents tend to have just a couple of children, thereby maximizing economic benefits and rising expectations.

Family control

During the subsistence farming period, in rural communities such as Drethia, "kou-mando" (being in command) was first the prerogative of the older male, and only second that of the older female. Since land was the only source of livelihood and this was in the hands of the older male, he had the power to abuse this right to his own interest. This he did many times according to local accounts. At that time, a young couple was expected to work on the father or father-in-law's fields in return for a modest board and lodging. By the 1980s, couples that were young in the 1950s expressed much resentment over this type of family control.

Given the entry of nonfamily-related economic activities in the full-scale tourism phase, this type of control appears to have diminished. Yet, there are cases where some of the *nouveau riche* have exercised similar types of control, leading to the same type of resentment, especially, but not always, from young brides.

224

Another position taken by grandparents, particularly maternal grandparents, relates to family assistance. During subsistence farming days, they used to help by working in the new couple's fields, as well as by taking care of the children while their parents worked. Sometimes, they hosted the new family unit until the young couple built their own house. Most orphans were also raised by these grandparents.

In the full-scale tourism phase, once again, it was mainly the grandmother who looked after the young children while her daughter worked in the tourist business. Grandparents, in their fifties and sixties, also assisted their children by actively participating in agricultural activities such as olive picking or potato planting and gathering (but they often complained that the young people were abandoning their fields). Although older people helped their children mainly in farming activities, some also aided them in their tourist business. Dishwashing or cleaning might be done by a young grandmother. A few older men in the village advertised their son's or daughter's *pension* to tourists entering the village, and often they would accompany tourists personally from the main street to the small hotel.

The influence of family members was not restricted to grandparents or vertical kin. It was also extended to lateral kin or such adult relatives as cousins, uncles and aunts, especially during the pre-tourism phase. High levels of interdependence among lateral kin meant that agricultural work was accomplished on time. This was extremely important, since there was no suitable equipment in the 1950s to help subsistence farmers. Cultivation was done by manual labor alone. This included plowing, planting, irrigating, and hoeing. In addition, the absence of fertilizers created more work. Most families could not afford to hire laborers from nearby villages. Only a few were able to hire local field hands as well as laborers from the area. Most local families took turns with lateral kin in working on each other's fields. Under this type of interaction, group solidarity and control was of great significance. Persons looked for social acceptability and approval in the lateral kin to which they belonged. Thus, influence and control by this group played a very important part of daily life, especially since most social activities were also practiced with lateral kin members.

During the 1960s, agricultural machinery was introduced in the village. Electricity also came to most homes. Roads were built to Iraklion and the nearby villages. Gradually, as more and more machinery was used in the fields, the value of manual work started to decrease. Help from relatives became less essential, especially for the vegetable and potato crops. Thus, the dependence on the help that relatives provided decreased.

The entrance of mass tourism made these changes even more dramatic in the village. Once the large hotels were built and the number of tourists increased, more and more local tourist businesses were established, while other locals started working for wages. The government helped in the establishment of these businesses with loans, grants, or other incentives. In this way, a small hotel, a restaurant, or a tourist shop was set up and operated by a husband–wife team, in collaboration with their children and parents. Therefore, lateral kin could not hold the central position they once did. Those locals who worked for others, usually the few, large tourist accommodation owners, also found themselves economically independent from lateral, and sometimes from vertical, kin.

Furthermore, given the nature of the businesses serving the tourist, social activities shifted from a concentration on extended family members to a focus on immediate family members and tourists. The seven-month tourist season every year exerts certain demands on the family that either owns or works in a tourist business, since all of the establishments are open seven days a week. Food and entertainment shops usually stay open until 2 a.m., hotels and pensions are open at all hours. Tourist shops close around midnight. Very little time is left to socialize with lateral kin, especially since economic ties have weakened considerably. Consequently, the influence and control once exerted by these relatives has diminished.

Marital arrangements and the dowry system

As in most subsistence farming communities in Crete, marriages in Drethia were basically a family affair, facilitated by local matchmakers. This was mainly due to the inheritance rights of land, the only source of livelihood. Both pairs of parents usually transferred partial land ownership to the new couple. Given the hard economic conditions of the time, in most cases, housing was not available. Thus the majority of couples shared accommodation with one set of parents, until they built their own house. During the 1950s, cash dowries were exceptional, usually provided by emigrants from the United States and Australia. Normally, young women received a dowry in the form of land and handmade necessities for the household.

The same marital patterns continued through the mid-1960s. Most marriages were arranged by families, although some couples did elope. In the early tourism phase, however, even though most marriages occurred between families of similar economic standing, offerings to the new couple began to assume the form of cash and housing.

Once tourism became the major economic activity in the community, parents owning tourist businesses either had the newly married couple work for them or made them partners. Helping the couple to establish its own business was another alternative. As a norm, the dowry given during the full-scale tourism phase assumed the form of housing, furnishings, and/or business transfer. More importantly, however, the parents of the newly married son gave him property as a dowry. Therefore, given the property transfer involved, families belonging to the first four socioeconomic groups (i.e. all those who are not just farmers or employees) had a great interest in choosing the marital partner for their child and so they would practice this as a norm. In contrast, parents of the last socioeconomic group were not as rigid. Although they would aim through the arrangements to increase their children's socioeconomic status, this was a serious challenge for them, given the limited supply of higher status candidates.

Marital ages

On the average, Drethians have a tendency to marry in their mid-twenties. The mean marital ages for the three phases differ only marginally: 24.7, 24.5, and 24.9, respectively. However, the observed decrease in the standard deviations (5.1, 3.2, 2.9) indicates that people in that community today tend to marry at ages closer to the mean marital age than they did in the past.

However, unlike the first phase, during the next two, marital ages for men and women spanned a wider range. As Figure 11.2 highlights, the gap between male and female marital ages widened in the full-scale tourism phase; in the late 1970s and early 1980s, females married younger than their male counterparts. The locals maintain that this pattern is a result of the courting relationships local males have with female tourists, as well as the ability of most local families to raise dowries much faster than they used to. This view is also supported by the media (including major Greek newspapers) and by the particular dynamics of these relationships which draw from the high female tourist to local male ratio during the tourist season.

The entrance of nonlocals who are seeking employment/business opportunities (most of them males) increases the native (Greek) male to female ratio. This situation leads to a higher demand for local brides whose marital ages, as a result, decrease. Although this demand is bound by the presence of foreign females during the tourist season, it is generally boosted by the attractive local dowries available. There is no conclusive evidence to substantiate this hypothesis. However, during the full-scale tourism phase, there are four times as many women as there are men emigrating from Drethia.

Looking at age differences between the spouses where men are older than women

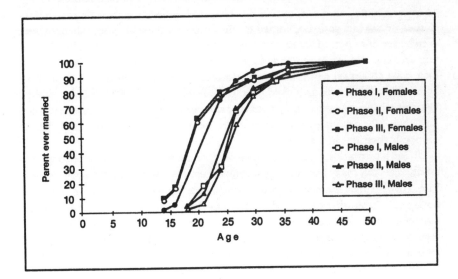

Figure 11.2 Marital ages in Drethia by sex

227

(the opposite is rare), the mean shifts from a 6.6 in the pre-tourism phase to a 7.5 in the full-scale tourism phase. This indicates that once more during the latter phase men tend to be older than their spouses, compared to men in the former phase.

Endogamy versus exogamy

In the pre-tourism phase, most of the locals married other locals, as seen in Table 11.2. This information is based on data from the community register in which year of marriage was used to categorize spouse under a phase. The outsiders who married locals, at that time, were mainly people from the nearby villages who were economically worse off than the Drethians. With the exception of one refugee, all nonlocals who married in Drethia were Cretans.

As the foundations of the new economic base were laid, more outsiders entered the local family structure. The number of locals marrying Drethians decreased, while non-Cretans from various parts of Greece married locals and established households in Drethia. The full-scale tourism phase is characterized by the entrance of non-nationals who married locals and became residents of the community. All of the European women who married locals were once tourists in Drethia.

The position of women

Under subsistence farming, the women in Drethia, supervised and controlled by parents and parents-in-law, worked in their fields with other members of their families. At home, they took care of the daily house chores such as milking goats, making cheese and bread, starting wood-fires, cooking, cleaning, and washing. Since neither consumer goods nor cash were available, all women worked in the evenings on the loom, making sheets, towels, blankets, rugs, tablecloths, napkins, material for clothes, or else they sewed or knitted for the family. Mothers of female children were even busier since part of the dowry comprised all such handmade necessities for setting up a household. As mothers, these young women were responsible for raising their children amidst all the demanding manual work in the fields and around the house. They either took their children into the fields, or left them in the care of older or single relatives. As wives, they were accountable to their husbands for every action they took. The paternal grandparent, more often than the husband, made all the economic decisions in the household.

Under tourism, local young women became participants in the emerging economic activities. Many became wage earners while some worked in family businesses. The typical female wage earner was either a cleaning woman in one of the larger hotels, or a dishwasher for a food/drink shop. As community records show, about a quarter of the local family businesses (usually medium-or small-size ones) were owned by married women. Since the men usually engaged in various other types of jobs, an even larger number of family businesses might be operated by married women and their children.

Women who operated tourist businesses, more than those who were wage earners,

Table 11.2 Spouse origin in Drethia by sex and three phases (1950–1982)*

| | Percent of spouses by phase, origin, and sex | | | | | | | | |
| Spouse origin | Pre-tourism (1950–1964) | | | Early tourism (1965–1973) | | | Full-scale tourism (1974–1982) | | |
	Males	Females	Total	Males	Females	Total	Males	Females	Total
Drethia	84.4	90.2	87.3	77.0	69.0	73.1	65.4	51.1	57.9
Nearby villages	8.7	4.0	6.3	10.5	7.3	9.0	6.2	6.7	6.4
Heraklion (city of)	1.7	0.6	1.2	2.7	5.5	4.0	6.2	11.1	8.8
Rest of Crete	4.6	2.9	3.7	7.1	5.5	6.3	13.6	15.6	14.6
Athens	0.6	0.6	0.6	1.8	1.8	1.7	1.2	1.6	3.5
Rest of Greece		1.1	0.6		10.0	4.9	6.2	1.1	3.5
West Germany					0.9	0.5	1.2	4.4	2.9
England								2.2	1.2
Rest of Europe								2.2	1.2
Asia Minor/mid East		0.6	0.3	0.9		0.5			
Total number	172.0	175.0	347.0	113.0	110.0	223.0	81.0	90.0	171.0
Total percent	100	100	100	100	100	100	100	100	100

* Assembled on the basis of data from the Community Register. Year of marriage is used in placing a spouse under a phase.

experienced an environment which demanded many more hours than any previous job in Drethia. Businesses serving the tourist are open between 14 and 24 hours per day, seven days per week, for seven months continuously. There are no weekends, holidays, or breaks for those who work in family businesses in Drethia. This way, they cut on costs by working all shifts themselves. This exhausting work schedule hits the married women hardest; they are responsible for both child and house care. Many of them do all of these jobs simultaneously, and if fortunate, they are helped by their parents.

Therefore, under tourism, in contrast to subsistence farming, the female wage earners (the larger group) are in a less advantageous position, given their almost powerless relationship with their employers. The self/family employed woman became her own exploiter in order to stay independent and out of debt.

Sexual freedom

Subsistence farming communities of the 1950s and early 1960s placed a lot of restrictions on the sexual behavior of young men and women. Even talking to each other, without the presence of others, was considered a sign of promiscuity. At dance parties, wedding receptions, baptism ceremonies, and church services, men and women sat separately. When parents permitted their daughters to dance, the latter were under strict supervision. Males were generally excused for their sexual lapses. Very often, however, they paid the price for their indiscretions by marrying the young woman, sometimes under the threat of her father's or brother's shotgun. According to locals, engaged couples were immediately placed in the bedroom by the elders. The sheets showing the loss of the female's virginity were proudly displayed for kin and nonkin. Such strict moral codes were subsequently eased in Drethia, in response to a number of factors including the inception of tourism.

Once tourism dominated the economic environment of the community, however, the sexual–moral codes were revised dramatically. Single female tourists constituted the main power to that end. The relationships between local males and female tourists involved a delicate and controversial issue. Some of the local men, often called "kamakia" or harpoons (metaphorically implying that the male was the harpoon and the female was the fish), systematically dated foreign women. In the full-scale tourism phase, there were about ten groups (or cliques) of single males between the ages of 16 and 30 who regularly dated female tourists. They constituted the majority of young males in Drethia; it was hard to escape this pattern because of strong peer pressure. This pattern is also attributed to the relatively stricter moral codes for local females who are not allowed to date. Dating foreign women might on occasion lead to marriage. According to the community register, about 9% of the women who married locals in the 1970s and early 1980s were foreign women who once were tourists in Drethia. Most locals viewed the high ratio of female tourists to local males as a problem, especially for married locals. During the seven-month tourist season, according to local estimates, this ratio is about 10:1. Thus, the persistence of strict sexual conduct of local females, *vis-à-vis* the relaxed relationships between local males and foreign women, has widened the gap of sexual code behavior for men and women in Drethia.

CONCLUSIONS

The complex combination of factors of change and adaptation in local contexts is missing from many models of tourism development. In the present paper, this combination of factors has been considered in relation to the issue of family change under tourism.

For communities like Drethia, factors of change involve outside actors such as the government, foreign agencies, and private entrepreneurs. They are the ones who have principally shaped the core of the new economic activities of tourism. For Drethia, this led to serious changes in land holding and wealth owning patterns where several outsiders now own the most valuable assets of the new capital base, leaving only limited room for locals to operate small businesses or work as wage earners. Thus, this paper supports the view that economic rather than ideological factors act as agents of family change.

Adaptation to social family life under tourism by rural family members reveals some moderate and some drastic modifications. To begin with, a decline in the influence of family control did not occur to the degree described by Stott (1978) for a Mykonian community. The power of parental authority depends on parental ownership of wealth. Marital arrangements as well as the dowry system continue to dominate the scene. Stott overlooked the fact that wages offered very little in comparison with the rising standard of living that had occurred in rural Greece since the 1950s.

Changes in the socioeconomic structure and the relationships between local males and female tourists have led to the widening of the gap between male and female marital ages. Drethia experienced an influx of outsiders entering to take advantage of the new opportunities. As a consequence, endogamy patterns changed. Moreover, a minority of foreign spouses brought a new identity to a once peasant family, a finding similar to that by Frey (1976).

Careful qualifications need to be made when describing women's independent employment. The demands made on women of the subsistence farming period underwent qualitative increases, with the onset of tourism. Finally, the increase in young men's and women's sexual freedom translates into a widening gap between male and female sexual codes in Drethia, which is consistent with Boissevain's (1978) findings.

In concluding, the following implication is drawn from this study. The literature on tourism development has not systematically explored the important interaction between factors of change and the local contexts for the case of the family. As a result, policymakers in tourism development may not have adequately incorporated this noneconomic aspect of tourism in their decisions. Future research on the topic would not only bridge this gap in the literature, but also do a great service to policy making.

ACKNOWLEDGEMENTS

Funding for this research was provided by SSRC. Drethia is a pseudonym for the rural Cretan community. I wish to thank Charles Tilly, Louise A. Tilly, and Karen Wilson for their valuable comments on an earlier draft.

REFERENCES

Boissevain, J. 1978 Tourism and the European Periphery: The Case of the Mediterranean. In *Underdeveloped Europe: Studies in Core Periphery Relations*, D. Seers *et al.*, eds. London: Harvester Press.

Boissevain, J., and P. Serracino-Inglott 1979 Tourism in Malta. In *Tourism – Passport to Development?*, E. de Kadt, ed. New York: Oxford University Press.

Cohen, E. 1984 The Sociology of Tourism: Approaches, Issues, and Findings. *Annual Review of Sociology* 10: 73–392.

Frey Haas, V. 1976 The Impact of Mass Tourism on a Rural Community in the Swiss Alps. Unpublished Ph.D. dissertation, Department of Anthropology, University of Michigan, Ann Arbor.

Goode, W. J. 1963 *World Revolution and Family Patterns*. New York: Free Press.

—— 1982 *The Family*. Englewood Cliffs NJ: Prentice Hall.

Greenwood, D. 1972 Tourism as an Agent of Change in a Spanish Basque Case. *Ethnology* 11: 80–91.

Hareven, T. 1982 *Family Time and Industrial Time*. Cambridge: Cambridge University Press.

Ismael, J. S. 1979 The Family and Social Change: A Historical Research and the Historical Alternative. *Journal of Comparative Family Studies* 10(1): 107–117.

KEPE (Centre of Planning and Economic Research) 1981 Regional Development Plan: Tourism Development Incentives. Internal Report of the Tourism Section, Athens.

Kousis, M. 1984 Tourism as an Agent of Social Change in a Rural Cretan Community. Unpublished Ph.D. dissertation, Department of Sociology, University of Michigan, Ann Arbor.

—— 1985 Proletarianization under Tourism: A Micro-level Analysis. Working paper No. 325, Center for Research on Social Organization, University of Michigan, Ann Arbor, April.

Levine, D., ed 1984. *Proletarianization and Family History*. New York: Academic Press.

Loukissas, P. J. 1977 The Impact of Tourism on Regional Development: A Comparative Analysis of the Greek Islands. Unpublished Ph.D. dissertation, Cornell University, Ithaca.

Ogburn, W. F. 1955 *Technology and the Changing Family*. Boston: Houghton Mifflin.

Schneider, J., and P. Schneider 1984 Demographic Transition in a Sicilian Rural Town. *Journal of Family History* 9(3): 245–272.

Shorter E. 1975 *The Making of the Modern Family*. New York: Basic Books.

Spagnoli, P. G. 1983 Industrialization, Proletarianization, and Marriage: A Reconsideration. *Journal of Family History* 8(3): 230–247.

Stone, L. 1981 Family History in the 1980s. *Journal of Interdisciplinary History* 12(1): 51–87.

Stott, M. 1973 Economic Transition and the Family in Mykonos. *The Greek Review of Social Research* 17: 122–133.

—— 1978 Tourism in Mykonos: Some Social and Cultural Responses. *Mediterranean Studies* 1 (2): 72–90.

Thornton, A., and T. E. Fricke 1987 Social Change and the Family: Comparative Perspectives from the West, China, and South Asia. *Sociological Forum* 2(4): 746–779.

Tilly, D. 1984 Demographic Origins of the European Proletariat. In *Proletarianization and Family History*, D. Levine, ed. New York: Academic Press.

Tilly, L. A., and J. W. Scott 1978 *Women, Work, and Family*. New York: Holt, Rinehart, and Winston.

12

THE PHILIPPINES*

The politicization of tourism[1]

Linda K. Richter

Welcome to the New Philippines
Department of Tourism Poster, 1987

"PEOPLE POWER, the Unarmed Forces of the Philippines" means, as the t-shirts proclaim, a new era in the Philippines. Forced from office after he massively rigged the 1986 Presidential elections, Ferdinand Marcos now lives in reluctant exile in Hawaii and Corazon Aquino, widow of his assassinated rival, now struggles with the Marcos legacy of political and economic decay.

The events that brought Aquino to power made dramatic street theater – unarmed millions massed to protest the Marcos election fraud, to protect defecting military, and to promote the Aquino victory. Even long-time Marcos friend, President Ronald Reagan, was compelled to airlift the Marcos entourage to Honolulu and to recognize the Aquino government.[2]

Aquino inherited a bankrupt nation with a negative growth rate, a $26 billion foreign debt, 70 percent of the population at or below the poverty line, and a country that lost between 10 and 20 billion dollars to the systematic plundering of the Marcos government.[3] She assumed the office of the president even as Marcos still cowered in his bullet proof palace. Following his exit, Aquino moved into a presidential guest house near the palace. There was no transition; records were in shambles. A new constitution, new institutions, and new leadership were desperately needed to replace those discredited by the 20 years of Marcos rule. Eighteen months later these were in place.

Under such circumstances one might assume that tourism policy would hardly be a salient issue for the government but Aquino, like Marcos, felt that tourism policy would prove a useful political weapon. To understand how and why the Aquino government has shaped current tourism policy it is necessary to trace in some detail the unprecedented use of tourism policy as a political instrument by the Marcos government. It is also important to understand the many ways in which tourism policy was utilized for national and personal objectives, because other governments,

* Reproduced with permission of University of Hawaii Press from Linda Richter (1989) *The Politics of Tourism in Asia*, Honolulu: University of Hawaii Press, pp. 51–81.

particularly authoritarian ones, have used and are using tourism for some of the same political advantages.

TOURISM BY DECREE: THE POLITICAL USES OF TOURISM UNDER PRESIDENT MARCOS

No regime has more blatantly used tourism policy for political leverage than that of ex-President Marcos of the Philippines. Although tourism contributed to many of the regime's political and economic objectives, it achieved those gains at enormous cost to the Filipino economy. As time went on, the insensitive development of tourism in the midst of deteriorating economic and social conditions spawned a counter-use of tourism – opposition violence against the tourist industry. Thus one finds in the case of the Philippines a microcosm of the political uses and abuses of tourism.

Background

Although 20 years ago the Philippine economy was second in Asia only to Japan, tourism development was quite a lackluster sector until the advent of martial law in 1972. It is indeed ironic that as a democracy the Philippines lagged behind all other Southeast Asian nations in tourist arrivals while under the martial law New Society, and later the authoritarian New Republic, it was billed as a tourist haven. It did not happen by accident. Within a month after Marcos declared martial law, claiming that the country was seething with subversion and violence, ambitious government plans to expand tourism were already being announced. "The New Society, if it lives up to the plans and promises, may come to be known in our history as that era when tourism was in flower."[4] To understand just how tourism administration was converted from a prosaic little bureau in a nondescript building into the flashy, high profile, skyline-changing, priority sector it would become requires understanding a bit of history of the Philippine political environment.

Between 1946, when the Philippines became independent from the United States, and 1965, when Marcos was first elected president, two loosely organized political parties had alternated in holding political power. The Philippines had a reputation as a freewheeling, sometimes corrupt but always lively democracy in the midst of the authoritarian nations of Southeast Asia. Its elections were always occasions for massive vote-buying on both sides, but the greed of the ruling party was to some extent checked by the free, often scurrilous, press, a strong opposition party and regularly held elections.[5]

President Marcos' first term in office more or less followed the characteristic pattern. In 1969 he became the first president since independence to be re-elected, in what was widely acknowledged to have been an unusually corrupt and possibly rigged election. The constitutional two-term limit for the president meant that Marcos was scheduled to leave office in 1973. During much of the president's second term he sought without success to have the constitution changed by encouraging a constitutional convention to develop a parliamentary system. That proved futile. At the same time, student protests, Marxist-encouraged agrarian revolts and the chronic (since Spanish times) violence of Muslim dissidents on the southern island of Mindanao made the Philippines seem unsettled at best. Critics would later argue that none of these events was particularly unusual for the Philippines. However, the president made them appear so. A terrorist bombing of a political rally (which was later discovered to have occurred at the president's behest) set the stage.

Using tourism to sell martial law

"Every visit is an endorsement of the continuation of the political, economic, and social stability achieved by ... martial law."[6] On September 22, 1972, the president declared martial law, dissolved Congress, closed most papers, abolished the vice-presidency, arrested political opposition, and began what Marcos labeled the "New Society." It was during this tumultuous period, under the scrutiny of world opinion and with all competing political institutions neutralized, that the government began its massive tourism program.

In an open political system, heavy dependence on patronage and public works tends to take the form of projects like schools, roads, dams, parks, and health clinics. Some are purely pork barrel in nature, but they are designed to be relevant to popular needs, and to sway public opinion and with it the vote. In a closed political system as the Philippines was between 1972 and 1986 different political needs surface. The winning political formula then becomes not votes but the support of the key elements in the domestic and international elite. Tourism is particularly well suited to assuaging elites and developing a clientele for authoritarian rule. Oriol Pi-Sunyer, in her study of Spain, documents the use of tourism in the 1930s as a way of countering hostility to fascist Spain.

> If Western governments were reluctant to support Spanish Fascism, Europeans could nevertheless be invited to Spain and, hopefully act as unofficial emissaries of the regime on their return home. Reports that the living was cheap (as it was for outsiders) and that the country was tranquil (as it appeared to be to the casual visitor) helped legitimate the system.
>
> For internal Spanish consumption, the hundreds of thousands of foreigners visiting Spain was offered as evidence that the dictatorship was accepted abroad. ... The controlled media constantly reported on the growth of tourism and explained at great length how this apparently inexhaustible source of wealth benefitted the Spanish people.[7]

The parallels with the Philippines grow even closer when one notes that the later expansion of Spanish tourism flourished and was in part linked to U.S. economic and military support, which was in turn a reflection of the American–Spanish bases agreement of 1953.

Similar contemporary examples using tourism for political credibility could be found in the two Koreas and the Republic of China. But it was President Marcos in the Philippines who in 1972 demonstrated that tourism could be developed to convey and create regime legitimacy in ways and to a degree not attempted before. At the time martial law was declared, Marcos had two critical foreign policy problems. First, he had to overcome the shock and dismay in some Western circles at the unexpected imposition of martial law in order to neutralize opposition to his leadership. Second, he had to assure that martial law would not jeopardize the flow of foreign capital investment into the country, encourage cuts in foreign aid, or erect new trade barriers to Philippine exports.

Internationally, Marcos' problem was his own political legitimacy as well as the continuation and enlargement of the economic base for his policies. He met these challenges with characteristic articulateness and political skill. The imposition of martial law was treated as an entirely constitutional, hence legitimate, response to an emergency situation created by communist subversion and communal violence. Moreover, he insisted that an integrated series of policies was required to effectively curtail the situation. Marcos thus served notice to the nation that martial law was not to be a 90-day phenomenon. At the same time that he was building a case that internal subversion required the imposition of martial law, Marcos also managed to assure such international economic brokers as Robert McNamara (then president of the World Bank) and others that the Philippines was an ideal investment site for both international aid and multinational corporate development.

Tourism, which had fallen off dramatically in the period immediately before and after martial law, was quickly seized upon as a means to refurbish the Philippines' and especially Marcos' image. Tourism had not been a priority industry prior to the establishment of martial law and had, in fact, done very poorly during the first term of Marcos' administration. Yet within eight months of the declaration of martial law, tourism was a priority industry eligible for a variety of tax incentives and customs concessions. The regime had set up its first Department of Tourism (DOT) by May 11, 1973.[8]

Jose Aspiras, former congressman and presidential news secretary, was given the task of selling the Philippines as a safe and delightful destination thanks to the achievements of the New Society. He approached his job with customary gusto and a well-developed public relations instinct. In days, tourism went from a once-a-month news item to a daily media blitz. The impression was fostered that no visitor had had a good night's sleep in pre-martial law Philippines and no hotelier would have a good one ever again without martial law.

Aspiras had a formidable ally in First Lady Imelda Marcos. She is rumored to have been responsible for the creation of a cabinet level post for Aspiras. Though a longtime associate of the president, Aspiras began his most rapid ascent after he reputedly was among those who saved Mrs Marcos from a would-be assassin.[9] This rumor gains credence from the fact that tourism had just been reorganized a few months before on the recommendation of a three-year study by the Presidential Commission on Reorganization. The new Philippine Tourism Commission was scuttled and with it tourism's longtime head, Gregorio Araneta, who was then given the position of undersecretary in the new DOT. Aranera contends that Aspiras' first choice was to be secretary of the Department of Public Information, but Marcos persuaded him that tourism would be a major responsibility in the New Society.

For some time, Imelda Marcos had nurtured expansive and expensive ambitions to see the Philippines, and particularly Manila, blossom into an international oasis for the luxury traveler. Her interest in the arts and in the general cultural life of the city had been ridiculed and frustrated by outspoken critics and a frequently hostile press. These constraints were gone after martial law began.

Nor was Marcos constrained in promoting tourism. Tourism, as a relatively new and heretofore harmless government function, had no real enemies. Unlike his key domestic program of land reform, with tourism policy the president had no need to perform amazing feats by juggling power interests. At the same time he could build a potentially important industry critically dependent on stability and relatively unconcerned about political freedom.

Tourists *per se* mattered less in the early years of martial law than did the publicity about tourism. The DOT launched an ambitious series of invitational visits to the Philippines for travel writers and tour operators, groups that could be depended upon not to bite the hand that fed them and who were not likely to be preoccupied with civil libertarian issues. In a friendly, beautiful country it was enough that the gun slingers were gone and no tanks patrolled the streets. To further the image of a peaceful, contented society, the DOT built a promotional campaign around the Philippines' most important asset – a cheerful, hospitable people. The slogan, "Where Asia Wears a Smile," was a particularly adroit choice for defusing criticism of life under the New Society.

Aggressive efforts were made to attract international gatherings of global appeal. Once these media events were secured, the mobilization required to assure their success began. The Miss Universe Contest in 1974 occasioned among other things such modern-day pyramid building as the construction of its venue, the huge Folk Arts Theatre, in an incredible 77 days of nonstop construction and at still unreported costs. The media event for 1975 was the "Thrilla in Manila," which pitted Joe Frazier against Mohammed Ali in their championship fight. Once again world publicity focused on the Philippines as journalists toured the country developing background stories for the spectacle. The government's tourism department promoted tourist events with a positively dazzling eclecticism that ranged from the "Miss Gay World Beauty Pageant" in Baguio to evangelist Rex Humbard's rally in Manila.

The IMF–World Bank Conference

But it was the International Monetary Fund–World Bank Conference in October 1976 that stimulated the most frenzied and politically motivated use of tourism. In late 1974, as soon as the Philippines learned that its bid had been accepted, all pretense at orderly and phased tourism development was abandoned. "In the rush to build the hotels, normal loan procedures were shelved . . . Irineo Aguirre (Director of the Bureau of Tourism Services) says that 'the Government never planned for so many hotels to be built at one time.' The hotel construction had been planned on a staggered basis . . . but then the IMF came along."[10] The IMF–World Bank Conference offered an opportunity for both regime recognition and personal aggrandizement. No expense was spared to assure that this most prestigious international conference would find in Manila a showcase of stability, prosperity, elegance, and beauty. The tantalizing prospect of hosting 5,000 VIPs, even for just a week, led to a rush to complete 12 luxury hotels within 18 months, though the tourism master plan had not expected such accommodation needs for at least a decade.[11]

The opportunities seized for private gain became clear later. By 1976 the First Lady had become the first governor of the super-political unit known as Metro Manila. As the highest authority over the 17 constituent cities, Imelda Marcos was clearly in a position to monitor development. Her financial interests and those of her friends were not neglected. Enormous convention facilities and exhibition halls were rushed to completion. Round-the-clock shifts were utilized to complete her own hotel, the $150 million Philippine Plaza, conveniently located on reclaimed land in Manila Harbor alongside the lavish convention facilities. "Ownership" is rather a misnomer, as the hotel was built without Philippine Tourism Authority (PTA) approval and with 100 percent government financing. Friends, including the tourism minister, also borrowed massive sums with little or no equity of their own involved. Although the Central Bank looked askance at such risky ventures, the deal was struck when the Government Service Insurance System agreed to guarantee the loans. Some entrepreneurs reputedly borrowed far more at the concessional interest rates than needed for the hotels and used the extra funds for speculation.[12]

In any event, the IMF–World Bank Conference is still, over a decade later, the reference point for grandiose tourism development among international travel industry specialists. It is hard to exaggerate what occurred.

The skyline of Manila, indeed, part of what was once Manila Harbor, was drastically altered by the rise of more than a dozen luxury hotels. Depending on one's sources of information, in the 1975–1976 period, between US$410 and $545 million of government money became directly tied up in hotel financing. The relative size of this commitment can be compared from several angles: it was between one-seventh and one-fifth of the government's total proposed 1976 expenditures of $3.05 billion.[13] It is more than the nation's total 1976 borrowing from the World Bank of $315 million. The Development Bank of the Philippines (DBP) alone spent a "staggering $229.29 million" through July 1976 on tourism projects.[14]

From a development standpoint, the expenditure on hotel financing alone is between 30 and 40 times the amount that the government has spent on public housing. It could, of course, be argued that comparing government expenditures with government loans is misleading because, theoretically, the latter will be repaid. But one could make a similar case for such things as agrarian reform loans and low-cost housing loans. The government was not so zealous in those sectors. Further, with hindsight we now know that the massive loans were not repaid. Tourism was subsidized by the Filipino citizen through taxation and social security payments. The government now owns most of these hotels.

Though a spokesperson for the Philippine Hotel Association acknowledged that the hotels built in 1976 would "not be economically viable for the next 15 years," the government continued to extend their credit.[15] As of 1983, the Development Bank of the Philippines had acquired 80 percent control of 70 tourism related accounts that owed the DBP 4.5 billion pesos. Though 1.5 billion pesos in loans were converted into equity in the hotels, the total owed now far exceeds the value of the hotels as collateral.[16] The government lost the taxpayers' money with every move.

Nor was the hotel building the extent of the government-financed tourism infrastructure. The reclamation of land in Manila Harbor was a very expensive project. Most of the subsequent building on the reclaimed land was related to tourism. The First Lady's Philippine Plaza Hotel has already been mentioned. Its waterfront location was envied by competing older hotels that found their once waterfront locations disappear as vast acres of reclaimed land emerged in front of them.

Other grandiose projects included the Philippine International Convention Center (PICC), estimated to cost $150 million, and the Philippine Center for International Trade and Exhibitions. The convention facility spurred the creation of Presidential Decree 867 creating a Philippine Convention Bureau, which was supported by taxing all hotels big and small. The small hotels resented the tax because they had no convention traffic. The larger ones resented the PICC because it competed with their own convention facilities.[17] Airport, port development, and accelerated road building projects were also developed with considerable attention to their impact on tourism.[18]

The IMF–World Bank Conference was apparently a huge political success, despite the fact that only 3,000 of the 5,000 anticipated participants came. Few delegates seemed to note the incongruity of a New Society that was supposedly aimed at redressing inequities spending many times more for the construction of luxury hotels than for public housing and land reform. As was intended, most were impressed by the stability and attractiveness of the society, the tremendous growth in international investments, and the obvious improvement in law and order, at least in the tourist belt.[19] Military and economic aid to the Marcos government increased, in both bilateral U.S.–Philippines terms and in terms of IMF–World Bank aid. To most of the 138 government and tourism industry officials interviewed in 1976 and 1977 the tourism development effort was an enormous political success for President Marcos, though few would argue that it made economic sense. Secretary of Tourism Jose Aspiras even went so far as to conclude that Filipino Martial Law was itself a tourist attraction.[20]

Balikbayan and Reunion for Peace

Although the international beauty contests, boxing events, and wooing of VIPs garnered excellent publicity for the Philippines, a different approach was needed to counter the bitter criticism of many expatriates abroad and the ever-lengthening lines outside the U.S. embassy of those seeking to leave the Philippines. The Marcos government was extremely sensitive to charges of political torture, "salvaging" (killing) of opponents, repression, and corruption by the regime because they threatened to alienate the more than 1.5 million Filipinos living abroad, three-quarters of whom reside in the United States. Through consulate activities of sometimes dubious legality,[21] through referenda giving overseas Filipinos preferential treatment over local Filipinos, and through tourism, the government sought to influence Filipinos abroad politically and the nations in which they resided. There was an economic rationale as well. Filipino remittances grew in importance even as the Philippine economy deteriorated. Moreover, criticism abroad threatened to jeopardize the foreign aid and investment upon which the government grew ever more dependent.[22]

The government response was the *Balikbayan* (homecoming) Program. Begun in 1973 as a special project to assuage concerns about martial law among overseas Filipinos, it continued as an elaborate project to subsidize the travel to the Philippines of Filipinos living abroad. Over a score of national departments and provincial and local agencies are involved in the development of the program, as well as embassies and consulates abroad. *Balikbayan* initially was built upon the close Philippine family ties and what is the longest Christmas season in the world: All Saints Day (November 1) to Epiphany (January 6). Hundreds of thousands of copies of "Invitation to a Traditional Philippine Christmas" were sent abroad. School children were assigned to invite relatives home for Christmas. Local governments were charged with developing local festivals and immigration, tax, and customs officials were instructed to exempt *Balikbayan* visitors from most requirements. The icing on the cake was the Department of Tourism's 50 percent discount of airfares and concessional rates on accommodations and shopping.[23]

Although the DOT initially contended that the *Balikbayan* Program was an economic success,[24] fervent supporters of the regime would later acknowledge that it was very costly but well worth it in public relations value.

> The government of course, has lost and is losing a rather substantial amount of revenue from the program. But the benefits derived by the country are not only from the foreign currency spent here . . . it is also an effective means of rebutting through actual experience the lies they have spread about this country in foreign lands.[25]

Though the Philippine people bore the economic costs of subsidizing the tourism of more affluent expatriates abroad, the regime benefited politically from the program. By 1978 nearly a million Filipinos and their families had taken advantage of the government program. Even so, the government took pains to inflate the figures further, because they so successfully belied the "you can't go home again" claims of Marcos' critics. All overseas Filipinos and their families, regardless of their present or past nationality, who visited the Philippines were considered *Balikbayans* for accounting purposes. The effect of this was to inflate the number of Filipinos coming home and deflate the number of American tourists, for most *Balikbayans* came from the United States. This accounting procedure also made it impossible to compare the number of overseas Filipinos who were visiting in the pre-martial-law period with those who have returned subsequently. Statistics for the former do exist but were never cited and are not comparable, for the earlier statistics noted only Filipinos born in the Philippines, not whole families.[26]

The favorable comments on the cleanliness, beauty, and order that prevailed in tourist areas provided positive feedback to both the world and the local press. Like tourists in general, *Balikbayans* rarely were interested in the political details of the government but were impressed with the relatively relaxed atmosphere and congenial surroundings. Yet they were more credible for political purposes than ordinary tourists because they were in a position to compare the New Society on a temporary basis with the Old Society they had known more closely and had opted to leave. Encouraged by the success of the *Balikbayan* Program, the government developed other specialized homecoming programs geared to influential expatriate groups. Presidential Decree 819, in October 1975, established the *Balik*-Scientist Program, which included incentives for practicing professional skills in the Philippines as well as regular *Balikbayan* privileges.[27] *Balik*-UP was set up in July 1978 to encourage visits from University of the Philippines alumni, the graduates of the most prestigious university in the country. As an indicator of the coordination and high-level support these programs enjoyed, *Balik*-UP was promoted jointly by the ministries of Tourism, Foreign Affairs, and Trade as well as Philippine Airlines.[28]

Toward the end of the Marcos era there was some curtailing of concessions to non-Filipino family members, but the basic political objectives of the *Balikbayan* Program continued to overshadow the economic costs of the program.[29] Actually, in many respects the economic costs have decreased. Since the government now owned many of the lavish hotels for which it underwrote the construction in 1976, it was eager to get whatever tourists it could or face added empty rooms and an increasing drain on government reserves.

The "Reunion for Peace," launched in early 1977, was another attempt to bring back to the Philippines groups who could not help but compare the atmosphere under martial law favorably with what they had known before. The program appears to have been inspired by a similar initiative in South Korea, another authoritarian nation in search of better press notices. The political formula is astute: promote the return of former World War II servicemen and their families from America, Japan, and Australia for a nostalgic tour of old battlegrounds and memorials as well as a panoramic view of the New Society. Concessional fares and touring discounts sweeten the package. Meanwhile, the entire project complemented the president's initiatives in presenting the Philippines as an independent and vigorous society intent on reconciliation and relations with a variety of powers. Even for this small program the government developed a committee composed of the heads of the Departments of National Defense, Development, Foreign Affairs, Local Government and Community Development, Highways, the National Historical Institute, and the Veterans Federation.[30]

Both *Balikbayan* and Reunion for Peace have been programs that were generally well received at home and abroad. The United Nations' praise of both programs as innovative and constructive has encouraged the adoption of similar programs elsewhere.

Until about 1980 one would have had to credit the regime's tourism policy with general political success. For example, tourism was used as the basis for forging closer political and economic ties within the Association of Southeast Asian Nations (ASEAN). Marcos was interested in strengthening ASEAN for several reasons. First, it was showing increasing signs of becoming an important counterweight to Japanese and American influence in the area. Second, it offered an international forum for the Philippines and for Marcos, where Marcos could enhance his global reputation as a statesman while representing the Philippines as a genuinely nonaligned nation. Third, by initiating patterns of regional cooperation in such sectors as tourism, he hoped to avoid ruinous competition that could abort the Philippine tourism effort. At the same time, the campaigns for ASEAN cooperative efforts encouraged the general political support of the other member nations. Harmony with such ASEAN powers as Malaysia and Indonesia also reduced the likelihood that those Muslim nations would actively support the Muslim rebels in the southern Philippines.

To further these objectives, the Philippine government worked tirelessly to develop closer communication and transportation links with ASEAN member nations. An air link was opened between the Philippines and Indonesia, thereby completing the air routes among ASEAN nations and facilitating regional travel and promotion. A permanent committee on tourism in ASEAN was established and plans were made to develop a regional passport, for free exchange of currency among member countries, and to ease customs and trade restrictions.[31]

So important had tourism become as an ingredient in Philippine foreign policy that for a while every government employee going abroad had to take a departure course on both the policies of the New Society and the major tourism attractions in the country. In early 1977, the Department of Foreign Affairs went a step further and recalled the ambassadors and consuls from North America for briefings on tourism and a ten-day tour of the Philippines.[32] Even city governments were expected to use their mandatory tourism councils to initiate visits and exchanges with other foreign countries.

In May 1977, the city of Manila hosted its first international conference, the Pacific and Asian Congress of Municipalities (PACOM). Manila has been the site of numerous international conventions, but this was ostensibly the first international convention hosted by the city. The demarcations between municipal and national government, however, were certainly blurred under martial law. National government figures and departments were prominent in the planning, but most of the costs were borne by the Manila city government. The city had given "voluntarily" over 1 million pesos ($134,700) a year to the DOT since martial law began.[33]

Appropriately, the subject of the conference was national and local partnership in the development of tourism. PACOM is an apolitical body of local government executives from countries in the Pacific Basin. It has few resources and little influence, being largely a creation of rather particularistic political needs. The organization was founded by Honolulu Mayor Frank Fasi, who until the Manila convention had remained PACOM president. PACOM offered Mayor Fasi a forum that nicely complemented the variety in his Honolulu constituency.

The conference was also important to the Philippines. By giving this diffuse collection of local officials the VIP treatment, not only was the Philippines likely to be remembered favorably, but attitudes toward Filipinos abroad might be positively affected. It was mentioned, for example, during the planning for the PACOM conference that many politicians in a position to help overseas Filipinos would be attending. Moreover, a few of the mayors coming would be overseas Filipinos who would themselves be important in improving the Philippine image abroad. Philippine planners were particularly hopeful that Mayor Fasi would be elected Hawaii's governor in the next election so that his favorable record of appointing Filipinos to government posts could continue at a higher level.[34] They were to be disappointed.

It was also to be a morale booster for the local Filipino mayors who would attend and later help one of their own, Manila Mayor Ramon Bagatsing, become without opposition the new president of PACOM. Other nationality groups did not bother to put up presidential candidates when it became obvious that out of a total PACOM membership of less than 500, 300 were new Filipino members. PACOM planners were ready should there be a challenge to Mayor Bagatsing's candidacy, but their contingency plans were not necessary.[35]

PACOM was a small conference with fewer than 500 attending, but its keynote address was given by the First Lady and its closing speech by the President. This is an indication less of the significance of the convention than of the unstinting efforts of the First Couple to garner influence and prestige for the New Society.

The conference was executed with a precision and attention to detail that was most impressive. Thanks to the convention center and new hotels, the facilities were impeccable. The costs were enormous, however, and only a fraction of the investment was recovered by registration fees and visitor spending.[36] As several delegations remarked, the Philippine conference would be a hard act to follow for those national groups who have to account to the public for their expenditures.

Opposition uses of international tourism

Only two major tourism conferences have clearly boomeranged politically. The first was the Eighth World Peace Through Law Conference, which was held in Manila in August 1977. The theme of the conference was international protection of human rights. It was perhaps too much for the Marcos regime to woo jurists as successfully as it had wooed boxing and beauty contest fans, but certainly a Herculean effort was made. Marcos announced the lifting of curfew and the ban on international travel as well as the release of about 1,500 prisoners, and intimated that local elections would be forthcoming. They were not. The presence of the conference in Manila also encouraged local campus dissidents to stage antigovernment demonstrations. These the government gingerly confronted, reluctant to show its strength with American conventioneers in the crowd and with a stellar international gathering in the city.[37]

The second, which failed more dramatically, was the American Society of Travel Agents (ASTA) Conference. The Philippines had accelerated its tourism initiative by 1980 as hotel overcapacity became a major liability and public criticism of tourism policy grew. Overseas offices multiplied and a new tourism offensive was signalled by the opening of the Philippine Embassy's Special Services Office in Washington, D.C., to coordinate tourism, convention, and cultural activities. The effort was dramatic enough that even the industry press took notice.

> The travel industry has been practically commuting to Manila in the past 12 months. The Philippine capital, with its superb meeting facilities and plethora of deluxe hotels, has managed to attract, like a giant Pacific magnet, a procession of travel industry conferences (PATA, the Pacific Area Travel Association, the World Tourism Organization) and this month welcomes the biggest annual industry event of them all: the ASTA World Congress.[38]

The American Society of Travel Agents convention was eagerly wooed and, once won, the Philippine government spared no expense in preparing a lavish conference for the more than 6,000 delegates expected at this important event. More than any other single tourism organization, this one influenced access and trade to the American and even the world market. This the government knew. But so did the President's opposition.

Despite elaborate security, no more than minutes after President Marcos opened the gala conference a bomb went off in the convention hall, injuring several delegates. The timing of the attack could not have been more adroit from the opposition's perspective, for it followed the president's ridicule of Western press accounts of the Philippines as unsafe.[39] The president was unhurt but politically stunned by his opposition's bold and successful breach of security. Moreover, the ASTA Conference was in chaos. The conference was cancelled despite Philippine government pleas. The Ministry of Tourism offered free stays, food, and travel to delegates who would remain, as it desperately sought to recoup its political standing. There were few takers.

With one bomb, the most important clientele group in global travel had been initiated into the state of domestic politics in the Philippines. Philippine tourist arrivals declined immediately by 10 percent and continued to decline throughout Marcos' tenure, despite the deteriorating exchange rate of the peso, which made the Philippines a very inexpensive destination for international visitors.

The tourist industry, developed in large measure by massive media promotion, had been undone by media events beyond the regime's control. Tourism had also been affected dramatically by the reports of general violence and political instability.[40] A case in point: the airport assassination of opposition leader, Benigno Aquino, as he returned from exile had an immediate impact on tourism.

> Reaction to the killing (and its consequences) among potential visitors to the Philippines can be illustrated by what happened in the convention market. Within days of Aquino's death, eight large international events scheduled for

244

late 1983 and early 1984 were cancelled. Revenue from an estimated 10,500 high spending visitors was lost.[41]

Even business travel, which had been more resistent to bad political news, declined as the economic situation deteriorated.

This was not the first violence directed at tourism. Selective acts of terrorism against tourists and the tourist establishment had been increasing as the Marcos family and friends became identified as the prime beneficiaries of the government's expensive tourism development program. In the mid-1970s there were kidnappings of two Japanese tourists on the southern island of Mindanao. Since Japan assumes an unusually paternal role toward its citizens abroad, this event had severe repercussions. Japan, by far the largest tourism market for the Philippines, had a tremendous impact on Philippine tourism. The Japanese issued a warning to Japanese tourists to avoid the southern Philippines and almost all did, for years.

A beefed-up military police and stiff sentences for crimes against tourists were the government's attempt to keep lawlessness to a minimum in Manila's tourist belt.[42] By the late 1970s and early 1980s, however, a new opposition strategy targeted the hotels owned by the Marcos family and their associates. A series of arson attacks began that severely damaged or destroyed several luxury hotels and a floating casino. A group calling itself the "Light a Fire" movement claimed responsibility for the fires.

In mid-October 1984, a famous hill city monument, the Pines Hotel in Baguio, went up in flames. The hotel was filled with American veterans on a Reunion for Peace tour. Two other hotels burned in as many weeks. After over 50 deaths and 500 injuries from the three fires, the luxury hotels were forced to search visitors and their luggage, a decision scarcely compatible with the carefree ambience the Tourism Ministry spent millions to promote.[43]

If one tried to imagine the mindset of militant opponents of the regime, generally denied open channels of dissent, subject to preventive detention, not allowed habeas corpus, and frequently the target of torture and disappearances, the attacks become more understandable. After 20 years of Marcos rule, the radical opposition had correctly concluded that the president would not peacefully relinquish control and that foreign governments would not precipitate action against him.

Tourism was at once a highly visible, salient example of "crony capitalism" and of the grandiose pretensions of the Marcos regime. The tourist infrastructure was largely owned by the First Family or those close to them. Moreover, terrorist victims of arson or bombs were likely to be foreigners or the most conservative strata of the Filipino elite. Thus a backlash among the mass of citizens was unlikely. In any case, because foreigners were involved, there was publicity abroad, if not in the Philippines. Moreover, tourism is an important export sector that can be killed at little effort or expense. A pattern of sporadic violence can cripple it.

The government found that its very zealousness in building a tourism infrastructure simultaneously created new pressures to draw tourists even as it has developed new vulnerability to opposition violence. In June of 1984, the government announced it was putting 200 street marshalls in Manila with instructions to shoot to maim suspected criminals. By the end of July 1984, over 40 people had been shot dead by the marshalls and others wounded. Such law and order efforts did not inspire great confidence among tourists but for a time it had a chilling effect on dissident activity in the capital.[44]

The domestic uses of tourism

Within the Philippines, the decision to rapidly expand tourism also had predominantly political objectives, though the rationale developed for explaining the importance of tourism was in terms of the economic benefits that would accrue from an accelerated program. Political legitimacy was no less important at home than abroad, and the spectacle of thousands of tourists visiting the Philippines supported the administration's claim to have earned the confidence and respect of the world. As in fascist Spain, tourist arrivals were hailed as personal endorsements of martial law.

Favorable comments of tourists and visiting officials were included in nearly every issue of the major newspapers since the creation of the DOT in 1973. Conventions, however, were a new variant on bread and circuses in terms of regaling the masses with lively details. Each was assured daily coverage and numerous feature stories.

The impression of an inordinate amount of newspaper space devoted to tourism was confirmed during my four-month content analysis of the three leading dailies, from January to May 1977. The three papers studied during the period were *Bulletin Today*, *Philippine Daily Express*, and the *Times Journal*, all Manila dailies. The first paper was owned by a friend of the president. The latter two were under the ownership of members of the First Couple's family.

By way of comparison, there were 12 stories on tourism for every one on agricultural policy, including the much heralded land reform. Another measure of tourism's prominence was the coverage devoted to it in the prestigious *Fookien Times Yearbooks* to which government and civic leaders contribute. In 1976, for example, tourism coverage (14 pages and cover) exceeded space devoted to agriculture, industry, or trade. This imbalance is even more pronounced when one compares the prominence, length, and accompanying photographs on the two subjects.[45]

Moreover, even mild criticism of tourism policy was dealt with more summarily than criticism of land reform implementation. Columnists who failed to be properly enthusiastic about government implementation of the tourism program lost their jobs. One such example, Frankie Lagniton, writing for the *Bulletin Today*, had his column cancelled without notice to him or his readers. Individuals in the DOT leadership felt he had been critical of their approach to attracting increased travel trade.[46]

Bernard Wideman of the *Washington Post* also ran foul of the regime after a number of articles in the *Far Eastern Economic Review* and the *Washington Post*. His unpopularity with the regime was based, among other things, on his writing about the massive hotel construction and financing. Wideman was ordered deported, but a hearing shortly before the World Peace Through Law Conference led to a personal determination by the president that he could stay. Not so lucky was Associated Press Manila bureau chief Arnold Zeitlin, who the preceding November, after the IMF–World Bank Conference, was given 24 hours to leave the country without a hearing.[47] The decisions made no mention of the conferences, of course, but it seemed that the different treatment accorded these journalists was, in part, occasioned by the timing of the conferences and by the regime's awareness that such publicity could not be helpful to the image desired.

Tourism implementation and administration

Political legitimacy was only one of the president's domestic goals. Promotion of tourism has also proved to be a very flexible tool for selectively investing in the country according to political priorities, both personal and regime-based. Initially, the tourism master plan identified eight large priority zones for development. These reflected the stated desire to diffuse the investment, and thereby the benefits from tourism, as widely as possible. The nine priority areas were northern Luzon, Manila Bay region, eastern Visayas, Bicol region, western Visayas, Lanao area, Davao provinces, Zamboanga, and Sulu.[48]

The implementation of tourism development bore little relationship to that schema, which had been the product of more politically detached technocrats.[49] In an effort to accelerate development it was decided that government financing of tourism initiatives would go primarily to four much smaller areas where there was already sufficient demand to make such ventures quickly profitable.[50]

Two of the areas selected fit the critical mass criterion – the Greater Manila area and the Laguna–Cavite areas across Manila Harbor. The third, Zamboanga, was an area of potential rather than present mass demand. It was located on the troubled island of Mindanao and was developed as an international gateway to the Philippines in what proved to be a futile attempt to redress the years of neglect in the southern Philippines. Of course roads, airports, and improved communications facilities can serve military and tourist requirements alike. In fact, some military personnel suggested that rebel sanctuaries might be most effectively undermined by turning their rustic settings into tourist attractions, thereby bringing jobs and development to the area.[51] By 1988, it appeared that the tourists rather than the rebels were in flight from the area.

The fourth focus of tourism development was the popular resort area of La Union, the home province of the then Secretary of Tourism and the heartland of President Marcos' Ilocano supporters. There are many tourist attractions in this area, so that although it does not have the critical mass of a city like Cebu, which was ignored in the five-year plan, La Union is, nevertheless, a reasonable center for tourism development.[52] Patronage is a time-honored tradition in the Philippines that martial law only embellished. La Union is one of the smaller provinces in the Philippines, but it has the largest of the 12 tourism field offices and is the only province with a subfield office, in this case in Secretary Aspiras' home town of Agoo. The La Union Field Office is itself a tourist attraction, sitting as it does on a high hill overlooking

Bauang, the provincial capital, and the ocean. The staff numbered 22 to 24, including 8 to 10 employees in tiny Agoo. The field office nearest in size was in the Philippines' second largest city, Cebu, where there was a staff of 16.[53]

La Union has also benefited from tourist-related infrastructure, including the Marcos Highway, which conveniently connects Baguio and Agoo. In the 1980s construction workers were also kept busy building a Mount-Rushmore-sized bust of President Marcos along a mountainside. Despite the financial crises endemic to the Philippines that worsened each month, there were always funds for the Ilocano region.

Filipino politicians historically have furnished bountiful patronage for their home or linguistic regions. None, however, had the 20-year opportunity of President Marcos. In the 1980s, the First Couple had their greatest personal tourism patronage opportunity. They redecorated and in large part re-created an entire Spanish town, in the sleepy town of Sarrat. The architecture was circa 1800 but the amenities were strictly jet set. The motivation was not entirely political, because it was in the president's home province and the initial flurry of construction was for his oldest daughter's wedding. Sentiment and politics are not incompatible, however. Just as the cabinet and army, particularly the officers, have come overwhelmingly from the Ilocano regions, generally, patronage also tended to flow to the region.[54] Even in 1988, over two years after Marcos was unceremoniously whisked into exile and after hundreds of stories verified his plunder of the country, he retained the loyalty of many Ilocanos.

Tourism under martial law became an elastic source of patronage not only in terms of the private sector but also in terms of government employees. Just as only a fraction of the government's budget was strictly within the Department of Tourism, many of the personnel working on tourist-related activities were attached to departments and agencies with nontourist-oriented primary functions such as city governments, cultural bodies, banking and investment institutions, transportation, and planning and budgetary bodies.

The Board of Travel and Tourist Industry (BTTI), the precursor to the DOT, employed about 250 people at its zenith in 1972. By 1977, over 1,000 were on the Department of Tourism payroll, with an additional 300 in the powerful Philippine Tourism Authority.[55] Given the growth in tourism, such increases may appear reasonable. Not so easy to rationalize were the six people assigned to the DOT library, which at that time had 58 items in its cardfile – mainly speeches – and a clippings file. The library had no magazine subscriptions and no books![56] The Marawi City Field Office had a staff of six, but only five tourists had come by late May 1977. They were ecstatic when this writer's family arrived, nearly doubling the arrivals for that year.[57]

An intensive examination done by the University of the Philippines College of Public Administration of the administration of the Department of Tourism and its comparison with other departments is not possible to detail here, but there was an unusual amount of slack in the Department of Tourism and the Philippine Tourism Authority even by Philippine standards.[58]

There were also some rather bizarre wage policies. Most government positions are subject to wage standardization, but certain positions are exempted if the work is of a highly technical nature. The Department of Agrarian Reform (DAR) could not hire lawyers because of abysmal salary scales, and because, for all the government's rhetoric, it refused to consider DAR lawyers "technical" and therefore exempt from pay ceilings. Thus, landlords could abort most land reform decisions by paralyzing litigation to which DAR could not adequately respond.[59] The Philippine Tourism Authority, on the other hand, had not only its lawyers but 70 percent of its staff of 300 declared "technical," so their salaries were much higher than most in the Civil Service.[60]

Most DOT personnel receive salaries based on the wage standardization scale. Nearly a third, however, were employed as casual or contractual workers, which allowed them to bypass the civil service examinations and position ceilings in the department. This means that such employees are considered temporary, but Aspiras saw to it that most of the benefits of permanent government service accrued to them.[61]

An indicator of the importance the president attached to tourism is the number of former presidential assistants and military men heading key tourism departments. In the field offices and in other central office positions could be found many former congressional aides, former politicians, and journalists supportive of Marcos. Despite the reforms attempted by the Civil Service Commission, all levels of the department exhibited the characteristic bureaucratic problems of nepotism and overstaffing.

The growth of tourism brought fresh hope to travel agents and other sectors of the tourist industry, initially hard hit by the ban on Filipino travel abroad. This ban was lifted, but in its place the Central Bank established an enormous travel tax that made travel abroad prohibitive for all but the most affluent.[62] Other Southeast Asian governments followed suit, despite ASEAN fears that such protectionism would only hurt all regional tourism. For most Filipinos, the domestic tourism slogan "See the Philippines First" is incontestable. All travel agencies were instructed by the government to develop and promote low-cost travel packages, but no statistics exist on the response to such package excursions or on domestic tourism in general.

The government was hopeful that domestic tourism could be a means of better integrating the country's scattered linguistic and cultural groups, of broadening the support for the New Society, and preserving the arts, culture, and antiquities of the Philippines. A promising start in this direction was taken in 1973 by the DOT and the Department of Education and Culture. Student group exchanges and local travel clubs were organized. Other small and politically interesting programs designed to develop local support for tourism were the DOT familiarization tours of the islands given to various farmers' groups and to selected local political leaders.[63] Unfortunately, most of the government's domestic tourism budget was axed in fiscal year 1977 in one of the DOT's rare economy moves. The disastrous economic decline and political instability since then kept both citizens and the government preoccupied with survival.

In an effort to control and direct the industry more effectively, the government sought to incorporate tourism into its general policy of corporativism.[64] After a lengthy campaign, Aspiras succeeded in getting all tourist-related organizations to band together in a single umbrella organization, which the department promised to utilize as an advisory body. The Tourism Organization of the Philippines (TOP) operated, however, more as a channel for communicating tourism policies than as a limited forum of interest-group opinion.

Academics were drawn into the support of tourism. The prestigious University of the Philippines College of Public Administration contributed to the administrative critiques of the Department of Tourism's personnel policies and members of its distinguished faculty have spoken before a variety of domestic and international forums regarding tourism's relationship to other sectors of the government hierarchy. The government moved early in its departmental history to institutionalize tourism on the campus through its economic and political support for the Asian Institute of Tourism (AIT). The AIT was expected to draw and develop professional talent for the tourism industry from all over Asia.[65]

So multifaceted had the support for the involvement of tourism in the nation's developmental perspective become that even the highest planning body, the National Economic Development Authority, played no appreciable role in regulating the direction or extent of the commitment. As one official observed, "Tourism is a given."[66]

The effect of the decision to concentrate on luxury tourism was extremely costly from a developmental standpoint. Huge amounts of foreign exchange earned by tourism went to imports needed for the industry. Most of these items are for five-star amenities; for example, over 98 percent of the luxury transport, limousines, and air-conditioned coaches must be imported from Japan. There was also an enormous gap between what the Central Bank actually received in foreign exchange and what the department, on the basis of polls, estimated that visitors spend.

The gap, in part, reflected three other problems: the tendency of many tour operators to "salt dollars abroad," the unobtrusive but thriving black market, and the tendency, still unstudied, for a significant percentage of foreign exchange to remain in or be remitted eventually to the tourist's own country. In an interview with owners of small hotels, the complaint was made that many imports end up on the black market or are clearly superfluous, like imported detergents.[67]

Unnecessary expenditures are particularly common when luxury hotels are managed by multinational chains that use a central purchasing system for all supplies for the chain. The problem of foreign exchange leakage is more characteristic of luxury travelers who are apt to stay, eat, and shop in the multinational, five-star hotels while on tours organized in the guest's home country. If, for example, a Japanese tourist flies with a prepaid Japan Airlines ticket to the Philippines, where he or she stays at the JAL-owned Manila Garden Hotel and takes Japanese-language tours in Japanese coaches, what portion of his or her overall vacation expenses accrues to the Philippines?

Tourism development for personal political objectives

At the outset, it was noted that in a system as centralized as the Philippine government under Marcos the distinction between international, national, and personal political uses of tourism was often blurred. This section, however, explores those specific instances in which tourism policy has been used directly to affect personal fortunes without any apparent long-range considerations.

It is sometimes said that in the Philippines there is no such thing as conflict of interest, merely convergence of interest. The development of the tourist industry tends to substantiate that observation. Many of the top policymakers and close friends of the president have large economic interests in tourism. As mentioned before, the First Lady's holdings included the Philippine Plaza Hotel. According to individuals interviewed, former Secretary of Tourism Jose Aspiras had numerous investments in Manila motels, as well as hotels in La Union and in Zamboanga. Marcos' friend, Roberto Benedicto, Philippine ambassador to Japan and former president of the Philippine National Bank, was a major backer of the Manila Holiday Inn and the Kanlaon Towers. Nemesio Yabut, former mayor of Makati, owned the Makati Hotel. Jose Manzanan, Dean of the Asian Institute of Tourism, owns the Pagsanjan Rapids Hotel, and his wife has several handicraft and souvenir shops in Manila. Former Secretary of Public Highways Baltazar Aquino's wife owns the Mayon Imperial Hotel in Legaspi.[68]

Not surprisingly, then, many of the new hotel owners secured government financing far in excess of the generous ceiling the president endorsed, though small inns and pension owners often could not obtain even standard bank loans. When Manila became severely overbuilt with luxury hotels constructed with cheap and seemingly limitless credit, the president's friends complained in their role as hoteliers that they must renegotiate their loans or they would be unable to meet their payments, leaving the country's banking institutions with 14 hotels to manage.[69] It has since happened.

It must be kept in mind that the surplus of rooms was not unforeseen. DOT planners were right on target with most projections of accommodation needs. It was not mere error or oversight, therefore, that made the PTA unwilling to adhere to its own forecast of what was economic in the face of what was politic. The regime motivations have already been discussed. Relevant here, however, is that through pursuing massive hotel construction, Marcos could make himself and several strong supporters more affluent, and in the process create a prima facie government rationale for increasing tourist promotion and development to fill up the hotels.[70]

The government renegotiated hotel mortgages and launched several waves of new initiatives to help the hotels. What appeared to be a vicious if deliberate cycle, however, was not that at all. The new repayment terms were in fact arranged by financial leaders who were in some instances linked directly or indirectly to the ownership of the hotels. Moreover, as one government official remarked in another context, "If you owe the bank 1 million pesos, you are at the mercy of the bank. But if you owe the bank 200 million pesos, the bank is at your mercy."[71] In this case the figure is over 4.5 billion pesos owed to the Philippines' major national credit institutions and guaranteed by the Central Bank.

An entirely different type of personal political interest involved the national carrier, Philippine Airlines (PAL). The government, in contrast with some other nations, did not own its own airline before 1978 but retained a 24 percent interest in it through the Government Service Insurance System (GSIS). Since 1964, PAL had been substantially controlled by Benigno Toda, who reputedly acquired PAL as a political concession from the pre-Marcos Macapagal administration. Since then, PAL's fortunes have followed rather closely the waxing and waning of friendship between PAL's owner, Benigno Toda, and President Marcos.

At the beginning of the martial law period relations were close. President Marcos announced a "rationalization" of the air industry in which PAL's two domestic competitors were absorbed and their routes taken over by PAL.[72] In January 1974, PAL became the sole carrier for the Philippines.[73] The airline's growth was steady despite a very small capital base of approximately P25 million (US$3.43 million). Relations between Toda and Marcos reputedly soured, however, sometime in 1975. The breach occurred when Toda presented the Marcos family with a US$2.4 million flight bill for the "his" and "her" jets used by the family on their many trips abroad. The details of the dispute were confirmed by a high-ranking government tourism official interviewed in mid-1977.[74]

A none-too-subtle campaign against the management of the airline followed. A series of newspaper articles attacked PAL's service and administrative record, and in all disputes and petitions that appeared publicly, PAL lost every round. Among the defeats was the loss of the right to run the prestigious Manila Hotel. This left PAL, which had planned most of the restoration of the historic hotel, without a hotel to promote, unlike most national carriers. The next defeat came in 1977 when PAL appealed unsuccessfully to raise rates on the domestic routes, then among the lowest in the world.[75] This petition occasioned a spate of anti-PAL publicity that was in marked contrast to the usual laudatory comments about various sectors of the travel industry.

The final defeat occurred when PAL sought to keep the government from further liberalizing its air rights policy. The controversy pitted PAL management directly against the powerful Secretary Aspiras and the rest of the tourist industry. In the name of promoting tourism and protecting the nation's huge investment in tourist infrastructure, Aspiras vigorously pushed for an increase in flights to and from the Philippines by foreign carriers, a liberalization in policy that PAL could not expect to match by reciprocal agreements.[76] Though there was an appearance of evenhandedness in the dialogue over air policy, there was little doubt as to which way the government would move.[77]

Toda, who evidently realized his mistake in crossing Marcos, had attempted as early as 1976 and again in 1977 to sell PAL to the government. The policy defeats of 1977 left him no options. Not only could he not promote a hotel, raise uneconomic fares, or compete against foreign air rights, Toda also could not obtain a loan from the Development Bank of the Philippines for the purchase of additional aircraft. The reason given by the government was PAL's small capitalization.

In October 1977, the government, which had twice turned down Toda's selling price, decided suddenly to implement a 1966 board decision ordering Toda to increase his capitalization from pesos 25 million (US$3.43 million) to pesos 100 million (US$13.72 million) – something he could not do. Roman Cruz, who succeeded Toda as PAL president and board chairman, insists that although the government and Toda could not agree on a selling price, Toda had waived pre-emptive rights and allowed the government to subscribe "wholly to a 900 percent increase in capitalization." In the process, Toda's control went from 74 percent to 8 percent, and two weeks later Cruz took control.[78]

Since then the economic facts of life have remained as intransigent as the political events have been mercurial. Most of the criticisms of PAL's performance went unmet, though Cruz found several foreign multinational concerns willing to finance increases in PAL's fleet. Moreover, Cruz discovered that PAL had to raise domestic fares, an argument not unlike the ill-fated Toda's.[79] PAL continued to lose money throughout Marcos' rule. Between 1981 and 1984, the airline, which had been profitable before takeover, lost $167.5 million. Vindictiveness proved very expensive.[80]

PAL's uneasy situation was resolved in a manner reminiscent of the expresident's financial dealings with other economic giants not firmly controlled by the regime until after martial law began. Exploring how these takeovers took place offers insight into the way economic policy tends to trail political objectives. The process proceeds cautiously, without any nationalistic rhetoric that might frighten foreign investment, and is implemented with minimum expense and maximum political leverage.[81]

TOURISM POLITICS AFTER MARTIAL LAW

Just days before the inauguration of President Reagan in January 1981, President Marcos "lifted" martial law. The New Society became the New Republic, but the political reality was that the Old Oligarchy of the Marcos family and friends was still in control. As National Assemblyman Orlando Mercado put it: Marcos "facelifted" martial law.[82] Obvious controls on the press were loosened; subtle ones such as control of newsprint were not. Decrees issued under martial law remained in force, and the president was authorized to issue new ones. Writs of habeas corpus were gone; preventive detention was not.

One dismayed critic put it this way, "At least under martial law there was an illusion that it was merely a transitory stage . . . now we are locked into a system where the President possesses all the same sweeping powers with no end in sight."[83] Perhaps the most striking thing about the lifting of martial law from the standpoint of tourism politics was that, there too, nothing really changed. Resorts were still being built.[84] Lavish conventions, film festivals, and other activities were still being promoted.[85] The Department of Tourism was still opening foreign offices, and the personal political and patronage uses of tourism continued unabated. It was rather like having a gala dance party on the *Titanic*.

The Philippines was broke but *Balikbayan* and Reunion for Peace continued. The New People's Army grew daily, but so did the mountainside bust of the President.

Malnutrition was on the rise, but luxury imports of food and vehicles for the tourist industry were exempted from taxation. Television public service announcements designed to alert some 50,000 estimated victims to the danger signs of leprosy were cancelled because they might cast a pall over visitors to the Philippines.[86]

Tourism development in the Philippines was not just a case of a government suddenly and belatedly deciding to cash in on the boom in world travel. If that had been the case the government had the data and the expertise to expand tourism prudently. The 1973–1977 DOT plan is a model of careful, phased, and diffused development.

It was not departmental error that led to the reckless edifice complex or allowed virtually unlimited levels of government financing to be diverted ostensibly to hotel construction but often to short-term speculation. The department simply acknowledged the parameters of its political authority. One could even argue that from a departmental perspective, the DOT stood to gain from the hotel situation: it emerged as a potential savior of the national finances.

Ralph Waldo Emerson once remarked that "a foolish consistency is the hobgoblin of little minds." But President Marcos was no dummy. Unfortunately, where tourism policy is concerned, pervasive irrationality in terms of the national interest may be quite ingenious and lucrative at the individual or regime level.[87]

What happened with Philippine tourism under Marcos makes sense only in terms of the political objectives and personal taste of the First Couple. Too often the critics attempt to educate decision makers on the hazards of over-building, chide them on their folly, and call for better planning and information. Such critics miss the point entirely. They assume that if decision makers only knew better such debacles would not happen. This is not so. The policy-making structure must include an insistence on cost-benefit analysis, competitive bidding, and adherence to supply–demand factors and encourage public input. Otherwise, the debacles in tourism policy are almost certain to be repeated. Obviously, no decision maker is a prophet. Mistakes or lucky guesses do occur. But one must not be misled by rhetoric designed to persuade rather than inform.

Political elites do not generally opt to maximize economic rationality, at least not for the entire society.[88] That has been true historically in the Philippines as well as elsewhere. Martial law did not make the president more altruistic or better equipped to pursue the economic needs of the nation. It simply allowed him to proceed unchecked by democratic institutions.

Optimizing political benefits is not uncommon in politics, be they democratic or authoritarian. However, in developing or de-developing nations such as the Philippines under Marcos, the weak political institutions are rarely a match for political elites, particularly when those elites are immune from democratic repercussions or insulated from investigative reporting.

The administration created a veritable "stage set" of modernization and elegance that few development specialists at home or abroad saw fit to challenge. President Marcos got increased military and economic aid, international recognition, personal wealth, and the loyalty of relatives and associates who parlayed government loans for development into wealth and prestige. By the time Marcos was forced into exile in

1986 the tourist industry was in disarray, political instability was growing, and the Philippines had the highest per capita debt in Asia. Ironically, the Philippines owes so much that the country has a kind of reverse leverage on international banking circles. Thus the international finance community is forced to restructure loans, much as the Philippine government was forced to do with the hotels, to avert further decline and maintain hope of future repayment.

Though the IMF and the World Bank have forced some fiscal austerity moves on the Philippine government, the real costs are being borne by the government's banking, credit, and insurance systems and by the public as a whole. In fact, the very depressed nature of tourism is now being used as an argument for still more tourist initiatives.

Tourism probably could have been a successful developmental tool. The Philippines is blessed with both outstanding scenic beauty and a rich cultural heritage. The Philippines needed an alternative to the volatile commodity markets in sugar, coconut, and pineapples. The original 1973 plan might well have matured into a more gradually developing, but less import-dependent tourism program.

Investment decisions should have weighed tourism's economic and social costs and benefits against what an infusion of credit might have meant to other revenue-producing or developmental sectors. The National Economic Development Authority, as the government's highest planning body, should have been prepared to do this, but because of the First Couple's priorities it did not.

To the government, the immediate primary value of Philippine tourism clearly was not its economic value. The policy is consistent and credible only when tourism is evaluated against the critical and immediate legitimacy needs of the Philippine leadership, made suddenly vulnerable by the imposition of martial law, and the domestic and personal political advantages derived from tourism development.

Tourism development bought time, good will, and influence at home and abroad at a time when all three were in dangerously short supply. The administration kept hoping it could continue to do so, but that was wishful thinking. The debacle of massive luxury tourism development was too obvious. The spillover effects of enormous inflation, housing shortages, energy and water shortages, and the shame of the gigantic proportions of mass prostitution (over 100,000 prostitutes in Manila alone), made tourism a political liability to the regime, a source of controversy, and an avenue for violence.[89] But leaders have a way of staying with the formulas that initially brought them success, long after they have proved their declining utility.[90] It was too late for a new scenario. The third act was ending.

TOURISM POLICY UNDER AQUINO

In 1986, President Corazon Aquino had at least two of the same political motives in promoting tourism that Ferdinand Marcos had immediately after martial law was declared. She needed first to reassure her international constituencies, such as Japan and the United States, that she was in control, with a stable situation, and that she was a legitimate leader who enjoyed popular support. Second, she needed to guarantee that there was no interruption in aid or investment occasioned by her assumption of power.

Unlike President Marcos in 1972, however, Aquino was coping with a ravaged and impoverished nation as well as two serious and sizeable insurrections – one communist-led, one Muslim. Further, she chose to pursue her agenda in a turbulent democratic environment rather than one in which all opposition had been crushed.

Aquino also had a stark economic motivation for furthering tourism. She inherited over a dozen luxury hotels the government had taken over when President Marcos and entourage abandoned them rather than pay back their loans. These constituted a formidable drain on government coffers. Similarly, resorts, the convention center, and the 17 private homes of ex-President Marcos also needed to be put to good use. So Aquino not only needed tourists for their ordinary economic utility, but she also needed to stop the nonproductive and costly government-controlled tourist infrastructure from continuing to be a drain on scarce resources.

Still she accurately observed that the best way to promote the Philippines was to reduce the misery, filth, and malnutrition that pervaded the beautiful islands. Soon after coming to power Aquino made several key decisions that dramatically changed tourism policy from what it was under President Marcos. First, she made it clear from the outset that, however politically useful tourism was, the Philippines' tourism policy would be designed to subsidize national development, not to support tourism *per se*. That meant that the Department of Tourism would have a modest budget and would have to live on it, rather than raiding other agencies and levels of government.[91]

This has meant that in practice tourist revenues have also been supportive of projects impacting on tourism but not specifically connected to the industry. For example, the "Tourist Belt" of Manila is an exciting and important venue for most international tourists. It is also filthy, congested, and smelly. Under the Aquino administration the DOT is helping the city of Manila with garbage collection, since both bodies realize that the safety, health, and appearance of the Tourist Belt reflect on their own mission and effectiveness.[92]

Second, she chose as leader of the DOT an individual who, though a friend of her late husband, had entrepreneurial talent and marketing experience. Jose ("Speedy") Gonzalez has a reputation for executive skills and integrity. This contrasts dramatically with Marcos crony Jose Aspiras, whose public relations zeal exceeded any administrative ability and who was also involved in some of the seamiest aspects of sex tourism.[93]

Under Aspiras, tourist brochures promised "a tanned peach on every beach," sex tours including those for pedophiles flourished, and his own notorious motels and massage parlors were exempted from the martial law curfew. Under Gonzalez it has been firmly announced that no government advertising exploitative of women will be tolerated, promotional exhibits at tourism conferences will be tasteful, and pedophiles will be prosecuted and protests lodged with their countries of residence.[94]

In fact one of Gonzalez's first acts as tourism secretary was to bring a tour of

Japanese women to the Philippines as part of a new media effort to sell the country as a wholesome destination rather than a prostitution den.[95]

Gonzalez is also attempting to deal with the massive overstaffing that occurred under Aspiras. In January 1987, a 50 percent reduction in staff was contemplated,[96] but as time went on interviewers found Gonzalez and his undersecretaries increasingly vague about how much cutback would be feasible and politically tolerable. He has discovered that, in a country with massive unemployment and low wages, a lean public organization is not as practical or politically advantageous as it was when he operated in the private sector.[97]

Gonzalez, perhaps more than most of Aquino's Cabinet, is completely in accord with the government's new push toward privatization. The Philippines has a history of adopting U.S. management and budgeting fads of which privatization is simply the latest. (Actually one might argue that President Marcos was quite innovative in thoroughly blurring the distinction between what belonged to the Philippines and what belonged to him. He personally "privatized" many of the nation's public resources.)

The desperate financial plight of the government makes privatization less an ideological decision than a pragmatic necessity. This is especially true of the DOT, which manages lavish properties of little public utility that were sequestered from the Marcos holdings. Until they can be sold, the DOT has converted 3 of the 17 homes the Marcos' owned in the Philippines for use as VIP lodges and luxury tourist quarters. Other properties built and managed by the Philippine Tourism Authority are also up for sale as the government struggles to raise cash.[98]

One building not for sale is Malacañang Palace, the traditional home of the chief executive since Spanish times. Malacañang is historically important and a veritable museum of Philippine treasures and the excesses of the Marcos era. Aquino chose not to live there in such opulence but to make it a tourist attraction and a center for ceremonial functions.

As such, it has been an enormous touristic and political success. Where once only the elite of the world could enter, now long lines form daily for the five peso (US25 cents) tours through the mansion. Others can take more individualized tours for P200 (US$10.00) and on Sunday it is absolutely free. Over US$50,000 a week are collected.[99]

Beyond the badly needed money and the exployment it furnishes guides and staff within and the vendors outside serving the crowds, the tour itself is useful as political education. Since the Marcos family fled at night, one of the particular fascinations of the tour is the portrait of their greed and insecurity that emerges from the things they left behind. No amount of charges and countercharges by Aquino and Marcos has the impact of actually seeing Imelda Marcos' 3,000 pairs of shoes, the hundreds of ball gowns, the vault with hastily ransacked jewel boxes, the toy Rolls Royce, the drawers of imported chocolates, and amidst it all the remnants of fear.

The bulletproof glass that enclosed immense verandas, the miniature hospital for the president (who never admitted he was ill), and the mattresses laid on the floor where all the entourage slept the last few days reveal the Marcos family concerns. So fearful were they of the revenge of Filipinos that they all slept in the one room with a secret exit that could take them to the river and away from the palace if it came under seige. It was a lifestyle of the rich and infamous that has no peer.[100]

The Aquino government also conducts a "Freedom Tour" that goes to the major sites of the 1986 "People Power" revolution that unseated Marcos. A pictorial of events is another popular attraction at the Philippine International Convention Center, built by Imelda Marcos to showcase the arts and entertain "the beautiful people."[101]

The government initially decided to keep the *Balikbayan* program for both political and economic reasons. Like Marcos, Aquino wanted to encourage Filipinos living abroad and those of Filipino ancestry to visit. They represent a potentially influential constituency of nearly two million.[102] Economically, they are also needed. Beyond their normal tourist expenditures *Balikbayans* also are a source of remittances and gifts. In fact *Balikbayans* are now allowed to bring special "*Balikbayan* boxes" of goods when they enter. Gift-giving to relatives in large extended families actually constitutes in many cases the *Balikbayan*'s greatest travel-related expense. Currently, there is a campaign on to get *Balikbayans* to do their special gift-shopping in Manila so the economy can get the benefit of their largesse and they in turn are spared the hassle of transporting gifts.[103] Though still encouraged to return, *Balikbayans* found some special discounts stopped in 1988 as an economy move by the government.

Tourism policy under Aquino has also attempted to diffuse benefits more broadly than was true under Marcos. Three initiatives are worth noting. First, the government during its negotiating has found the Muslim insurgents interested in getting more tourism opportunities for their community. The government is receptive to encouraging more investment and doing more tourist promotion of Muslim areas, but it needs to be assured of tourist safety. It was insurgents in the past who by their kidnappings and terrorism of tourists destroyed earlier initiatives on Mindanao.[104]

A second program that is already being implemented is an effort to help families get involved in tourism, earn some money, and still provide basic accommodations for both domestic and international tourists. By early 1987 the homestay program, as it is called, was operating in several communities where tourist infrastructure is scarce and yet where local festivals attract both foreign and domestic visitors.

> The main idea behind the program is to make travel affordable to domestic tourists and give the foreigners the chance to experience Philippine culture among Filipino families. Participants in the program are families (in the festival areas) who can accommodate at least one paying guest ... provided with ... a guest room, clean toilet and bath, meals at reasonable prices, [and] clean surroundings.[105]

The only problem thus far is neither supply nor demand but over-generous hospitality: the host families are simply spending too much on their guests to make a profit! The DOT is now attempting to instruct host families in how to achieve a balance between their sociability and their need to develop a profitable clientele.[106]

A third initiative has been to set up in key cities several courtesy centers with clean restrooms and other basic facilities for the traveling public. Anyone familiar with touring in developing nations will realize that these are the kinds of no-frill amenities that demonstrate that the Aquino government's priorities are at least in good order.[107]

Whether the Philippines will be able to build a solid tourist record, however, is still open to question. The Aquino government has had to retrieve tourism policy from its disastrous and expensive excesses and recast it according to the needs and budget of the impoverished nation. This it has done with flair, taste, and imagination.

However, tourism remains ultimately dependent on political stability and peace for its successful development. To the extent tourism is no longer a source of instability and a focus of opposition, much progress has been made. Ultimately, however, Aquino must face the economic despair that has fed the insurgencies and encouraged widespread property crime or steady development of tourism is impossible.[108]

The Philippines has been the scene of many "miracles" since the 1986 presidential election provided the catalyst for a new political dynamic. If the past administration made "martial law, Filipino style" into a tourist attraction, then "People Power" and the re-emergence of democracy should be at least as attractive.

NOTES AND REFERENCES

1 This chapter is a revised, updated, but very abridged composite of several studies I have done on Philippine tourism. For more detailed political analysis of tourism organizations and their administrature milieu, see my book, *Land Reform and Tourism Development: Policy-Making in the Philippines*. My article, "The Political Uses of Tourism: A Philippine Case Study," examines at greater length the political needs of the Marcos regime served by tourism. This chapter is based on 138 interviews and other research done in the Philippines in 1976 and 1977, while I was on a Fulbright Research grant. During that time I was a Visiting Research Associate of the University of the Philippines, College of Public Administration. In 1987 more interviews were conducted at the Department of Tourism and among interest groups affected by tourism policy.

2 Reagan, as governor of California, had been a friend of the Marcos' since the opening of the Philippine Cultural Center in Manila in the early 1970s. The First Couple had invited many celebrities but Governor Reagan was the only prominent American politician to accept, perhaps as a way of bolstering his presidential ambitions with foreign contacts. It would be over a decade before he realized that such notorious friends were a political liability.

3 See Carl Lande, ed., *Rebuilding a Nation: Philippine Challenges and American Policy*.

4 Leticia Magsanoc, "The View from the Tourist Belt."

5 Carl Lande, *Leadership, Factions and Parties: The Structure of Philippine Politics*.

6 Jose Aspiras, "Tourism in 1973 . . . A Success All the Way."

7 Oriol Pi-Sunyer, "The Politics of Tourism in Catalonia," p. 61.

8 The Department of Tourism became the Ministry of Tourism in 1978 after the president had developed his own martial law variant of a parliamentary system. Its internal dynamics and leadership were not changed by the terminology any more than martial law was affected by parliamentary labels. The labels reverted to department and secretary following the ratification of the 1986 Constitution. In discussing the tourism body the term

"department" will be used, and "secretary" will be used to designate its head throughout the chapter.

9 Interview with Undersecretary of Tourism, Gregorio Araneta, June 8, 1977.

10 Bernard Wideman, "Overbuilt, Underbooked."

11 Macrina Leuterio Ilustre, "Metro Manila Opens Its Arms to the World's Top Financiers."

12 Wideman, "Overbuilt, Underbooked."

13 Department of Tourism, *Accomplishment Report, 1976*, p. 38.

14 *Asia Travel Trade*, October 1976, p. 65.

15 *Bulletin Today*, September 6, 1983, p. 19.

16 *Bulletin Today*, September 9, 1983, p. 19.

17 Interviews with hoteliers, November 1976–July 1977. See also Presidential Decree No. 867 and *Asia Travel Trade*, March 1983, p. 34.

18 *Asia Travel Trade*, October 1976, pp. 50–51.

19 Department of Tourism, *Accomplishment Report*, 1976, p. 14.

20 Confidential interviews in Manila, January 1977.

21 Primitivo Mijares, *The Conjugal Dictatorship of Ferdinand and Imelda Marcos*, pp. 120–121.

22 Linda K. Richter, "Philippine Policies Toward Filipino Americans."

23 Department of Tourism, "Invitation to a Traditional Philippine Christmas."

24 Interview with Vic Portugal, Research and Statistics Division, June 1977.

25 Commentary by Teodoro Valencia, *Filipino Reporter*, March 13–19, 1981, p. 12. The *Filipino Reporter* was a pro-Marcos paper published in New York.

26 Interview with Vic Portugal, June 1977.

27 Department of Tourism, *Accomplishment Report*, 1976, pp. 42–43.

28 *Manila Journal*, July 23–29, 1978.

29 Official, Philippine Consulate in Hawaii, October 29, 1984.

30 Presidential Letter of Instruction, No. 331, October 29, 1975.

31 *Times Journal* (Manila), January 10, 1977; "The Number 1 Market," *Far Eastern Economic Review*, January 2, 1977, pp. 42–44; *Datafil*, 16–30 November 1976, p. 1931, reports tourist arrivals from the ASEAN region were up 98 percent over 1975.

32 *Bulletin Today*, March 14, 1977.

33 Interview with mayor of Manila, Ramon Bagatsing, May 26, 1977

34 Confidential interviews, March–June 1977.

35 Ibid.

36 Interview, June 1977. I was most fortunate to have been included in planning sessions during the two months before the PACOM conference and I was an observer throughout the meeting.

37 Rodney Tasker, "Campus Confrontation."

38 Judy Bredemeier, "Convening Where Asia Wears A Smile," p. 27.

39 Robert E. Wood, "Ethnic Tourism, the State, and Cultural Change in Southeast Asia," p. 371.

40 *International Tourism Quarterly*, No. 4, 1982, p. 11.

41 *Asia Travel Trade*, May 1984, p. 39.

42 *Bulletin Today*, May 26, 1982.

43 *Honolulu Star-Bulletin*, November 12, 1984, section A-9.

44 Linda K. Richter, "The Philippines," in *Colliers Encyclopedia Yearbook 1984*. The communists are also finding the rampant prostitution associated with tourism to be a major irritant to exploit. See also Francisco Nemenzo, "Rectification Process in the Philippine Communist Movement."

45 *Fookien Times Yearbook, 1976*, p. 6.

46 Interview with personnel of *Bulletin Today*, February 1977.

47 Rodney Tasker, "Journalist Vindicated."

48 Department of Tourism, *Accomplishment Report, 1973–1975*, p. 4.

49 Interviews. See also the recommendations in William L. Thomas, "Progressive Rather than Centralized Tourism: A Recommendation for Improving International Tourism in the Philippines."

50 Department of Tourism Planning Service, *The Tourism Investments Priorities Plan, 1977–1981, Final Revision.*

51 *Bulletin Today*, December 2, 1976.

52 Interview with Gregorio Araneta, June 8, 1977.

53 Interviews in La Union and Agoo in May 1977.

54 Araneta contends that his non-Ilocano background worked against him and for Aspiras in the selection of a Secretary of Tourism. Interview, June 1977; see also "The Wedding of the Year."

55 Interviews with Araneta and the personnel managers of the Department of Tourism and the Philippine Tourism Authority, 1977.

56 I read all their resources in two days, but often just read other things in there just to see if there was some function assigned to them I had missed. There was not.

57 Interview with Marawi City Tourism Coordinator Mitz Ramusen, May 1977.

58 See Linda K. Richter, *Land Reform and Tourism Development.*

59 Ibid.

60 Interview with Colonel Cacdac, General Manager of the Philippine Tourism Authority, June 7, 1977.

61 Department of Tourism, *Accomplishment Report, 1976*, pp. 6–7.

62 *Business Day* (Manila), September 20, 1977, p. 2.

63 Department of Tourism, *Accomplishment Report, 1976*, p. 21.

64 For background on the government's corporativism policy see Robert Stauffer, "Philippine Corporativism: A Note on the New Society."

65 Interview with Dr. Eryl Buan, Associate Dean of the Asian Institute of Tourism, January 18, 1977.

66 Interview with an official working with tourism policy in the National Economic Development Authority (NEDA).

67 Interview with hotel manager, 1977.

68 There was some controversy about whether highway personnel and machinery were used to build the lengthy driveway for the Mayan Imperial. Wideman, "Overbuilt, Underbooked"; and various interviews in Manila, June 1977. By 1987 other scandals had emerged concerning the hotel.

69 *Business Day*, May 26, 1977, p. 1.

70 *Stars and Stripes*, November 26, 1977.

71 Orlando Sacay, *Samahang Nayon*, p. 63.

72 Matt Miller, "Will PAL Remain National Airlines?"

73 *Pacific Travel News*, February 1974. See also *Bulletin Today*, December 30, 1973, p. 6.

74 *Asia Travel Trade*, January 1977.

75 *Bulletin Today*, August 6, 1977, p. 17. See also *Manila Journal*, September 4–10, 1977.

76 *Business Day*, July 13, 1977, p. 1.

77 *Bulletin Today*, July 15, 1977.

78 Miller, "Will PAL Remain National Airlines?," p. 10.

79 *Manila Journal*, January 22–29, 1979.

80 *Asia Travel Trade*, May 1984, p. 11.

81 Mijares, *Conjugal Dictatorship*, pp. 200–202.

82 Orlando Mercado in a speech at the East–West Center, Honolulu, October 26, 1984.

83 *Newsweek*, July 13, 1981.

84 *Asia Travel Trade*, March 1983, p. 34.

85 Linda K. Richter, "Tourism by Decree."

86 *Contours*, Vol. 1, No. 4, 1983.

87 Ralph Waldo Emerson, "Essay on Self-Reliance," p. 70; Linda K. Richter, "The Hobgoblin Factor in International Tourism."
88 Harry G. Matthews, *International Tourism, A Political and Social Analysis.*
89 Richter, "Tourism By Decree."
90 James Barber, *The Presidential Character.*
91 Interview with Undersecretary of Tourism Narzalina Lim, February 2, 1987.
92 Ibid.
93 Ibid.
94 Ibid.
95 Margot J. Baterina, "Promoting Tourism on a Small Budget."
96 Interview with Undersecretary of Tourism Narzalina Lim, February 2, 1987.
97 Baterina, "Promoting Tourism."
98 Interview with Undersecretary of Tourism Narzalina Lim, February 2, 1987.
99 Interview with Director of Malacañang Palace, February 5, 1987.
100 Personal observation, January 1987; see Belinda A. Aquino, *The Politics of Plunder*, for an excellent analysis of the dynamics and techniques by which President Marcos amassed his illegal Philippine fortune.
101 Tourism brochures.
102 Interview with Undersecretary of Tourism Narzalina Lim, February 2, 1987.
103 *Philippine News*, May 1987.
104 *Philippine Panorama*, May 24, 1987.
105 Baterina, "Promoting Tourism," p. 16.
106 Interview with Undersecretary of Tourism Narzalina Lim, February 2, 1987.
107 Baterina, "Promoting Tourism," p. 16.
108 Claude A. Buss, *Cory Aquino and the People of the Philippines.*

Part VI

TOURISM AND SOCIAL CHANGE

13

GENDER AND ECONOMIC INTERESTS IN TOURISM PROSTITUTION*

The nature, development and implications of sex tourism in South-east Asia

C. Michael Hall

INTRODUCTION

Sex tourism is one of the most emotive and sensationalised issues in the study of tourism. Although it is in prevalent in Western capitalist nations, sex tourism has generally come to be associated with 'Western' (including Japanese), usually male, visits to the Third World. Otherwise known as tourism prostitution, sex tourism can be defined as tourism for which the main purpose or motivation is to consummate commercial sexual relations (Graburn, 1983; Hall, 1991a). Sex tourism has become substantial fare for Western media over the past decade, a factor which has fed into often sensationalist reports of tourism prostitution in Thailand or the Philippines. However, much of the recent attention given by commercial media to sex tourism has not arisen because of any new-found concern for the disempowering nature of tourism prostitution, rather it has emerged because of the spread of AIDS. Therefore, current concern with sex tourism does not reflect a discovery of relationship between gender issues and tourism development, but is instead often regarded as a health concern (Ford and Koetsawang, 1991).

Sex tourism is a major component of international travel to South-east Asia and has been given both overt and covert encouragement by government as a source of foreign exchange. For example, the Philippine Women's Research Collective (1985, p. 8) reported that sex tourism is the third largest source of foreign exchange for the Philippines. Although exact figures are impossible to obtain, one estimate suggests that 70–80 per cent of male tourists travelling from Japan, the United States, Australia, and Western Europe to Asia do so solely for purposes of sexual entertainment (Gay, 1985, p. 34). While Gay's estimate is almost certainly on the high

* Reproduced with permission from Vivian Kinnaird and Derek Hall, 1994, *Tourism: A Gender Analysis*, New York: Wiley, pp. 142–63.

side, especially given contemporary concerns over AIDS, it is likely that 'sexual entertainment' is a major motivating factor for many male travellers to the region.

This chapter will present an account of the various dimensions of sex tourism in South-east Asia. The chapter is divided into several sections. First, there is a discussion of the methodological difficulties encountered in the study of sex tourism. This is followed by an overview of the development of sex tourism. Thirdly, this chapter will note the various dimensions in growth of sex tourism in two South-east Asian countries: South Korea, and Thailand. Finally, the chapter will acknowledge that sex tourism can only be understood within a framework which addresses the varieties of structural inequality that occur within the South-east Asian region.

STUDYING SEX TOURISM

The study of sex tourism is difficult because of a number of factors:

1 the seeming blindness of many tourism researchers to actually acknowledge that a link exists between sex and the tourism industry, particularly in respect of sex as a motivating factor for travel and the sexual relationships between hosts and guests;
2 the extreme difficulties to be had in conducting research on tourism prostitution, which is typically an illegal informal activity, often with substantial crime connections, and which may place both researcher and subject at risk from brothel owners, criminal gangs and syndicates, and police and politicians who wish to keep the subject hidden and away from public examination;
3 the lack of common methodological and philosophical frameworks with which to explain the complex web of gender, productive, reproductive and social relations which surround sex tourism.

BLINDED BY THE LIGHT

With few exceptions, there is an apparent blind spot or natural reticence among academics to acknowledge the relationship between tourism and sex (Cohen, 1982; Crick, 1989; Hall, 1991b). Nevertheless, if an effective analysis of the proposition that 'tourism is prostitution' is as true of metropolitan tourist resorts of the First World as it is of the Third World (Graburn, 1983, p. 441), then such studies are necessary. For example, despite claims by Mathieson and Wall (1982, p. 149) that there is little evidence to confirm or deny a positive link between prostitution and tourism, there has been a long history of this relationship (Turner and Ash, 1975). Just as Miami has become synonymous with vice, the combination of sun, sea, sand and sex has become associated with tourism resorts in general (Hall, 1991b). Tourist promotion often focuses on the more licentious attributes of the tourist and highlights the erotic dimensions of a tourist destination:

> Tourism promotion in magazines and newspapers promises would-be vacationers more than sun, sea, and sand; they are also offered the fourth 's' – sex.

Resorts are advertized under the labels of 'hedonism', 'ecstacism', and 'eden-ism' ... One of the most successful advertizing campaigns actually failed to mention the location of the resort: the selling of the holiday experience itself and not the destination was the important factor.

(Baillie, 1980, pp. 19–20)

Most observers who have examined the motivations of tourists have overlooked the very explicit messages purveyed in tourist advertisements. Despite the self-confessed selling of hedonism by the tourist industry, it is interesting to note that several studies of tourist brochures and images have failed to identify the pervasive use of erotic images in promoting travel destinations (Buck, 1977; Dilley, 1986). Furthermore, the 'red light' districts of many cities are tourist attractions which draw the sensation-seeking tourist as participant or observer.

CONDUCTING RESEARCH

There is a substantial lack of systematic research on sex tourism. This is because of the often informal and illegal nature of tourism prostitution, and the general un-willingness of police, government authorities, and politicians to acknowledge its ex-istence. Given the circumstances surrounding sex tourism it is therefore extremely difficult to measure the scale of sex tourism at any given location and obtain inter-views with prostitutes and other sex industry workers and their customers (Wihtol, 1982). Furthermore, Burley and Symanski (1981, p. 239) suggest that widespread disdain for prostitutes inhibits researchers and hinders the reliability of informants' accounts.

The analysis of sex tourism is also complicated by the several forms which tour-ism prostitution takes (Hall, 1991a):

1 Casual prostitutes, who move in and out of the pursuit according to financial need. This may be regarded as incompletely commercialised and the nature of the relationship between partners is ambiguous (Cohen, 1982).
2 Prostitutes ('callgirls', 'callboys') who operate through intermediaries and who visit tourists in their accommodation.
3 Sex workers who operate out of clubs and brothels, which are often regarded by authorities as a mechanism for the containment of prostitution (Symanski, 1981).
4 Bonded prostitutes working in brothels who have often been sold in order to pay debts or reduce loans. Bonded prostitution is a form of slavery in which the body of the prostitute is treated solely as a commodity to be bought and sold without the prostitute's consent.

In addition, it should be noted that prostitution is often 'spatially, economically and socially differentiated between locals and tourists, a point that is often forgotten by many researchers who tend to blame all prostitution on tourism' (Hall, 1991a, p. 65).

FINDING COMMON GROUND

A major problem in studying sex tourism, as with any analysis of gender issues, is the considerable divergences in approaches towards feminist and gender studies. Most theoretical developments in the study of women and gender have emerged from the studies of women in the First World, not from Third World studies (Brydon and Chant, 1989). Indeed, Barrett (1986, p. 8) has described that a feminist theoretical framework remains 'fragmentary and contradictory'. Nevertheless, research into the nature of sexual politics and the interrelated structures of gender relations may shed some insights on sex tourism.

Sexual politics examines the links between sexuality and power. Gender and sexuality are socially constructed at both the personal level of individual consciousness and interpersonal relationships and the social structural level of social institutions. The control of sexuality through, for example, such mechanisms as formalised tourism prostitution, is a key component of domination. Sex tourism serves to establish certain erotic images of Asian women in Western society which are then perpetuated within the social institutional structures of the destination. For example, Davidson (1985, p. 18) reported a Frankfurt advertisement which stated, 'Asian women are without desire for emancipation, but full of warm sensuality and the softness of velvet'. More recently, South-east Asian airlines such as Thai Airlines, Singapore Airlines and Cathay Pacific have portrayed 'submissive' Asian women in their promotional material, with Singapore Airlines running a campaign of 'Singapore Girl – you're a great way to fly'. The control of sexuality harnesses erotic power for the needs of larger, hierarchical systems by controlling the body and hence the population (Foucault, 1981). Therefore, there is a need to examine the processes by which power as domination occurs on the social structural level – namely, the institutional structures of racism, sexism, and social class privilege – and annexes the basic power of the erotic on the personal level (Collins, 1990).

Connell (1987) identified three interrelated structures of gender relations: first, the division of labour; second, power, particularly as connected to masculinity; and third, 'cathexis' – 'the construction of emotionally charged social relations with objects [that is, other people] in the real world' (p. 112). The three structures contribute to the institutionalisation of a 'gender order' which described the historically constructed pattern of power relations between men and women at the societal level. From this perspective, which relies heavily on notions of structuration, gender relations are stable to the extent that groups constituted in the network of gender relations have interests in the conditions for cyclical rather than divergent practice (Connell, 1987, p. 141; Crompton and Sanderson, 1990, p. 19).

Gendered perspectives on sex tourism which place significant emphasis on the exercise of power in gender relations will have substantially different insights on the conditions which allow sex tourism to prosper. For example, the economic justification provided for sex tourism in Fiji in the early 1970s by Naibavu and Schutz (1974, p. 6) argued that prostitution as a fully localised industry provided income for unskilled female workers for most of whom no other employment would be available.

It required no foreign capital investment, but attracted significant foreign exchange with minimal leakage. By contrast, a gendered perspective (ISIS, 1984, p. 4) views the acceptance of the inevitability of prostitution as being vested in the assumption that access to sexual relations is a male right, and that viewing prostitution as woman's choice reduces all women to the lowest and most contemptible status within male-dominated society.

> Prostitution is both an indiction of an unjust social order and an institution that economically exploits women. But when economic power is defined as the causal variable, the sex dimensions of power usually remain unidentified and unchallenged.
>
> (Barry, 1984, p. 9)

Any oversimplified utilitarian justification for prostitution may obscure the gender and class interests associated with tourism prostitution. According to Thanh-Dam (1983, p. 536), prostitution is differentiated in response to the processes of 'capitalist development and conditioning labour relations, demand and supply'. In contrast, feminist researchers such as Hawkesworth (1984) and Rogers (1989) argue that prostitution is a direct result of the patriarchal structure of society. The reality for many of the women involved in the sex tourism industry lies somewhere in between. The roots of disempowerment are found not only in economic status and within the sphere of production but also in reproduction and social and cultural structures. As Connell and other writers on gender issues have indicated, the explanation for the exploitation of women through tourism prostitution lies in the interweaving of class relations and gender relations. The present chapter therefore draws upon empirical evidence and attempts to synthesise the interpretations and arguments used by authors working from different theoretical perspectives in arriving at an understanding of sex tourism in South-east Asia.

THE DEVELOPMENT OF SEX TOURISM IN SOUTH-EAST ASIA

The economic and social problems of Asian women tend to be viewed by Western feminists as stemming from the patriarchal nature of local cultures (Hall, 1991a). Yet there is not an undifferentiated Asian model of women and gender relations, although the state as a structure is dominated by patriarchal interests (Agarwal, 1988). Each region within South-east Asia needs to be examined in terms of its own particular set of gender, class and social relations. However, certain generalisations about the role and position of women within South-east Asia and the development of sex tourism can be made.

The growth of the newly industrialised economies within South-east Asia through the relocation of multinational export-processing manufacturing forms to the region has led to the highest rates of female employment in the Third World. In each country of South-east Asia women constitute at least 26 per cent of the labour force, and 36–45 per cent in the Philippines, Laos, Burma and South Korea (Seager and Olson, 1986). The transfer of export-processing plants to South-east Asia from Japan,

North America and Europe has led to a demand for young, unmarried, relatively educated women who are assumed to have the manual dexterity and docility needed for the tedious, repetitive and monotonous work. The new international division of labour has perpetuated existing gender disparities in wages, except in situations where labour market shortages have significantly improved the status of women in the workforce (Phongpaichit, 1988). Despite the high rates of female labour force participation,

> women's 'status' in Southeast Asia is not necessarily 'better' than in regions
> where their involvement in remunerated work is lower, since many of the ac-
> tivities engaging women are those which reinforce patriarchal power struc-
> tures, such as prostitution and sexual services in Thailand and the Philippines.
> (Brydon and Chant, 1989, p. 43)

Ong's (1985, p. 2) argument that the new trade in the labour and bodies of Asian women is more rooted in corporate objectives of profit-maximisation than in the persistence of indigenous values is only partially correct. The new international division of labour and the spread of consumerism to the Third World has had substantial impacts on gender, racial, and social relations in South-east Asia, but rather than supplant indigenous values, it has built on them. Indeed, as noted above, it was the indigenous set of gender relations which helped to provide a female labour force for various multinational enterprises and which has assisted in the development of sex tourism. In spite of high rates of remunerated employment among women in east Asia, religious and cultural ideals emphasise that their primary responsibility is to home and family (Brydon and Chant, 1989). Women who have fled rural poverty, only to be forced into prostitution by urban unemployment, are victims of a double standard (Claire and Cottingham, 1982). Brydon and Chant (1989, p. 44) point to the fact that economic development has had an equally important role to play in reinforcing women's secondary status. When women move into the labour force their work is often of a highly exploitative nature, whether in a multinational-owned factory or brothel.

Sex tourism may be conceptualised as a series of linkages between a legally marginalised form of commoditisation (sexual services) within a national industry (entertainment), dependent upon, but performing a dynamic function within, an international industry (travel) (Thanh-Dam, 1983, p. 544). The post-war development of the new international division of labour has meant radical restructuring of the economies of South-east Asia through their closer integration within the global economy. The influx of rural women to urban centres in order to support village families has encouraged the marginalisation of female participation in the labour market (Hall, 1991a). Both in the light manufacturing industries of the electronics revolution and in the sex industries, women have been integrated into the global economy. Indeed, Ong (1985) argued that women who have lost jobs in the industrialised sector have been forced to seek work in hotels and brothels. However, the new international division of labour should be seen as only one stage in the development of sex tourism in east Asia.

In east Asia the institutionalisation of international sex tourism commenced with the prostitution associated with American and Japanese colonialism and militarism and has now become transformed through the internationalisation of the regional economy (Hall, 1991a). Prostitution is illegal in most South-east Asian countries, but laws regarding tourism prostitution are poorly enforced by authorities in the region's patriarchal societies (Sentfleben, 1986).

Prostitution in South-east Asia clearly existed before the arrival of tourists. One of the ironies of the current Japanese involvement in sex tourism is that the Japanese used to export their own prostitutes *Kara-Yuki San* to their colonies. *Kara-Yuki San* were bonded Japanese women who were sent abroad to serve as prostitutes in ports frequented by Japanese merchants and soldiers (Hawkesworth, 1984; Matsui, 1987a). However, since the 1920s when the Japanese government issued the Overseas Prostitution Prohibition Order, and with the prohibition of legal prostitution in Japan in 1958, women from the former colonies 'are now imported into Japan as prostitutes' (Graburn, 1983, p. 440). In addition, the nature of sex tourism varies from country to country according to different gender, cultural and social factors. For instance, in a report on child prostitution Rogers (1989) states that in Thailand child prostitution is 90 per cent female, in Sri Lanka 90 per cent male, and in the Philippines, young boys account for 60 per cent of child prostitutes.

The evolution of sex tourism in South-east Asia has generally gone through four distinct stages (Hall, 1991a). The first stage is that of indigenous prostitution, in which women are already subject to concubinage and bonded prostitution within a patriarchal society which regards such sexual relations as acceptable and normal. The second stage is that of economic colonialism and militarisation in which prostitution is a formalised mechanism of dominance and a means of meeting the sexual needs of occupation forces. For example, the American presence in Taiwan from the Korean War through to the end of the Vietnam War provided a major stimulus for tourism prostitution centred on Shuang Cheng Street in Taipei. Similarly, until the closure of the bases, the 12,000 registered and 8,000 unregistered hostesses in Olongapo City provided the major source of sexual entertainment for the United States military personnel based at Subic Naval Base and Clark Air Force Base in the Philippines (Claire and Cottingham, 1982, p. 209). In this stage both the indigenous and occupier's set of gender relations are used as a justification by government and occupying forces for economic or military enforced prostitution, thereby encouraging the dependency of host regions on the selling of sexual services as a means of income generation. The commoditisation of sexual relations is marked in the third stage by the gradual substitution of overseas military forces by international tourists. For example, Australian criminal elements made substantial investments in the sex industry in Angeles City near Clark Air Force Base in the Philippines prior to it being closed, in order to offer sexual services to Australian tourists as well as American servicemen. Bacon (1987, p. 21) reported that 'Australians now have a financial interest in more than 60 per cent of the 500 bars and 7,000 prostitutes around the base'. At this third stage, sex tourism now becomes a formal mechanism by which authoritarian governments can further national economic goals.

271

The fourth, and current, stage for most of the nations of South-east Asia is that of newly industrialised nation status. Despite the increased material standard of living for many areas, tourism prostitution is still widespread. The series of stages in the development of sex tourism have established and reinforced a network of gender conditions which may take many years to dismantle, particularly given their continued maintenance by patriarchal elite structures. The following overview of sex tourism in South Korea and Thailand indicates the four stages of development of sex tourism.

SOUTH KOREA: *KISAENG* TOURISM

The term *kisaeng* originally referred to females who served a similar social function to the *geisha* in Japan. The word is now, however, synonymous with Japanese-oriented tourism prostitution in Korea. 'No less than 100,000' tourist service girls were estimated to be operating in South Korea in 1978 (Korea Church Women United, 1983, p. 2), although this figure excluded the large number of unregistered sex workers. Gay (1985) estimated that some 260,000 prostitutes were operating in South Korea, the majority of whom originated from economically marginal rural areas. *Kisaeng* girls were required to obtain identification cards from authorities in order to enter hotels. While at the government 'orientation programme' for sex workers, which was a prerequisite for issuance of the card, women were told that their carnal conversations with foreign tourists did not prostitute either themselves or the nation, but expressed their heroic patriotism (Gay, 1985, p. 34). The orientation programme was regarded as similar to that given by the Japanese to the 100,000 Korean women who were forced to serve in the 'women's volunteer corps' as prostitutes to the Japanese troops during World War II (Yoyori, 1977; Matsui, 1987b). Indeed, the relationship between sex tourism and the use of 'comfort women' by the Japanese armed forces during World War II is now expressly made by Korean women's groups. The South Korean Church Women's Alliance adopted the 'comfort women' issue, not so much as a major cause in its own right, but rather in the hope that it could be a vehicle to generate feelings against Japanese sex tours (Hicks, 1993, p. 34). According to Korea Church Women United (1983, p. 27):

> The only difference we can find . . . is the circumstances under which they are conducted – the one during the Japanese invasion days was in a war effort; the other, currently in a sovereign state.

Kisaeng sex tourism is dependent on the Japanese. Following the severing of Japanese diplomatic ties with Taiwan in 1973 there was a massive increase in Japanese tourism, predominantly male, to South Korea (Kikue, 1979). (Conversely there has been a reduction in Japanese sex tourism to Taiwan.) *Kisaeng* sex tourism has taken three major forms:

1 The establishment of *kisaeng* houses (brothels) in the major urban areas geared towards tourists.

2 *Kisaeng* act as companions to Japanese businessmen and travel with them during their visit to Korea. This role is regarded as an integral component of the conduct of Japanese business overseas and may be likened to a contemporary equivalent of the 'comfort women' role that Korean women were forced to take during World War II.

3 *Kisaeng* has been exported to Japan, with at least one report of 'musical talent' being sent to 50 Japanese *kisaeng* houses operating in Osaka (Korea Church Women United, 1983, p. 46).

The authoritarian nature of successive South Korean governments has played a major role in the commoditising of women through *kisaeng* tourism. Existing sets of gender relations, reinforced through periods of colonisation by the Japanese, have been consciously maintained by government policies which have exploited women's bodies for national economic gain. The 1988 Seoul Olympics saw South Korea put on a modern façade for the world but despite the promotion of cultural and natural attractions by the government, *kisaeng* tourism is still a major factor in attracting Japanese male tourists to the country. Korean women's groups have attempted to draw attention to Japanese sex tourism in recent years, most successfully by highlighting the 'comfort women' issue. However, the Korean government appears at pains not to damage its economic relationship with Japan and provides little support for the women's movement. The continued maintenance of the authoritarian, patriarchal institutional structure by the present Korean elite can only continue to perpetuate the existing set of gender relations which allows *kisaeng* tourism to exist.

> The tidal wave to modernize this land has brought along in its wake the spurring on of tourism – a form of 'prostitution tourism' which is condoned and encouraged, for the sake of foreign exchange, by the powerful in our country. This depersonalization of the sex act and through it the dehumanization of Korean women, takes sex from being a soul–body relationship to one of a 'subject–object' relationship, in which the Subject is the man who buys and the Object is the woman who is sold. This deliberate action of earning foreign exchange by selling the flesh and souls of our women for sexual exploitation, is not only stripping the human dignity from Korean women, but is also an affront to all humanity.
>
> (Korea Church Women United, 1983, p. 1)

THAILAND: 'LAND OF SMILES' OR 'THE BROTHEL OF ASIA'

Substantial debate has surrounded the indigenous gender relations of Thai society. Several authors have argued that women have enjoyed relative equality (Davidson, 1985), while others have noted that women have long been subordinated, with the patriarchal Buddhist culture maintaining both class divisions and the exclusion of women from political power (Phongpaichit, 1982; Thanh-Dam, 1983; O'Malley,

1988). Indeed, the importation of elements of Brahminical culture into Thai society has allowed concubinage and polygamy to be legitimised and has created a ready-made framework within which sex tourism can be acceptable to government and local elites.

The promotion of sexual services is an important element in the marketing of Thailand to tourists. It seems likely that in no other country has tourist motivation been so explicitly linked to sex (Cohen, 1988, p. 86). In 1957 there were 20,000 prostitutes, by 1964 400,000, and between 500,000 and a million in the early 1980s (Phongpaichit, 1982; Taylor, 1984; Gay, 1985; Hong, 1985). Not all the prostitutes cater to sex tourists. Indeed, Cohen (1982, p. 409) argued that the women working with *farangs* (foreigners) are in many ways the 'elite' among the prostitutes: they earn significantly more than those working with Thais, enjoy greater independence, and are rarely controlled by pimps or pushed into prostitution against their will. However, according to Richter (1989, pp. 98–9), Cohen's empathetic and empirical approach describes the lifestyle of only a tiny minority of female prostitutes. Most have no such protective ambiguity. Furthermore, many Thai women become 'rented wives' (*mia chao*), somewhat similar to the *kisaeng* of Korea, who often return to tourist generating regions, particularly Germany, Holland, and Japan, where they may suffer linguistic and social isolation while often being forced to perform sexual services (Skrobanek, 1983).

The economically marginal rural areas of the north-east and northern provinces of Thailand have been the major source for child and female prostitutes, with many families and villages economically dependent on remittances from the prostitutes (Wereld, 1979; Phongpaichit, 1981, 1982; Cockburn, 1989). The rapid and uneven development of Thailand has been closely viewed as integrating militarisation, tourism, and industrialisation as institutionalised systems of female exploitation (Ong, 1985). Phongpaichit (1982) has argued that the economic condition of northern Thailand stems from the hegemonic power of Britain and the United States which have encouraged export-oriented development on the Central Plain at the expense of the north. The northern provinces are structurally disadvantaged within the Thai economy and in the absence of alternative income sources, a vested interest and dependency upon a continuation of the sex industry is created (O'Malley, 1988, p. 110), providing an assured supply of workers for the sex industry based in the nation's urban and industrial centres.

Until the end of the 1980s the Thai government placed great emphasis on sex tourism as a means to earn foreign exchange, to the extent that ministers often openly advocated tourism prostitution as a means of job creation (Mingmongkol, 1981; Gay, 1985; Barang, 1988). Successive Thai governments therefore continued to perpetuate existing gender relations and exploitation through support of the sex tourism industry. The last five years have seen public statements from the Thai government on the control of sex tourism, but the reasons for this probably lie more within concerns for AIDS and the image of the country overseas than in any intrinsic changes in social attitudes.

In 1989 the Thai Public Health Ministry started campaigning against prostitution and the promotion of Thailand as a sex tour destination. The primary reason for the campaign was the recognition that sexually transmitted diseases such as AIDS could pose major problems for Thailand's rapidly growing tourism industry. A Ministry survey of AIDS indicated that about 3,000 prostitutes in Thailand were carrying the AIDS virus and a World Health Organisation report estimated that the number of infected people in Thailand was between 45,000 and 50,000, compared to an official figure of 14,000 (Corben, 1990, p. 7). In tourist centres such as Chiang Mai, tests have suggested that one of every two prostitutes in the region carries the virus (Robinson, 1989). According to the Thai Deputy Public Health Minister, Suthas Ngernmuen (in Robinson, 1989, p. 11):

> Thailand's profitable tourist industry has been an inhibiting factor in promoting AIDS awareness . . . More than two-thirds of the overseas visitors entering Thailand are single men, and medical officials avoided publicising the appalling AIDS statistics for fear of damaging the country's healthy tourist business . . . But it is long past time for the government to change Thailand's image as a sexual paradise.
>
> We should promote tourism in more appropriate ways, and campaign more against AIDS.

The Thai government has considered a number of options to curb the spread of AIDS, including operating testing programmes for certain visitors, and distributing condoms in hotels and at the airport. Part of the AIDS programme includes 're-habilitating' and skilling prostitutes, while the Defence Ministry is providing an AIDS education programme for servicemen. The Public Health Ministry has proposed the issuing of health cards to brothel workers which would indicate the holder's personal background and the results of tests conducted for sexually transmitted diseases, including AIDS (Corben, 1990). The issue of three-monthly health cards is only an indirect measure of controlling AIDS and does relatively little to deter the sex tourist. Instead health cards may only further attract visitors to certain brothels or locations and may also allow the possibility of corruption in order to obtain cards which give workers in the sex industry clean bills of health. If the Thai government is to be serious about controlling AIDS a major emphasis has to be given to replacing Thailand's image as a sex tour destination and, more significantly, providing alternative economic and social support mechanisms for those who are forced to use prostitution as a means of employment (Hall, 1993). However, Thailand's tourism authorities are extremely sensitive to reports on the causes and consequences of the rising incidence of AIDS and other sexually transmitted diseases in Thailand (Corben, 1990, p. 9). After a damning examination of the sex trade and the major role it plays in Thailand's tourism industry by the *Far Eastern Economic Review*, the Governor of the Tourism Authority of Thailand (TAT), Dharmnoon Prachuabmoh, attacked the *Review* and argued that any effective measures adopted by the Thai government to curb the spread of AIDS were acceptable, and that the welfare of the Thais, not the tourist dollar, was top priority (Corben, 1990, p. 9).

The AIDS scare has already impacted on some Thai tourist destinations (Hall,

1993). Visitor arrivals by road from Malaysia were estimated to have dropped by over half during 1989 at the southern town of Hat Yai, renowned for its bars and massage parlours (Asia Travel Trade, 1990). Similarly, AIDS has tarnished the image of Pattaya, and this, associated with environmental problems, is believed to have contributed to a sharp decline in the number of visitors to the resort area in the later 1980s.

In order to attract more female tourists and to counteract Thailand's image of sex tourism and AIDS, the chairman of the Tourism Authority of Thailand and a leading anti-AIDS campaigner, Minister Mechai Viravaidya, planned a Women's Visit Thailand year campaign in 1992. According to the Minister:

> We want women to come particularly from countries where some of their men have come here on sex tours . . . We want them to see what their men get up to and how they have exploited uneducated women and children. We want their women to come and see the good Thai women and encourage Thai women to stand up to the brutality and disrespect they have suffered. More action must come from Thai women themselves, otherwise the country will still be seen as the brothel of the world.

> (Kelly, 1991, p. 44)

The Minister's stance led him to be criticised severely by some members of the tourism industry and accused of a conflict of roles (Hail, 1992). It is readily apparent that some sections of Thailand's tourism industry are still keen to promote sex tourism because of its financial benefits. However, the long-term health implications, through the spread of AIDS and other sexually transmitted diseases, and the social impacts of sex tourism are enormous and represent a potential time-bomb for Thailand's economic and social development. In order to confront the implications of sex tourism for Thai society and economy, the government must overcome official corruption and deep-rooted cultural attitudes towards sex and the role of women. Such a task will not be easy, but unless the government takes firm and decisive action, not only will the broader tourism industry be damaged, particularly that geared towards the family market, but the human base of Thai economic development will be undermined. However, as long as the gap between the city and the rural areas continues to widen, real living standards in the country remain low, and the pre-existing set of gender relations continue to be maintained by successive military regimes, the women of the rural north will continue to furnish Thailand's sex industry with its raw material.

CONCLUSION

Sex tourism is an integral part of the tourism industry of east Asia. It has developed for a number of reasons. The indigenous set of gender relations in many parts of east Asia provided the gender base upon which sex tourism could thrive. However, the post-war militarisation of the region plus the integration of the area in the international economy has provided a further dynamic element in the perpetuation of pre-existing gender and social relations, including institutionalised racism. The significance of economic marginality and racial inequality as a causal factor in prostitution is witnessed in the origins of prostitutes. For example, the majority of tourism prostitutes in Taiwan are not Han Chinese but instead come from the island's aboriginal population, which is Polynesian–Malayan in origin and which lives in marginal rural areas (Sentfleben, 1986). Similarly, hospitality girls working in the ago-go bars in the tourist nightlife belt of Ermita in the Philippines come mainly from low-income families in economically marginal rural areas (Wihtol, 1982). According to the Philippine Women's Research Collective (1985) many women expend almost a quarter of the total income to support family members in Manila or send remittances back to the home province, a situation which is replicated throughout the region.

The desire for economic development by many of the region's newly independent nation states has led to dependence on tourism as a source of foreign exchange (Center for Solidarity Tourism, 1989; Harrison, 1991; Hall, 1993). Indigenous gender relations have allowed many women to be sold as commodities in order to achieve this goal, whether it be through sex tourism, as cheap labour in multinational manufacturing enterprises, or as migrants. For example, following the campaign against sex tours by Japanese and Filipina women, numbers of Japanese tourists to Manila decreased while the proportion of Filipina women travelling to Japan increased (Matsui, 1987a, p. 32). Economic relations both within the new international division of labour and between the First and Third Worlds have only exacerbated gender inequality. A direct analogy has been drawn between prostitution in the Third World and that in metropolitan resort centres, where the women (and men) may be largely drawn from economically and socially disadvantaged sections of the population whose position illuminates forms of 'internal colonialism' commonly found in stratified, industrial societies (Graburn, 1983, p. 442).

The stability of the gender relations which allow the maintenance of sex tourism is a reflection of the interests of the groups which constitute the gender order. The 'insidious tourist first attitude' (Philippine Women's Research Collective, 1985, p. 36) of many governments of eastern Asia has actively promoted the images of subservient women waiting on the needs of the (male) tourist. To change this set of circumstances will therefore require fundamental change both within the host countries and in the international tourism industry.

East Asian sex tourism has further institutionalised the exploitation of women within patriarchal societies although its more overt forms do appear to be on the decline. However, the reasons for this probably relate more to the threat of AIDS to the visitor rather than widespread concern for the sex worker or fundamental changes in gender and economic relations. Indeed, modern mass tourism and its accompanying images only serve to promote unequal gender relations in which women are subordinate to male interests. Tourism will continue to be a mainstay of the region's economies. However, while tourism prostitution may decline, the intrinsic inequality of host–guest relationships in mass tourism can only continue to perpetuate the current set of gender relations. (Mass) tourism is sex tourism.

REFERENCES

Agarwal, B., 1988, Patriarchy and the 'modernising' state: an introduction, in Agarwal, B. (ed.), *Structures of patriarchy: state, community and household in modernising Asia*, Zed Books, London, pp. 1–28.

Asia Travel Trade, 1990, AIDS problem menaces tourism, *Asia Travel Trade*, **22** (November): 56–7.

Bacon, W., 1987, Sex in Manila for profits in Australia, *Times on Sunday*, 19 April: 21, 24.

Baillie, J. G., 1980, Recent international travel trends in Canada, *Canadian Geographer*, **24**(1): 13–21.

Barang, M., 1988, Tourism in Thailand, *South*, December: 72–3.

Barrett, M., 1986, *Women's oppression today: problems in Marxist feminist analysis*, Verso, London.

Barry, K., 1984, *Female sexual slavery*, New York University Press, New York.

Barry, K., Bunch, C., Castle, S. (eds), 1984, *International feminism: networking against female sexual slavery*, International Women's Tribune Centre, New York.

Brydon, L., Chant, S., 1989, *Women in the Third World: gender issues in rural and urban areas*, Edward Elgar, Aldershot.

Buck, R., 1977, The ubiquitous tourist brochure: explorations in its intended and unintended use, *Annals of Tourism Research*, **4**(4): 192–207.

Burley, N., Symanski, R., 1981, Women without: an evolutionary and cross-cultural perspective on prostitution, in Symanski, R., *The immoral landscape: female prostitution in western societies*, Butterworths, Toronto, pp. 239–73.

Center for Solidarity Tourism, 1989, Impacts of tourism in the Philippines, *Contours*, **4**(2): 29.

Claire, R., Cottingham, J., 1982, Migration and tourism: an overview, in ISIS, *Women in Development: a resource guide for organisation and action*, ISIS Women's International and Communication Service, Geneva, pp. 205–15.

Cockburn, R., 1989, The geography of prostitution, part I: the east, *The Geographical Magazine*, **60**(3): 2–5.

Cohen, E., 1982, Thai girls and farang men: the edge of ambiguity, *Annals of Tourism Research*, **9**, 403–28.

Cohen, E., 1988, Tourism and AIDS in Thailand, *Annals of Tourism Research*, **15**: 467–86.

Collins, P. H., 1990, *Black feminist thought: knowledge, consciousness, and the politics of empowerment*, Unwin Hyman, Boston.

Connell, R. W., 1987, *Gender and power*, Polity Press, Cambridge.

Corben, R., 1990, Thailand takes another step to curb AIDS, *Asia Travel Trade*, **22**(June): 7–9.

Cottingham, J., 1981, Sex included, *Development Forum*, **9**(5): 16.

Crick, M., 1989, Representations of international tourism in the social sciences: sun, sex, sights, savings, and servility, *Annual Review of Anthropology*, **18**: 307–44.

Crompton, R., Sanderson, K., 1990, *Gendered jobs and social change*, Unwin Hyman, London.

Davidson, D., 1985, Women in Thailand, *Canadian Women's Studies*, **16**(1): 16–19.

Dilley, R. S., 1986, Tourist brochures and tourist images, *Canadian Geographer*, **30**(1): 59–65.

Ford, N., Koetsawang, S., 1991, The socio-cultural context of the transmission of HIV in Thailand, *Social Science of Medicine*, **33**(4): 405–14.

Foucault, M., 1981, *The history of sexuality*, vol. 1, *An introduction*, Penguin, Harmondsworth.

Gay, J., 1985, The patriotic prostitute, *The Progressive*, **49**(3): 34–6.

Graburn, N., 1983, Tourism and prostitution, *Annals of Tourism Research*, **10**: 437–56.

Hail, J., 1992, Thailand: a new approach, *Asia Travel Trade*, **23**(May): 24–31.

Hall, C. M., 1991a, Sex tourism in south-east Asia, in Harrison, D. (ed.), *Tourism and the less developed countries*, Belhaven, London, pp. 64–74.

Hall, C. M., 1991b, *Introduction to tourism in Australia: impacts, planning and development*, Longman Cheshire, South Melbourne.

Hall, C. M., 1993, *Introduction to tourism in the Pacific: development, impacts and markets*, Longman Cheshire, South Melbourne.

Harrison, D., 1991, International tourism and the less developed countries: the background, in Harrison, D. (ed.), *Tourism and the less developed countries*, Belhaven, London, pp. 1–18.

Hawkesworth, M., 1984, Brothels and betrayal: on the functions of prostitution, *International Journal of Women's Studies*, 7(1): 81–91.

Hicks, G., 1993, Ghosts gathering: comfort women issue haunts Tokyo as pressure mounts, *Far Eastern Economic Review*, 18 February: 32–6.

Hong, E., 1985, *See the third world while it lasts*, Consumers Association of Penang, Penang.

ISIS, 1979, *Tourism and prostitution, International Bulletin*, 13, ISIS, Geneva.

ISIS, 1984, Prostitution: who pays?, *Women's World*, 3: 4–5.

Kelly, N., 1991, Counting the cost, *Far Eastern Economic Review*, 18 July: 44.

Kikue, T., 1979, Kisaeng tourism, in ISIS, *Tourism and prostitution, ISIS International Bulletin*, 13: 23–6.

Korea Church Women United, 1983, *Kisaeng tourism, a nation-wide survey report on conditions in four areas Seoul, Pusan, Cheju, Kyongju*, Research Material Issue No. 3, Korea Church Women United, Seoul.

Mathieson, A., Wall, G., 1982, *Tourism: economic, physical and social impacts*, Longman, London.

Matsui, Y., 1987a, The prostitution areas in Asia: an experience, *Women in a Changing World*, 24(November): 27–32.

Matsui, Y., 1987b, Japan in the context of the militarisation of Asia, *Women in a Changing World*, 24(November): 7–8.

Mingmongkol, S., 1981, Official blessings for the 'brothel of Asia', *Southeast Asia Chronicle*, 78: 24–5.

Naibavu, T., Schutz, B., 1974, Prostitution: problem or profitable industry?, *Pacific Perspective*, 3(1): 59–68.

O'Malley, J., 1988, Sex tourism and women's status in Thailand, *Loisir et Société*, 11(1): 99–114.

Ong, A., 1985, Industrialisation and prostitution in southeast Asia, *Southeast Asia Chronicle*, 96: 2–6.

Philippine Women's Research Collective, 1985, *Filipinas for sale: an alternative Philippine report on women and tourism*, Philippine Women's Research Collective, Quezon City.

Phongpaichit, P., 1981, Bangkok masseuses: holding up the family sky, *Southeast Asia Chronicle*, 78: 15–23.

Phongpaichit, P., 1982, *From peasant girls to Bangkok masseuses*, International Labour Office, Geneva.

Phongpaichit, P., 1988, Two roads to the factory: industrialisation strategies and women's employment in south-east Asia, in Agarwal, B. (ed.), *Structures of patriarchy: state, community and household in modernising Asia*, Zed Books, London, pp. 151–63.

Richter, L. K., 1989, *The politics of tourism in Asia*, University of Hawaii Press, Hawaii.

Robinson, G., 1989, AIDS fear triggers Thai action, *Asia Travel Trade*, 21(September): 11.

Rogers, J. R., 1989, Clear links: tourism and child prostitution, *Contours*, 4(2): 20–2.

Seager, J., Olson, A., 1986, *Women in the world: an international atlas*, Pan, London.

Sentfleben, W., 1986, Tourism, hot spring resorts and sexual entertainment, observations from northern Taiwan – a study in social geography, *Philippine Geographical Journal*, 30: 21–41.

Skrobanek, S., 1983, The transnational sex-exploitation of Thai women, MA thesis, Institute of Social Studies, The Hague.

Symanski, R., 1981, *The immoral landscape: female prostitution in western societies*, Butterworths, Toronto.

Taylor, D., 1984, Cheap thrills, *New Internationalist*, 142: 14.

Thanh-Dam, T., 1983, The dynamics of sex-tourism: the cases of South-east Asia, *Development and Change*, 14(4): 533–53.

Turner, L., and Ash, J., 1975, *The golden hordes: international tourism and the pleasure periphery*, Constable, London.

Wereld, O., 1979, Sex tourism to Thailand, in ISIS, *Tourism and prostitution, International Bulletin*, 13: 9–12.

Wihtol, R., 1982, Hospitality girls in the Manila tourist belt, *Philippine Journal of Industrial Relations*, 4(1–2): 18–42.

Yoyori, M., 1977, Sexual slavery in Korea, *Frontiers, A Journal of Women's Studies*, 2(1): 76.

14

INTERPRETATIONS OF TOURISM AS COMMODITY*

G. Llewellyn Watson and Joseph P. Kopachevsky

INTRODUCTION

Tourism is a complex sociocultural experience, one that cannot be properly understood except in relationship to other units of the whole in human society: the family, the economy, class structure, ideological constructions, the physical environment, and most importantly, in terms of what MacCannell (1976) labels a "semiotic of capitalist production" and what Baudrillard (1981) calls "the political economy of the sign."

It is inaccurate to claim that "sociological attempts to provide accounts of tourism have so far produced little work that is interesting" (Allock 1988: 35). Tourism is undertheorized, but it is not atheoretical. Even a less-than-exhaustive review of the existing literature on tourism attests to the fact that social scientists have, over a considerable period of time, been seriously engaged in the development of theoretical frameworks and research strategies appropriate to the systematic study of tourism.

In this respect, Cohen (1972, 1979a, 1979b, 1988b), Redfoot (1984), Pearce (1978), Smith (1977/1989), and Hamilton-Smith (1987), among others, have all developed typologies of tourists and tourism that are interesting and heuristic, and which at once reveal that there is no such thing as a universal tourist, or a universal tourist experience, but rather numerous varieties of tourists and experiences. As well, such works show that tourism is a complex phenomenon implicating in social relational sense many spheres of social life. These efforts are further complemented by Jafari's (1990) typology of tourism literature.

Similarly, one can find conceptualizations of tourism as a sacred crusade or pilgrimage for authenticity (MacCannell 1976); a sacred journey or pilgrimage (Allock 1988; Graburn 1977, 1983); play (Cohen 1985; Graburn 1977); a form of colonialism and friendly conquest (Krippendorf 1989); a type of ethnic relations (van den Berghe 1980); a form of imperialism (Crick 1989; Harrison 1985; Nash 1977); an acculturation process (Cohen 1984); cultural commodification (Greenwood 1977); an agent of social change (Greenwood 1972); a form of migration (Cohen 1972); a form of leisure (Nash 1981); and so on. Finally, MacCannell's (1976) adaptation of

* Reproduced with kind permission of Elsevier Science Ltd, Pergamon Imprint, Oxford, England, from *Annals of Tourism Research* 1994, 21: 643–60.

Goffman's dramaturgy to tourism, his semiotics, not to mention his description of different types of tourist space, are all inspiring for tourism studies.

The diversity and richness of these conceptions alone provide compelling testament that tourism has occupied scholars over a considerable period. Even so, some who have longstanding involvement in tourism research acknowledge that this recent attention given the subject notwithstanding, the following are true: "differences in perspective lead to alternative ways of conceptualizing tourism" (Dann and Cohen 1991); research efforts in this field, while growing, are still small (Jafari 1990: 39); tourism was, until recently, "neglected as a social phenomenon" (Cohen 1972: 164); and "research on tourism by social scientists has not so far been very sophisticated theoretically" (Nash 1992: 222).

This brief essay aims to extend, and perhaps enliven, the discussion of one significant aspect of tourist relations: commodification. Adopting a macro-sociological approach, this essay offers a set of sociological interpretations, and a provisional exploration of tourism as one activity within modern consumer culture that is shaped by the essential logic of capitalism, namely commodification. This paper is not so concerned, then, with the actual process of tourism, as with the ways in which the tourist pursues multiple commodities, including behavior, and experience-as-commodities, while each in turn is regarded as a commodity.

INTERPRETATIONS OF TOURISM COMMODITY

The study of tourist behavior is paradigmatic, and is indeed an integral part of the study of consumer behavior and consumer culture as new configurations of capitalist modernity (Featherstone 1991). As in the modern culture of consumption generally, touristic consumption is "sign-driven" and media-driven, subject to the dictates of commodity exchange and consumption patterns.

Since the discussion in this paper is couched in the theoretical context of societal structures, and in terms of interpreting tourism as commodity, some attention is given to insights derived from Marx, not because Marx was a student of tourism, but because he was "the first to trace the emergence and historical development of the commodity form as the structuring principle of capitalist society" (Best 1989: 23). This paper understands tourism as an extension of the commodification of modern social life.

But this position does not share totally the overgeneralized view that the modern tourist is a metaphor for the shallowness and inauthenticity that are endemic in the commoditized inauthentic quality of modern life (Boorstin 1964; Redfoot 1984: 291, 303). What is recognized is a complex social process whereby with the growing commodification of modern life, "leisure time," which is proclaimed and expected to be an escape from routine work, in turn often becomes another routinized, packaged, commodity, thereby failing to be anything like a carefree, relaxed alternative to work. As will be argued, with the growing commodification of symbolic forms, and their virtual incorporation into, and monopolization by, mass communication, tourism as social behavior becomes more and more dislodged from the spontaneity and free choice that is regularly assumed to define the tourist and tourism experience.

The analysis that is briefly sketched in this paper sheds light on the common claim that in modernity people's lives seem to be a series of traps spun by forces over which they have little control (Mills 1959); and the more recent claim that the contemporary *tourist* is "the quintessential, if tragic modern person searching for authenticity in an increasingly meaningless world" (Hummon 1988:181), a sentiment echoed in Baudrillard's aphorism (cited in Featherstone 1987) that "we live everywhere already in an aesthetic hallucination of reality."

One aim of this essay, then, is to identify in what ways tourism confirms some of these descriptions, and to critically assess, or at least put another interpretation on, the conventional wisdom that the tourism journey represents that segment of tourists' lives over which they have maximum choice and control. When commodification is introduced into the discussion of tourism, it seems less convincing that, as a Graburn (1983:13) puts it, "tourism is that epitome of freedom and personal choice characteristic of Western individualism."

Commodification

Commodification is the process by which objects and activities come to be evaluated *primarily* in terms of their *exchange value* in the context of trade (Best 1989; Cohen 1988a: 380; Dupré 1983), in addition to any *use-value* that such commodities might have. This primacy of exchange conveys the gist of Marx's original formulation on the subject. But, as will be seen, when one conceptualizes tourism as commodity, this meaning has to be extended to include at least one other value as well: sign-value.

It was Marx who, in the first volume of *Capital*, outlined that a commodity has two qualities, or what he termed "powers": the power to satisfy some human or material need (use-value); and the purely symbolic power to command other commodities in exchange (exchange-value). In discussing these qualities, Marx commented that "a commodity is a mysterious thing, simply because in it the social character of men's labor appears to them as an objective character stamped upon the product of that labor" (Marx n.d.: 76–77). Exchanging products as commodities is thus simply another method of exchanging labor, the "exchange-value" being expressed in a single abstract equivalent, money. All these exchanges help to veil the social roots of the commodity.

According to this conception, the mystery of the commodity derives from its *hidden* form; and what is hidden is the very thing that determines its value (namely, human labor). Three key points immediately come into focus at this stage of the discussion. First, all commodities are the products of human *labor*. As Marx put it, all commodities are only definite masses of congealed labor-time; "every commodity contains the same kind of labor, human labor in the abstract." Second, human labor is the original source of the value that products have, and that which continues to influence added value as the commodity is serially traded. Third, such products become *commodities* the moment they are introduced into the flow of trade, and are exchanged (or are exchangeable) for other commodities. Among modern "products," one must now include human *services* that facilitate the abstract exchange of commodities. In this context, tourism experience is among such "products"; "experience" and their delivery become commodities. Other services that exemplify this exchange are those that occur on the truly mysterious social institutions of capitalism (stock exchanges) which perfect the *abstract* exchange of major commodities. Similar exchanges are facilitated in tourism commodities: transportation, hospitality, entertainment, even gazing.

Further, commodification is thus synonymous with all modern societies, not just capitalist ones, since all human societies must produce their own material conditions of existence. Only under capitalism does commodity production *take primacy* to the degree where the worker also becomes a commodity. But in the post-barter condition of the present day, almost all human products and efforts have been rendered into commodities. Thus, one can just as easily buy tourism souvenirs and other commodities in Moscow's huge department store in Red Square, in the bazaros of Havana, as well as in Bloomingdale's in New York or Harrod's in London, or be exposed to multiple symbolic signs in any of those cities. The crucial difference between the first two and the latter is which social group in the society in question effectively controls and manipulates commodity exchange and the political economy of the sign. The question, in short, turns on who has the freedom to commoditize and manipulate the signs and images which are so central to modern consumer culture. This question is addressed later in this paper.

MacCannell (1976) has provided important clues to, and has gone one better than Marx in, broadening the concept of commodity for today. For, according to Mac-Cannell, all tourism attractions are cultural experiences. As culture commoditized, the value of such experiences has nothing to do with Marx's labor theory of value, but rather is a function of the quantity and quality of experience they promise. Thus, beyond Marx's simple dichotomy between use-value and exchange-value, MacCannell, anticipating Baudrillard (1981), introduces one other type of value embedded in modern commodities: sign-value. This new value is a manufactured signifier. Tourism, as commodity, is best grasped (especially within capitalism) as an expression of the "semiotics of capitalist production" (MacCannell 1976: 19–23; cf. Culler 1981).

Appadurai (1986), too, has introduced some crucial considerations into the analysis of commodities that show not only how complex this subject is, but also the quasi-chameleonic nature of the commodity. Thus, given contexts (such as dealings with strangers, or auctions, or bazaar settings) can serve to alter the flow of commodities in ways that may well be considered atypical. In Appadurai's words, "commoditization lies at the complex intersection of temporal, cultural and social factors" (1986: 15).

The theoretical elaborations by MacCannell and Appadurai force one to recognize that people now live in a world in which tourism and tourist experience are major components. Such a world is one in which image, advertising, and consumerism – as framed by style, taste, travel, "designerism," and leisure – take primacy over production *per se*, and in which commoditization is shaped and honed by specific, influential groups in society utilizing a mixture of social, cultural, and political resources. Indeed, Rojek (1985) suggests that, given the complex relationship between base and superstructure today, leisure activity seems to include social relations that have little to do with the economic base. Introducing the idea of sign-value then takes researchers closer to the reality of the current period, where modern men and women seem far more fascinated by, and interested in, the "spectacle" – the chaotic flow of signs and simulated images so carefully purveyed by the mass media – than by any supposed use-value of commodities.

Commodity fetishism and reification

Given the "powers" and mystery inherent in commodities, according to Marx, certain implications follow with respect to social relations. He, therefore, developed the concept of commodity fetishism and the economic valorization of symbolic forms, in order to demonstrate the curious fact that as soon as human labor takes on the commodity form, it displays a variety of "theological whimsies," and a certain amount of "magic" and necromancy surrounds the products of labor (Marx n.d.: 80–81). This happens simply because when products are rendered into commodities, their abstract being is symbolized in the medium of pure exchange. In assigning symbolic attributes to objects, one endows them with a life of their own, and, as in religious fetishism, the objects appear, and are regarded, as things having an existence and control over and above human choice. This is the meaning of reification, clearly implicit in Marx's discussion of commodity fetishism, and subsequently elaborated by "critical humanists" such as Lukács (who actually provided the term), Jean-Paul Sartre, Karl Korsch, and the Frankfurt School.

The concept of the commodity, therefore, cuts across the phenomenon of reification. Indeed, Lukács' conception of reification can best be initially grasped as a commodification theory in programmatic form. Reification is a process whereby the structures and practices intentionally built, piece by interlocking piece, are subsequently "externalized" so that one begins to act as if, and genuinely comes to believe, that such structures are unchangeable, natural, and inevitable. Under such conditions – the conditions of capitalist consumerism – the social relations people enter into appear as things governed by natural laws and eternal verities, as inevitable rather than historically contingent, thus making it difficult to recognize human agency and deeds. This condition not only affects consciousness, but social theory as well, and explains why it is often said that all reification is a forgetting. Lukács argued that reification was the immediate reality that dominated capitalist culture as an ideological phenomenon (1971: 93–94), related to the commodity form and ultimately to the preservation of the status quo. It amounts to the naturalization of the present.

Any serious discussion of tourism as commodity takes one directly into this complex social process, and reveals that reification is a way people have of manufacturing necessity: real social relations between people in time mimic the socially necessary appearance of relations between things. Briefly, reified social relations turn on a dynamic in which the *social* character of activity, objects, and experience are transmuted into "things," or as Marx put it "a fantastic form of a relation between things" (n.d.: 76–77) that seems to have non-social properties, and putative mysterious powers. This apparent fetishism is a form of mystification. It effectively disguises human relations as object relationships (Gouldner 1985: 244) and encourages a submission to an abstract entity, not unlike the symbolic projection associated with the anthropology of religion.

The religious metaphor helps to interpret how, in the world of commodities, a serious cultural rift occurs between the subjective agent and his/her products, so that the human agent experiences a forceful, external domination by *by the created products* of his/her hands. In brief, the commodity "appears as personified expressions of human characteristics and relationships" and "the productions of the human brain appear as independent beings endowed with life" (Kline and Leiss 1978; Marx n.d.: 77). It is clear that many tourism attractions are analogous to the religious symbolism of, in Durkheim's phrase, "the religious life," where believers themselves assign sacredness to ordinary objects and or experience, then willingly submit themselves to their own designations, attributing to them supra-human powers (Durkheim 1912/1965). It is equally clear that what Urry (1990) refers to as the "tourist gaze" involves the symbolic transmutation of many ordinary objects, places, and experiences into sacred ones.

Therefore, briefly, Marx recognized that the commodity was a deceptively simple sort of thing which necessarily veils itself with what are illusions, not least because for it to exchange, the labor which went into its actualization has to be stripped of its social and historical character. It thus *appears* as an abstract, but powerful symbol, mystified by the exchange process, to eventually assume the fetishistic appearance of a thing that gives one no clues as to its social roots.

Sign manipulation

Tourism, like other commodities, is packaged for exchange by advertising, much of which appeals to people's deepest wants, desires and fantasies (often sexual), and is anchored in a dynamic of sign/image construction/manipulation. As an integral part of modern culture, advertising's main function is twofold: one, to serve as a discourse about objects, symbols and ideas, and as a template for erecting monuments to consumption and self-indulgence; and, two, to persuade people that only in consumption can they find not only satisfaction, but also mental and physical health, social status, happiness, rest, regeneration and contentment (Goldman and Wilson 1983; Leiss and Jhally 1986). It is a powerful symbolic message, if totally vicarious, and in this regard tourism attractions and commoditized experiences are clearly analogous to many types of religious symbolism that make claim to aid well-being. True valorization, rife in tourist activities, occurs when symbolic forms are ascribed a "value" to become, in Durkheim's terms, external and constraining.

Observing modern tourism, especially mass tourism, provides content for many of these points. While modern tourists typically hanker for leisure as a respite from deeply alienating, routine work, and presumably as an effective counterbalance to the pressures of work, in the very nature of the commodification of modern tourism, the sense of alienation seems to follow the travelers into their touristic adventures.

The alienation takes the form of restricted choices framed by a "staged" consumption pattern – a fetishism if you will – in which the singular theme of possession and display takes priority over all other social behavior.

This apparent tyranny of the commodity is most readily revealed in the type of tourism in North America and elsewhere that is prepackaged, such as in bus tours, ski packages, and shopping tours. The buses follow a tightly-scheduled set of stops at pre-contracted points that offer maximum opportunity for routinized activities such as consumption at "factory outlets" (a shibboleth for factory "seconds") and of souvenirs and "gazes." The stops are meant to facilitate commodity exchange as part of the very definition of the tourists' travel and experience. It is, nonetheless, a predefined experience, severely limited and constrained by the monetary dictates of other people whose prime reward is not the extrinsic experience of "fun," but the intrinsic reward of making money.

Clearly, when one speaks of commodification in tourism, this must be in terms much broader than mere objects, and include services, activities, and experiences. Thus, what Urry (1990) calls the "tourist gaze" should be regarded as a tourist experience that includes the "consumption" of signs, symbols, cultural experiences (some purely artificial), *and* the actual "gaze," a subset of the tourism experience. As some leisure research specialists have noted, for many touristic experiences, a "great vacation" is going to a place where opportunities to spend money and watch others perform dominate their time. A good time, it is suggested, has come to be associated with spending and consumption, and pleasure may be measured by spending (Goldman 1984; Kelly 1986: 460).

In the process of commodification, human relations become objectified as relations between *things*, and money is hoisted as the universal "doubly abstract" medium of exchange, the primary measure of value, if not the symbol of general alienation, having the chief function, as Marx put it, to supply commodities with the material for the expression of their values – the commodity of commodities. Simmel, like Marx, developed the insight that money constitutes a general equivalent of all values, the purest form of the tool (Simmel 1907/1978: 210). Simmel concluded that the money economy, coupled with the extensive division of labor of the modern period, transforms cultural forms into external, autonomous *things* or objects. For him, this was to be interpreted as the fragmentation of modern life, and alienation from one's own capacity to transcend contingent reality, a trend symptomatic of the eventual objectification of social relationships (Simmel 1907/1978: 296–297, 448–461). This key insight into the sociology of culture clearly has an affinity with Lukács' studies of reification. Ironically, modern sociology lacks a theory of money, except for Parsons' many isolated references to money (1961, 1967, 1975), references which, however, do not constitute a full-fledged theory of money, and which have not been elaborated upon by other scholars.

But in tourism as in other areas of social experience, consumers are structurally entwined in the seemingly interminable process that begins and ends with the same thing: money and exchange-value (Best 1989: 26). The fetishism of commodities clearly constitutes a reproductive political system operating under conditions of what can best be termed the terror of consumerism, and, as analyzed by Marx, what was only one instance of the fate of all cultural contents, corrupted and homogenized in their subsumption to the rule of exchange-value (Best 1989: 28).

In all of this, the "secret" of the commodity is played out through imagistic advertising that specializes in reproducing and expanding the commodity sign as well as the commodity form. Modern advertising, clearly, is the most powerful instrument of commodification, creating its own mystery in *form*, not text. In a class society, some groups, and not all groups, are well positioned to ascribe symbolic value to experience. Mass advertising is a private tool of the powerful class labeled the "captains of consciousness" (Ewen 1976). With it they manipulate the commodity sign and all commodities in the interest of profit maximization.

The culture of capitalist modernity is driven by mass consumption, which in turn relies on mass advertising for interjection of life-style imageries and an individualistic, stylistic self-consciousness. Leiss (1983) has shown that the pattern of advertising in Canada indicates a marked decrease in product information, and an increase in imagery and form. The same pattern is evident in advertising targeted for potential tourists, for example in both the print and electronic media. This same pattern holds true for other societies approaching "post modernity." At the end of the system of privatization and administered consumerism is what can best be described as a new form of tyranny, the "inevitable consequence of the structural control over one's life, work, consumption, lifestyle, and social relations" (Langman and Kaplan 1981: 90) and, one might add, one's leisure, including "staged" sports. As Young has argued (1986: 19), commodity sports present people with a modern metaphysic for daily life. It redeems, in a false and trivial manner, alienated conditions of work. It provides alienated solidarity in a conflict-ridden society. Its super-masculine model of play offers to redeem an alienated sexuality; and its aesthetics and metaphysics provide an envelope into which to insert a message vesting desire into possession of material goods rather than in primary social relations.

The social relations of tourism consumerism

With large scale commodification in tourism, as in other aspects of consumer behavior, those who have the economic and ideological power to shape institutions and consciousness to suit their ends also succeed in one other way. They are able to structure and standardize taste, and codify value priorities both in work and in leisure. Indeed, one of the major theses of scholars who analyze leisure from a neo-Marxist perspective, including the Frankfurt School, is that leisure relations are relations of power, and that mass leisure can properly be regarded as one of the main dynamics of capitalist ideological productions, part of the culture industry's technique of mass deception, social acquiescence, and passive conformity (Kellner 1979; Kelly 1986; Rojek 1985, 1987; Young 1986).

Mass tourism can also be interpreted as relations of power, and is equally susceptible to the standardization and homogenization of all massified experiences, including both at the ideological level and at the level of consumption. To the degree that this happens, individual thought, choice, and action are destroyed. Thus, as Cohen notes, caught in the staged "tourist space" from which there is no exit, modern mass tourists are denied access to the back regions of the host society where genuine authenticity can be found, and are presented instead with "false backs" (Cohen 1984: 373, 1988a: 373). Thus, MacCannell (1973, 1976) may very well be right in claiming that the modern tourist is a pilgrim in search of authenticity. But in reality the desire for such authenticity is severely thwarted by and through the antics of commodification and the consumer culture.

Interestingly, MacCannell admits that "a tourist experience is always mystified." He also points to the fact that the tourist rarely, if at all, achieves the sought after experience of authenticity, a fact that has less to do with the quality of the actual quest, and more with the manipulations undertaken by many tourism establishments. By conceptualizing some tourist relations as asymmetrical relations of domination, one can begin to recognize how the tourist is attracted by the semiotics that he/she does not control and cannot manipulate at will, and in the end seems damned to varying degrees of inauthenticity both in objects and in experience. For the latter, Greenwood (1977: 136–137) has shown how commodification of local culture in the Basque region of Spain amounted to a perversion of the local people's cultural rights; and Turner and Ash (1975: 197) insist that tourism is inimical to the integrity of cultural identity.

None of this is necessarily condemning tourism offhand, or claiming that it is irredeemably malevolent; neither is commodification *per se*. There are, in fact, studies that disclose counter-examples (Deitch 1977; McKean 1977) indicating that tourism is full of contradictory possibilities and potentials. Whether or not a given tourist situation or experience leads to cultural impoverishment and underdevelopment, or their opposites, is an empirical question, to be settled by actual, concrete investigation of people's lived experiences. In any event, the question of tourism being potentially a "special form of subversion" and a "kind of friendly conquest" (Krippendorf 1989: 55, 56) cannot be summarily dismissed.

The problem of mass trinketization, debased imitations of local culture, and the cheap commercialization of culture for touristic consumption has long been of interest to anthropologists. Graburn (1976), for instance, argued that tourists, though not as a deliberate objective, encourage a market for junky, inexpensive, often mediocre souvenir art forms. But Graburn also recognized the built-in dialectic of touristic folk art and craft. They represent a "mixed blessing" for the craftspeople and their culture generally, for while the labor expended to produce such items is potentially lucrative, the more the artists/craftspeople, out of necessity, cater only to tourists, the more the artists tend to remove those elements of style and content that might offend the unknown buyer. If that happens, the artisan "finds himself in danger of surrendering control of his product" (Graburn 1976: 32).

This dilemma is typically compounded by the availability of cheap, flimsy, *imported* simulations of native artifacts, plus other paraphernalia that make portable souvenirs. Most popular tourist destinations – Atlantic Canada, Disney World, the Caribbean, the Far East – abound with such commodities. Their appeal to tourists lies in the fact that they are cheap, because they are mass-produced. They also symbolize the tourists' experience; when bought and taken home for self or friends, they indeed are tangible evidences of travel, a representation of a gaze or experience. By virtue of the exotica associated with the touristic trip, together with consumer ethic, nothing short of a duty to acquire such commodities prevails, underlining once more the "mysterious power" of the commodity.

In tourism research, the relationship between people and things takes on special significance, for as Csikszentmihalyi and Rochberg-Halton have argued in their fascinating book *The Meaning of Things* (1981), things, even simple things, like souvenirs that simulate the real, carry symbolic meaning as signs of status. Some objects express qualities of the self. In some cultures, this could be represented by ownership of a hunting spear; other objects confer status (e.g. money – the most abstract form of status symbols); and still others operate as symbols of social integration (e.g. totemic items as explained in Durkheim's classic study of religion). They are, of course, crucial as a medium of gifting, where "what counts" is not necessarily the use-value but the sentiment attached. The point is, human beings can and do cultivate the meaning that objects, not to mention service and experience, eventually acquire. But the meaning is attached at the level of exchange, not at the level of production. This is what gives artifacts bought in "holy" places, such as Bethlehem on the West Bank, their multiple values: sign-value, exchange-value, and use-value. For all the average consumer cares, they may all have been produced by hard-core atheists.

The asymmetry in relations

Interpreting tourism as commodity discloses that it is a structure of social relationships that creates two categories of people: those who demand, and those who supply – those who serve, and those who are served. As such, it is not hard to see how feelings of superiority and inferiority develop in tourism relationships, and why it is the locals who more often than not must adapt to tourists' wishes, demands, and values, and not the other way around (Krippendorf 1987: 45, 101–103). Such relationships need not be, but they are, asymmetrical. The closer one looks at tourism as a social activity, the clearer it appears that it inherently creates dependency relationships (de Kadt 1979; de Vries 1978, 1981; Harrison 1985; Johnston 1990; Krippendorf 1987; Ritchie and Goeldner 1987). It also became clearer that working under the tourist gaze (Urry 1990: 68–81) places particular strains on the social relations between hosts and guests, if not between people and things.

Part of the strain stems from the fact that tourism involves simultaneously certain measures of nearness and distance, making it the archetypal social relationship that displays structural ambivalence. This nearness/distance dichotomy is provided in Simmel's classic essay on "the stranger" who, as Simmel says, may develop all kinds of charm and significance. But as long as he/she is considered a stranger in the eyes of the other, he/she is not an "owner of soil." Thus, the *special* proportion of nearness and distance involved in stranger/other relationships, and some reciprocal tension, produce the particular relation to the tourist/stranger (Simmel 1950). In Simmelian terms, the tourist is the quintessential stranger, and it is this status, together with the ephemerality of tourists' sojourns, which make relationships with them so different from other types.

Tourism provides for special types of social relationships that are a consequence (result) of "forced" interaction between host and guest, where the guest is a "stranger," and the interaction is predicated on the principle of commodification. These social relations are hinted at in the interactions described in common phrases, such as "to make the tourist feel at home" and "providing good service." But this situation is further compounded by the existence of other questionable tourism commodities, the type usually available in abundance in the shops in tourist areas and at airport shops. As these objects are mass-produced (typically in Hong Kong, Taiwan, Korea) and imported for the tourism market, local craftspeople and artists are placed at an economic if not a cultural disadvantage, which in turn adds to the strains in hosts/guests relationships.

The traditionally strong cultural obligation to purchase and possess such ostensibly fake commodities attests to the element of social make-believe in what Baudrillard (1975) calls the "postmodern situation." Such "playacting" is part of the larger fetishism of commodities in daily life. This is why MacCannell (1973) is able to argue that pseudo-events result from the social relations of tourism to the degree that such relations are premised on commodity as the key unit of capitalist society. Likewise, MacCannell can speak of "alienated leisure" (1976: 57–76, 1973: 58) because, in the modern machine-directed, commodity-driven consumer society, a psychological and anthropological distance is created between the producer and the product (Best 1989). Accordingly, alienated consumers tend to regard the purchase of commodities or the act of purchasing them as an end in itself. Tourists are not exempt from this lure. Nowadays, the culture of compulsive consumerism is so well heeled that the promotion of consumption as a way of life is generally taken as the cure for the spiritual desolation of modern life.

This type of consumerism, after all, is what the hedonistic, commercialized orgy called the "Spirit of Christmas Shopping" is all about, leading these writers' colleague Thomas Trenton to suggest that "Christmas is now the celebration of capitalism." To get into the spirit is to accept the suspension of one's critical judgment and voluntarily submit to the power and seductive influence of multiple commodity exchanges, whence the freedom to consume a plurality of images and objects is equated with freedom itself. This measure of alienation, ironically, is what makes tourism such good business, economically. The standardized consumption of mass-produced commodities that has come to characterize these times effectively mystifies class relations and brings about a seemingly natural identity of interests between the different social classes. But it also helps to create the conditions for the extension of the market for commodities, revealing another of Marx's *aperçu*: "production is also immediately consumption" (Marx 1973: 90, 91).

The reproduction of social life

Relations of power hold the potential for social domination. Certain social relationships, cultural manipulations, and ideological reifications shape individual character structure and collective consciousness to insure the continued dominance of the capitalist system (Langman and Kaplan 1981: 87). The end result of such manipulations of human desire, best illustrated in mass mediatized consumer messages and signs, is to project all needs and desires and social experience onto the cosmos of commodities.

Tourist behavior is no exception to the logic of capitalism. It is subject to the same cultural logic of late capitalism, and to the structural imperatives that regulate political economy for the purpose of the reproduction of the capitalist system. The private interests also safeguard the legitimizing ideologies (Featherstone 1991; Kellner 1988; Thompson 1990), including the ideology that particular forms of tourism are part of the very definition of modernity, even if that modernity is conspicuously fragmented, wasteful, alienating, and superficial. The psychological and social consequences of a culture of consumerism are premised, on the one hand, on a revamped individualism (as revealed in the free-wheeling hedonism and conspicuous consumption of many tourists) and, on the other hand, by a routinization and conformity that clearly detracts from and undermines genuine personal choice or self-actualization in the use of leisure time.

Mass touristic activities are definitely constrained by the prepackaging and staging of activities and experiences, because profitable commodification requires the standardization of inputs, production, clients, and consumption. This is played out daily in "tourism spaces" such as Disney World and the Epcot Center in Florida. Observing the thousands who queue for miles in roped-off trails waiting (sometimes for hours) to gain entrance to highly-staged attractions, one is bound to ponder to what extent the tourists are exercising genuine self-realization, personal autonomy, and freedom from external controls. There is something of a paradox. Such activities are voluntarily entered into and are hobbled by the dictates of mass commodification and consumerism. In MacCannell's terms, such activities constitute "alienated leisure" in which people lose control over the context in which their non-work time is socially structured.

What develops in tourism consumption, as in other areas of life, is a consumer culture administered, in Marcuse's (1965) telling phrase, by "repressive tolerance," that amount of ideational domination and mystification necessary to ensure the "dull compulsion of economic relations." There is yet another paradox here, for capitalist culture contains the clear message that, as Urry (1990: 5) notes, if people in modern society do not travel, they lose status; travel is the marker of status. Yet, tourist brochures and other media messages subtly encourage people to voluntarily surrender spontaneity, allowing more and more of the tourism experience to fall under the spell of commodification.

The often unrecognized consequence of all this is a form of internalized domination that preserves capitalism and its overriding aim: commodification for the maximization of private profit. Given the fact that one of capitalism's distinctive features is its ability to take on deceptive phenomenal forms that veil its social relationships, few in capitalist consumer culture actually comprehend that entrenched consumerism constitutes domination at two levels: the ideological–structural level and the psychological–interactional or individual level. While an antiemic standpoint is not adopted here, the fact remains that consumer culture proceeds through people's obliviousness to the reality that it privatizes the many social experiences of daily life, so that they are never perceived as *social*. This is why, as Langman and Kaplan put it, the mystified nature of class relations under late capitalism makes it easy for a small minority to achieve great material wealth and comfort while the majority of the population considers this inequality legitimate (1981: 88, 90).

Similarly, Thurot and Thurot (1983: 178 and *passim*) argue that the style and imagery of French tourism advertising is symptomatic of structural asymmetries, notably the ideological conflict between social classes. Perhaps, then, modern tourism, while not always matching Boorstin's description as frivolous activity, pseudo-events of capitalist society, nevertheless confirms Max Weber's melancholy conclusion that in modern society, a "highly reduced view of freedom and happiness was the most that intelligent people could expect from life" (Rojek 1985: 61).

Tourism, by its very nature, is shaped by a very complex pattern of symbolic valuation; and this takes place in a structured social context over which tourists themselves have no immediate control. The essence of modern capitalism is the remanufacture of images, many of which effectively obscure the injuries of class, race, and sex. This, at the same time, entrenches mythologies and dominant ideologies that legitimize and naturalize social structure. Research and analysis undertaken by students of culture, mass media, and ideology amply confirm this proposition (Gouldner 1985; Kellner 1979, 1988; Thompson 1990) as an integral part of the increasing commodification of reality at all levels in modern capitalist societies.

The question of what has been called "commodity-sign" (Baudrillard 1981; Featherstone 1991: 83–94, 1990: 7, 1987: 57–58; Goldman and Wilson 1983: 122; MacCannell 1976), or product symbolism, is germane to the discussion of tourist behavior, because perhaps in no other type of consumer behavior is the attribution of symbolic meaning to products more pronounced. From the glamorous and romanticized advertising of the tourism vacation itself, to the return from the trip with trinkets, mementos and artifacts, it is the symbolism embedded in many of these products, experiences, and their delivery that constitutes their appeal, and that furnishes strong justification for their purchase and consumption as touristic commodities. Those who sell such symbolically-endowed commodities can purvey fakeness without compunction, since what is provided is "psychological utility" (Goldman and Wilson 1983: 122), not the practical utility of the goods. What is being sold is not just the direct use of the commodity, but its symbolic significance as a particular ingredient of a cohesive life-style.

The structural determinants of the tourism experience are such that most modern tourists, and certainly mass tourists, do not, on their touristic pursuits, exercise as much freedom and personal choice as it is often assumed. There can be little doubt that a good part of what restricts their freedom to structure their free time as they wish can be identified in the actual commodification that transforms their experience into a thorough objectivism, thereby curtailing spontaneity, free choice, autonomy, and self-determination. As a result, "a major source of modern terror is the limited ability of anyone to comprehend the complexity of the division of labor" (Best 1989: 44; Langman and Kaplan 1981: 106). What one must seek to grasp are the ways in which those who labor, as well as those who *rest* from their labors, are phenomenologically enmeshed in this new terror. Perhaps a crucial lesson to be learned is that tourism is no exception to the general rule of commodification, a 20th-century process that is nothing if not all-pervasive.

CONCLUSIONS

This essay has raised some theoretical points relevant to the structural understanding of tourism. In particular, it has highlighted the fact that modern tourism is best understood in the context of the commodification process and contemporary consumer culture. It is argued that tourism is an extension of the commodification of modern life, and as such can be structurally grasped only when due consideration is given to the objectifying power of the commodity form. A Marxian ontology of labor, coupled with recent theorizing on the political economy of the sign (or the semiotics of capitalist development), yield a critical phenomenology of men and women. It reveals that as men and women instrumentally produce their objective world, this same world ironically appears alien and tyrannical, not because of its intrinsic objectivity, but because of the very way in which it was produced. The frenzied commodification of images in the modern media contributes to ideological domination and alienated leisure, including many tourism experiences.

Until the corrosive power of modern consumerism is given a critical macro analysis, and until tourism is understood not just in the narrow sense of recreational travel, but in the more general framework of commodification, effective social policies in tourism can scarcely be framed.

REFERENCES

Allock, John B. 1988 Tourism as a Sacred Journey. *Society and Leisure* 11:33–48.
Appadurai, Arjun, ed. 1986 *The Social Life of Things: Commodities in Cultural Perspective*. Cambridge: Cambridge University Press.
Baudrillard, J. 1975 *The Mirror of Production*, St. Louis: Telos Press.
—— 1981 *For a Critique of the Political Economy of the Sign*. St. Louis: Telos Press.
Best, Steven 1989 The Commodification of Reality and the Reality of Commodification: Jean Baudrillard and Post-Modernism. *Current Perspectives in Social Theory* 9:23–51.
Boorstin, D. 1964 *The Image – A Guide to Pseudo-Events in America*. New York: Atheneum.
Cohen, Erik, 1972 Towards a Sociology of International Tourism. *Social Research* 39:164–182.
—— 1979a A Phenomenology of Tourist Experiences. *Sociology* 13(2): 179–201.
—— 1979b Rethinking the Sociology of Tourism. *Annals of Tourism Research* 6:18–35.
—— 1984 The Sociology of Tourism: Approaches, Issues, and Findings. *Annual Review of Sociology* 10:373–392.

—— 1985 Tourism as Play. *Religion* 15: 291–304.

—— 1988a Authenticity and Commoditization in Tourism. *Annals of Tourism Research* 15: 371–386.

—— 1988b Traditions in Qualitative Sociology of Tourism. *Annals of Tourism Research* 15: 29–46.

Crick, Malcolm 1989 Representations of International Tourism in the Social Sciences: Sun, Sex, Sights, Savings and Servility. *Annual Review of Anthropology* 18: 307–44.

Csikszentmihalyi, Mihaly, and Eugene Rochberg-Halton 1981 *The Meaning of Things: Domestic Symbols and the Self*. Cambridge: Cambridge University Press.

Culler, Jonathan 1981 Semiotics of Tourism. *American Journal of Semiotics* 1: 127–140.

Dann, Graham, and Erik Cohen 1991 Sociology and Tourism. *Annals of Tourism Research* 18: 155–169.

Deitch, Lewis I. 1977 The Impact of Tourism upon the Arts and Crafts of the Indians of the Southwestern United States. In *Hosts and Guests: The Anthropology of Tourism*, Valene L. Smith, ed., pp. 173–192. Philadelphia: The University of Pennsylvania Press.

de Kadt, Emanuel 1979 *Tourism: Passport to Development?* New York: Oxford University Press.

de Vries, Pieter 1978 Towards an Anthropology of Tourism. *Canadian Review of Sociology and Anthropology* 15: 478–484.

—— 1981 The Effects of Tourism on Marginalized Agrarian Systems: West Indian Perspectives. *Canadian Journal of Anthropology* 2: 77–83.

Dupré, Louis 1983 *Marx's Social Critique of Culture*. New Haven: Yale University Press.

Durkheim, Emile 1912/1965 *The Elementary Forms of the Religious Life*. New York: The Free Press.

Ewen, Stuart 1976 *Captains of Consciousness: Advertising and the Social Roots of the Consumer Culture*. New York: McGraw-Hill.

Featherstone, Mike 1987 Lifestyle and Consumer Culture. *Theory, Culture and Society* 4: 55–70.

—— 1990 Perspectives on Consumer Culture. *Sociology* 24: 5–22.

—— 1991 *Consumer Culture and Postmodernism*. London: Sage.

Goldman, Robert 1984 We Make Weekends: Leisure and the Commodity Form. *Social Text* 8: 84–103.

Goldman, Robert, and John Wilson 1983 Appearance and Essence: The Commodity form Revealed in Perfume Advertisement. *Current Perspectives in Social Theory* 4: 119–142.

Gouldner, Alvin 1985 *Against Fragmentation: The Origins of Marxism and the Sociology of Intellectuals*. New York: Oxford University Press.

Graburn, Nelson H. H. 1977 Tourism: The Sacred Journey. In *Hosts and Guests: The Anthropology of Tourism*, Valene L. Smith, ed., pp. 17–31. Philadelphia: The University of Pennsylvania Press.

—— 1983 The Anthropology of Tourism. *Annals of Tourism Research* 10: 9–33.

Graburn, Nelson H. H. ed. 1976 *Ethnic and Tourist Arts: Cultural Expressions from the Fourth World*. Berkeley CA: University of California Press.

Greenwood, Davydd 1972 Tourism as an Agent of Change: A Spanish Basque Case. *Ethnology* 11: 80–91.

—— 1977 Culture by the Pound: An Anthropological Perspective on Tourism as Cultural Commoditization. In *Hosts and Guests: The Anthropology of Tourism*, Valene L. Smith, ed., pp. 129–138. Philadelphia: The University of Pennsylvania Press.

Hamilton-Smith, Elery 1987 Four Kinds of Tourism. *Annals of Tourism Review* 14: 332–344.

Harrison, Paul 1985 *Inside the Third World*. Harmondsworth: Penguin.

Hummon, David M. 1988 Tourist Worlds: Tourist Advertising, Ritual and American Culture. *Sociological Quarterly* 10: 179–202.

Jafari, Jafar, 1990 Research and Scholarship: The Basis of Tourism Education. *Journal of Tourism Studies* 1: 33–41.

Johnston, Barbara R. 1990 Breaking out of the Tourist Trap. *Cultural Survival Quarterly* 14: 2–5.

Kellner, Douglas 1979 TV, Ideology and Emancipatory Popular Culture. *Socialist Review* 9(3): 13–53.

—— 1988 *Critical Theory, Marxism and Modernity* Baltimore, MD: The Johns Hopkins University Press.

Kelly, John R. 1986 Commodification of Leisure: Trend or Tract? *Society and Leisure* 9: 455–475.

Kline, Stephen, and William Leiss 1978 Advertising, Needs, and "Commodity Fetishism". *Canadian Journal of Political and Social Theory* 2: 5–30.

Krippendorf, Jost 1989 *The Holiday Makers: Understanding the Impact of Leisure and Travel.* Oxford: Butterworth/Heinemann.

Langner, Lauren, and Leonard Kaplan 1981 Political Economy and Social Character: Terror, Desire and Domination. *Current Perspectives in Social Theory* 2: 87–115.

Leiss, William 1983 The Icons of the Market-Place. *Theory, Culture and Society* 1(3): 10–21.

Leiss, William, and Sut Jhally 1986 *Social Communications in Advertising: Persons, Products and Images of Well-Being.* Toronto: Metheun.

Lukács, Georg 1971 *History and Class Consciousness.* Cambridge, MA: MIT Press.

MacCannell, Dean 1973 Staged Authenticity: Arrangements of Social Space in Tourist Settings. *American Journal of Sociology* 79: 589–603.

—— 1976 *The Tourist: A New Theory of the Leisure Class.* New York: Schocken Books.

Marcuse, Herbert 1965 Repressive Tolerance. In *A Critique of Pure Tolerance*, Robert Paul Wolff, Barrington Moore and Herbert Marcuse, eds., pp. 82–123. Boston: Beacon.

Marx, Karl 1973 *Grundrisse: Foundations of the Critique of Political Economy.* Harmondsworth: Penguin.

—— n.d. *Capital: A Critical Analysis of Capitalist Production* (vol. 1). Moscow: Progress Publishers (original publication, 1887).

McKean, Philip Frick 1977 Towards a Theoretical Analysis of Tourism: Economic Dualism and Cultural Involution in Bali. In *Hosts and Guests: The Anthropology of Tourism*, Valene L. Smith, ed., pp. 93–107. Philadelphia: The University of Pennsylvania Press.

Mills, C. Wright 1959 *The Sociological Imagination.* New York: Oxford University Press.

Nash, Dennison 1977 Tourism as a Form of Imperialism. In *Hosts and Guests: The Anthropology of Tourism*, Valene L. Smith, ed., pp. 33–47. Philaldelphia: The University of Pennsylvania Press.

—— 1981 Tourism as An Anthropological Subject. *Current Anthropology* 27: 461–481.

—— 1992 Epilogue: A Research Agenda on the Variability or Tourism. In *Tourism Alternatives: Potentials and Problems in the Development of Tourism*, Valene L. Smith and William R. Eadington, eds., pp. 216–225. Philadelphia: The University of Pennsylvania Press.

Parsons, Talcott 1967 *Sociological Theory and Modern Society.* New York: The Free Press.

—— 1975 Social Structure and the Symbolic Media of Interchange. In *Approaches to the Study of Social Structure*, Peter M. Blau, ed., pp. 94–120. New York: The Free Press.

Parsons, Talcott, Edward Shils, Kaspar Naegele, and Jesse Pitts 1961 *Theories of Society: Foundations of Modern Sociological Theory.* New York: The Free Press.

Pearce, D. G. 1978 Tourist Development: Two Processes. *Travel Research Journal* (1): 43–51.

Redfoot, Donald L. 1984 Touristic Authenticity, Touristic Angst, and Modern Reality. *Qualitative Sociology* 7(4): 291–309.

Ritchie, J. R. Brent, and Charles Goeldner 1987 *Travel, Tourism and Hospitality Research: A Handbook for Managers and Researchers.* New York: Wiley.

Rojek, Chris 1985 *Capitalism and Leisure Theory.* London: Tavistock.

—— 1987 Freedom, Power and Leisure. *Society and Leisure* 10: 209–218.

Simmel, Georg 1907/1978 *The Philosophy of Money.* London: Routledge.

—— 1950 *The Sociology of Georg Simmel.* New York: The Free Press.

Smith, Valene L. 1989 Eskimo Tourism: Micro-Models and Marginal Men. In *Hosts and Guests:*

The Anthropology of Tourism, (2nd edn), Valene L. Smith, ed., pp. 55–82, Philadelphia: The University of Pennsylvania Press.

Thompson, John B. 1990 *Ideology and Modern Culture*. Oxford: Polity Press.

Thurot, Jean Maurice, and Gaétane Thurot 1983 The Ideology of Class and Tourism: Confronting the Discourse of Advertising. *Annals of Tourism Research* 10: 173–189.

Turner, L., and J. Ash 1975 *The Golden Hordes: International Tourism and the Pleasure Periphery*. London: Constable.

Urry, John 1990 *The Tourist Gaze: Leisure and Travel in Contemporary Societies*. London: Sage.

van den Berghe, Pierre 1980 Tourism as Ethnic Relations: A Case Study of Cuzco, Peru. *Ethnic and Racial Studies* 3: 375–392.

Young, T. R. 1986 The Sociology of Sport: Structural Marxist and Cultural Marxist Perspectives. *Sociological Perspectives* 29: 3–28.

Part VII

TOWARDS A 'NEW' SOCIOLOGY OF TOURISM

15

SOCIOLOGY AND TOURISM*

Graham Dann and Erik Cohen

INTRODUCTION

Sociology studies the values, attitudes, and behavior of human collectivities, whether such groups comprise just a few persons or whole nation states. Preoccupation with the former is reflected in micro-perspectives. Emphasis on the latter is characterized by macro-theoretical approaches. While both positions focus on the individual in society, they differ in their respective points of departure. One locates the power of defining social situations in human actors, as they mold, and in turn are shaped by, interaction. The other points to a central value system which, either by constraint or consensus, normatively prescribes and sanctions role attitudes and behavior.

Historically, sociology emerged at the macro-level in the aftermath of the French Revolution. In response to the perceived social chaos associated with the popular demand for liberty, equality, and fraternity, it offered an alternative image of society, one based on the concept of order. Its initial conservative ideology, grounded in a philosophy of idealism, was strangely combined with a seemingly progressive view which advocated the scientific study of society. This "grand theory" approach concentrated on examining the evolution of societies from an alleged original state of homogeneity to increasing heterogeneity and social differentiation. Whereas most adherents viewed such evolutionary development as progress, a few saw it as the erosion of authentic community values, and hence as a turn for the worse.

Although Emile Durkheim (1858–1917) can be classified among these early evolutionists, he is probably better known for his contribution to the sociology of religion. In this respect, he offered a theory of change predicated on differing forms of social cohesion as they emerged through a general process of secularization.

Perceived deficiencies in the foregoing perspectives were counteracted by another "grand theory," one based on conflict. Under the influence of Charles Darwin (1800–1882), a number of thinkers attempted to show that evolution was grounded, not on harmony and equilibrium, but on the clash of interests. Later, under Marx (1818–1883) and Engels (1820–1895), evolution was replaced by revolution, and interests became identified with antagonistic classes struggling for domination.

The micro-level reaction came from Germany, notably in the figure of Georg Simmel (1858–1918). He decided that sociology could only advance by isolating the

* Reproduced with kind permission of Elsevier Science Ltd, Pergamon Imprint, Oxford, England, from *Annals of Tourism Research*, 1991, 18: 155–69.

forms, rather than the content, of elementary interaction, in such a way that they could be generalized.

Subsequently, Max Weber (1864–1920) incorporated a kindred idea, that of the sociological method of Verstehen. Since his concept of the "ideal type" was based as much on motivation as on external behavior, it helped provide explanations which were "causally adequate and adequate at the level of meaning" (Weber 1968: 99). At the same time, by adopting such an approach, Weber could maintain sociology in a half-way position between the natural and cultural sciences. From this methodological standpoint, he was able to advance the study of such problems as the differential impact of religion on the rationalization of various institutional domains, especially that of modern capitalism.

Following Weber, a number of twentieth century micro-perspectives appeared which focused their attention on the meaning component of social action. Whereas symbolic interactionists concentrated on the ways such meanings were communicated and individuals were socialized, phenomenologists preferred to direct their attention to the structure and social construction of intersubjective reality. In turn, ethnomethodologists attempted to explore the ground rules by which individuals made sense of the world around them.

The micro-theorists were in due course accused of over-emphasizing the subjectivity of human experiences. An effort was hence made to replace or supplement their interpretative schemes by a supposedly more objective treatment of social reality as a system. Under functionalism, which was influenced by the work of some early anthropologists, human needs were related to a series of hierarchically ordered, interlocking, personality, social and cultural systems. This marked a return to grand theory and to the perspective of the early theorists.

However, there were those who rejected this holistic and integrative approach for its alleged inability to tackle social change. At the same time, they urged a more individualistic approach based on "social action." Others essayed a middle path by exploring such possibilities as functional conflict and micro-functionalism.

Although the foregoing scenario of the dialectical development of sociological theory is far from complete, it does at least highlight the origins of the numerous perspectives in contemporary sociology. It also indicates that the choice of topic treated by various sociologists is often predicated on their theoretical approach. However, there are those, the present authors included, who prefer to be more eclectic in their stance, not wishing to be tied down to any specific theoretical line, but rather open to the prime selections of a variety of offerings.

TOWARDS A SOCIOLOGY OF TOURISM

Just as there have been complex developments in general sociology, so too has there been a proliferation of approaches to applied sociology. "Sociologies of" extend across all institutional domains and even include sociology itself. One of these areas which has come into gradual prominence in the last two or three decades is the emergent field of a "sociology of tourism" (e.g. Cohen 1972).

302

It follows from the above that there is no single sociology of tourism, just as there is no single sociology of education or of the family. Instead, there have been several attempts to understand sociologically different aspects of tourism, departing from a number of theoretical perspectives. Although no established approach to tourism has developed with its own unique blend of theory and method, there is a growing number of researchers who are willing to treat tourism as a recognized target domain requiring sociological understanding and explanation.

The "sociology of tourism" requires contextualization. Here the consensus seems to be that tourism cannot be treated in isolation, but has to be seen as nestling within wider applied domains. While it remains problematic as to what these particular domains should be, the outlining of three positions, not necessarily exclusive, should help clarify matters.

There are those researchers, for instance, who argue that the "sociology of tourism" should be located within the parameters of the "sociology of migration," since "touring" essentially denotes movement to another place. Furthermore, the push and pull factors, which respectively dispose and attract persons to migrate towards greener pastures, are often apparently analogous to those encountered in tourism.

However, the analogy is far from complete. In the first place, emigration abroad is but a subset of the total phenomenon of migration, just as international tourism forms only a part of all tourism. Second, emigration is a more permanent phenomenon than tourism, and requires decisions which relate to such domains of life responsibility as education and employment. Tourism, on the other hand, is often viewed as a series of transitory events where responsibility is placed in abeyance or suspended. Third, although it is conceded that there can be a fantasy element in both emigration and tourism, it is likely to be more pronounced in the latter situation. Finally, some emigrants can be classified as expatriates or even as "permanent tourists." Yet it should be noted that these display quite different configurations of attitude and behavior from other more temporary tourists.

Other researchers maintain that the "sociology of tourism" should be contextualized within the "sociology of leisure." Indeed, the definition of a tourist as someone at leisure who travels (Nash 1981) is appealing for its simplicity. However, this definition is too broad and inexact to capture the distinguishing characteristics of the tourist-traveler. Moreover, the manner in which sociologists define leisure very much depends on their varying political, ideological, and social backgrounds.

Consequently, another group of researchers can be found which prefers to stress the travel dimension of tourism. Cohen's (1972, 1974) early work conceives the tourist as a traveler and devises typologies along a continuum of familiarity and strangeness, ranging from the conventional mode of travel of mass tourists to the novelty mode of the adventurer or wanderer. Cohen is at pains to point out that, contrary to popular misconception, all tourists are not the same. Since they differ with respect to attitude, motivation, and behavior, schemes have to be devised to encompass such variation. The realization that such typologies are still only heuristic, rather than explanatory devices, denotes that the "sociology of tourism," lacking powerful theoretical and analytical equipment, is still very much in its infancy.

303

SOCIOLOGICAL TREATMENT OF TOURISM

In his review of the field, Cohen (1984) points out that the sociological treatment of tourism originated in Germany in the work of von Wiese (1930), and was further advanced by Knebel (1960). In the English speaking world, the late 1950s and 1960s witnessed the emergence of two opposing camps. One comprised critics such as Boorstin (1964), who portrayed the tourist as a cultural dope manipulated by the establishment. The other included writers like Forster (1964) who attempted to document the phenomenon empirically and without prejudice.

In the early 1970s, the pendulum swung once more in the direction of the critics, notably in the works of Turner and Ash (1975) and Young (1973). However, beginning with MacCannell's (1973, 1976) paradigm, a more sociologically profound and fruitful approach to the field was initiated, in that he, more than any other, sought to rebut those portraying the tourist as a superficial nitwit by placing tourism in the context of a quest for authenticity, a topic which will be treated later.

Cohen (1984) also maintained that work on the sociology of tourism could be classified into four main issue areas: tourists themselves, interaction of tourists with locals, the tourism system, and tourism impacts. At the same time, he concluded that the "sociology of tourism" had failed to integrate theory with method. True, there had been some progress in the conceptualization of tourists and tourism, but this had not been substantiated by systematic empirical research. Similarly, while there were many piecemeal empirical studies, these were often deficient in theoretical insight. As a result, one was often left with either sociographic data of little relevance or unsubstantiated theoretical speculation. Many of these points are reflected in this article, but in order not to go over identical ground, they are placed within a different framework – that of the previously articulated general sociological perspectives. Several examples of such theoretical approaches are introduced and discussed here.

Developmental (evolutionary and cyclical) perspectives

Tourism has been treated at the macro-level or "from above" in several studies. In the tradition of grand theory, the emphasis in these works is placed primarily on the manner in which tourism has become institutionalized, rather than how individual participants are affected by it.

Many of these studies adopt an evolutionary approach. Thus Knebel (1960), for instance, contrasts early and late forms of tourism along a continuum of polarized variables. So too do chroniclers of the social history of tourism when they describe the transition from the aristocratic grand tour to contemporary versions of mass tourism. Even the critic Boorstin (1964) employs such a framework when he speaks of the development of tourism from traveler to tourist, by nostalgically suggesting that the way "we were" was far better than the way "we are."

For many, the principal institutional development is that tourism has undergone a process of industrialization (Hiller 1976) and internationalization (Lanfant 1980). According to these views, tourism usually begins on a small scale through the efforts of entrepreneurs. Later it becomes "nationalized" and then internationalized, as powerful external economic forces seek to exploit the touristic potential of the destination on a scale which surpasses local resources.

Other exponents prefer to adopt a cyclical, rather than a unilinear, evolutionary model of development. The approach was first proposed by geographers in terms of a resort cycle. The idea was subsequently taken up on a more general sociological level by Machlis and Burch (1983), who made one of the few recent attempts to propose a general theory of tourism in cyclical terms in order to integrate it with wider sociological concerns. Their theory is highly complex and speculative. It is, however, marked by the virtues of novelty and imagination, as well as a welcome attempt to suggest empirical measurements for their theoretical arguments. In fact, Machlis and Burch follow along several lines the process of structural change produced by the industrialization and internationalization of tourism and its implications at the experiential level (that of "meaning"). Maintaining that after the peaking of a touristic destination, there comes an inevitable period of decline, they examine in detail the characteristics of both the upward and downward trend of the cycle. Additionally, they postulate covariation between the cycles of structure and those of meaning, so that the nature of the touristic experience changes according to the stage of the cycle. In this respect, they reinforce the position of earlier commentators such as Hiller (1976: 102), who, while convinced of the "escapist" motivation of modern mass tourists, believes that tourism "replicates the essentials of the industrial process." Hence, he claims (1976: 104) that "the travel product is a commodity to be sold like any other." By implication, the tourist experience becomes assimilated to other experiences of modern consumerism. "Escapism," or any other related touristic experience, can be purchased like any other commodity. Others extend the "commoditization hypothesis" to include hospitality, culture, and sex.

As far as an evolutionary model of development is concerned, there is an unwarranted assumption about the unilinearity of the process. Writers presume that internationalization will inevitably be the last stage of the natural evolution of tourism. However, a contrary model has also been proposed (Cohen 1979a) which is applicable to situations of "sponsored" touristic development. Now internationalization lies at the very beginning of the process, as for example when a multinational corporation establishes a large hotel in virgin territory. In this case, the locals penetrate the industry only at a later stage, once they have realized the opportunities for the provision of small-scale auxiliary services made possible by the original large-scale development.

The cyclical model, on the other hand, has so far received insufficient empirical support. Moreover, Machlis and Burch's theoretical approach, while innovative, is nonetheless over-deterministic. Proponents of the "commoditization hypothesis" also appear to have succumbed to an assumption of inevitability. In fact, their position about the loss of meaning of commoditized artifacts and events seems not to be universally applicable (Cohen 1988b).

Neo-Durkheimian perspectives

Three major Durkheimian themes have been employed in tourism research – those of anomie, the sacred, and collective representations. The notion of anomie was combined with Veblen's (1899) concept of status enhancement in a motivational study of tourists to Barbados (Dann 1977). It was maintained that anomie, which reflects the general normlessness and meaninglessness prevalent in tourism generating societies, requires investigation at the pre-trip level of the tourist's experience in order to appreciate more fully the subsequent stages of a vacation. What had hitherto been rather loosely described as escape from the tedium of a dull 9 to 5 routine, into a situation free from cultural constraint and responsibility, was now placed on a firmer theoretical and empirical footing with the presentation of profiles of anomic tourists. Surprisingly, few other researchers continued this line of inquiry into the sociopsychological investigations of tourists' attitudes and behavior in post-industrial societies.

On the other hand, a great deal of intellectual excitement was generated by MacCannell's (1973, 1976) portrayal of *The Tourist* as emblematic of modern individuals who seek authenticity in times and places other than those of their allegedly contrived environments. Seen in this light, tourism becomes a sacred quest responding to their deepest longings, a form of secular pilgrimage. When this is combined with Goffman's (1959) "front–back" dichotomy, tourists are said to attempt to penetrate the false fronts of staged tourism settings in order to reach the back region of authenticity. That they are not always successful is not so much due to their own superficiality (Boorstin 1964) as to the manipulated structural features of tourist space, which can often be mistaken for the genuine article and lead to "touristic false consciousness."

Arguing even more along Durkheimian lines was Graburn (1989), who adapted the profane–sacred distinction in relation to time in tourism. Whereas profane time refers to the "ordinary" time of everyday life, Graburn maintained that touristic time is "nonordinary," and hence similar to "sacred time" in religious settings. Implicitly, tourism again becomes a form of modern pilgrimage.

Turner (1973), though not a Durkheimian, also believed that tourism was a form of pilgrimage. However, he locates the Center, the pilgrim's (or tourist's) goal, not within the spatial or symbolic boundaries of society, but rather within the antistructural liminal recesses of the Other; that Center is the repository of society's most sacred values. It is here that persons come into contact with the sources of their existence and experience their humanity in its unconstrained fullness as they enjoy "communitas" with their fellow beings. Turner differs from Graburn, however, in viewing structure and antistructure as complementary, and in seeing the journey as compensatory, as well as functionally re-integrative. Though the touristic pilgrimage to the Center-Out-There is essentially secular, tourist trips can come close in spirit to the religious odyssey, even if they are recreational journeys to playful pilgrimage centers such as Walt Disney World (Moore 1980).

Interestingly, scholars who follow Turner's inversionist model differ from those with an evolutionary or cyclical approach by arguing that the quality of touristic experience is unrelated to the institutional structures of tourism. Thus, Moore (1980) discovers that, although visitors to Disney World are certainly aware of the ludic nature of the blatantly contrived and commercialized attractions, nevertheless, they enjoy the experience of liminality. Their experience appears to resemble that of American vacationers, who, according to Gottlieb (1982), seek compensation for the drabness of everyday life by playfully becoming either a "king" or a "peasant for a day."

Another Durkheimian theme, developed mainly by MacCannell, is that the differentiations of modern society are symbolized by a variety of touristic attractions. Here sightseeing becomes a ritual whereby tourists, by paying homage to attractions, overcome the discontinuities of modernity by incorporating its fragments into a unified experience. Tourists thus become the cultural equivalent of the traditional pilgrims visiting holy places. Their visits, in turn, serve to ratify the status of the attractions or sights as objects of ultimate value through a process of sacralization. Since the first stage of this process is the signaling of attractions by markers, it is possible to elaborate the tourist–sight–marker model into a semiotics of attractions, a promising avenue of research in itself.

In spite of his innovative contributions, MacCannell has not escaped criticism (e.g. Schudson 1979). Moreover, his arguments are not always borne out by empirical data. There are also some theoretical ambiguities in his model of the tourist. The first concerns the *level* of analysis. Given that it is avowedly structural, it seems to suggest that *all* tourists are personally impelled by a quest for authenticity predicated on the late modern situation (Cohen 1979a). The second relates to the status of attractions in a modern world which has expanded also to include ("museumize" in his terminology) pre-modern features. Here, MacCannell does not seem to resolve the basic problem of demarcating the Center, since it is unclear whether the vast range of cited attractions collectively represents the structured Center of society or its unstructured opposite – the Other. Finally, as Schudson (1979) notes, such natural attractions as deserts, beaches, and jungles have entirely escaped MacCannell's attention.

These reservations apart, there is certainly no doubting the tremendous impact of MacCannell's work, since he, more than any other sociologist, has placed the problem of tourism squarely into the mainstream of the sociological study of modernity. In this sense, he has acted as a springboard for others. His insights have engendered a variety of theoretical contributions, such as Cohen's (1979a, 1979b) attempt to differentiate types of touristic experience and tourist spaces, and Pearce's (1982) further development of the latter. Most importantly, MacCannell's offerings have helped in a general way to reorient the study of tourism by encouraging others to take tourists' aspirations and conduct seriously (Cohen 1988a).

Conflict and critical perspectives

There exists in tourism a potential tension between the industry and its clients, between a highly routinized and impersonal establishment and the quality of expected touristic experience. Critical theories emphasize this disparity. They point to the contradiction between the routinized industry and the extraordinary expectations of the customers it woos, even to the extent of staging promoted attractions in order to satisfy the expectations it has itself aroused. For this reason, contemporary tourism becomes the butt of satire and caustic criticism (Boorstin 1964; Turner and Ash 1975).

But perhaps more than in any other formulation, the contradiction is best described in the notion of tourism as "commercialized hospitality" (Cohen 1988b). Social exchange between hosts and guests, based on principles of hospitality characteristic of the gift, now becomes largely replaced by economic exchange and the profit motive, often masquerading behind a phoney front of friendliness or even servility.

A deeper structural criticism of modern tourism emanates from dependency and world-system theorists. As the modern tourism establishment develops into a huge international complex of airlines, hotels, travel agencies, transport companies, and the like, there is allegedly an increasing domination by the center (major industrial companies whence the tourists originate) over the periphery (less developed tourist-receiving areas) (e.g. Turner and Ash 1975). Thus, a dependency syndrome emerges. Tourism becomes identified as a form of imperialism or of metropolitan dominance in a neo-colonial setting, in which the natives, particularly of Third World countries, are systematically exploited (e.g. Young 1973).

The conflict approach validly challenges the myth of tourism as an agent of universal brotherhood and world peace. However, that approach itself can be criticized for being too one-sided and deterministic, while disregarding the dynamics of tourist systems and variations within them.

Functionalist perspectives

The functionalist tradition in sociology is represented in the study of tourism by works which regard the phenomenon as a social system. While there are certain similarities with the evolutionary approach, here the accent is less on growth and change and more on the interdependency of the various subsystems and their contributions to the functioning of the whole. Goals and purposes are thus defined as deriving from a central value system which endows individual parts with meaning.

The "needs" approach in psychology is also congruent with a functionalist framework, since needs can both be hierarchically arranged and tend to be satisfied according to priority. The needs themselves can also be seen as having specific functions. Thus, Mayo and Jarvis (1981) speak of such sociopsychological needs as curiosity and adventure, which require satisfaction before equilibrium can be restored to the personality system. Others refer to recreational "motives" in a similar manner. So too do satisfaction studies which seek to measure statistical differences between the expectations of tourists and the reality of their destination. The functionalist perspective is thus particularly appropriate for a marketing approach to the analysis of tourism and its products, since packages can be examined according to their propensity to satisfy identified (or created) needs.

However, the functionalist approach, while denoting *how* the tourism system operates, fails to identify the deeper underlying factors which structure it and inform its dynamics. Hence, it has to be combined with other approaches in order to gain a fuller picture of the reality it seeks to investigate.

Weberian perspectives

Weber was the first sociologist to articulate clearly the argument that meaning, and hence motivation, lies at the core of all sociological understanding. Several students of tourism have come to appreciate that a grasp of tourists' motivation, as derived from their worldview, lies at the heart of the phenomenon. Moreover, an understanding of those push factors which dispose a person to travel in the first place is imperative for a further discourse about destination choice and the impact of visitors on the host community (Dann 1977).

Yet, Lundberg (1980), for example, while stressing the importance of answering the question "why do people travel?," at the same time admits that not much progress has been made towards an answer, and laments the fact that there is no well established theory of human motivation on which to base tourism studies. Pearce (1982: 53) echoes the same point:

> In this writer's opinion a general motivational theory for tourist behaviour needs to fulfill ... this requirement of emphasizing the self directing, autonomous, non-deterministic aspect of motivation.

According to Pearce (1982: 49), such a theory is still in its infancy. Even self-actualization or attribution theories have their limitations, given that tourists are likely to provide favorable and unreflective accounts of their vacations. In this connection, Dann (1981) suggests that it is also necessary to tighten up the terminology of motivation research. He also points to the methodological difficulties involved in persuading tourists to reflect on their real motives for travel.

Apart from Barthes' (1984: 74–78) essay on the Blue Guide, where he speaks of the Helvetico-Protestant morality and its puritanical quest for nature, few have apparently developed Weber's concept of the Protestant Ethic in relation to tourism. Yet, there is a rich potential for the application of such ideas, not only to tourists, but also to their hosts. For example, differing historical patterns of conquest and colonization have yielded a variety of influences on Caribbean and Latin American host societies, ranging from a predominantly spontaneous Catholic–calypso–carnival mentality, to a more Protestant and sober business oriented approach, amenable to the production of less spontaneous, staged touristic events.

Another promising avenue of investigation is suggested by Weber's analysis of the work ethic. It should be possible to make a motivational distinction, not only between those who "work" at their tourism (as they trudge, guidebook in hand, from one site to another) and those who go out to play for "fun," but also to explore differences between a "modern" (Protestant oriented) and a "post modern" (anti-Protestant, "greening") attitude to tourism.

Formalism, phenomenology, and ethnomethodology

In this presentation, examples of formalism, phenomenology, and ethnomethodology are brought together, since all three perspectives typically emphasize the philosophical dimensions of meaning by adopting an initial micro-approach, focusing on the individual in society.

Georg Simmel (1858–1918), the protagonist of sociological formalism, and Alfred Schutz (1899–1959), the founder of sociological phenomenology, contributed, each in his own way, some important seminal ideas to the study of tourism, through their respective concepts of the "stranger." Unfortunately, these ideas have found too little application in tourism research, even though Cohen's (1972) early typology is based on Schutz's view that strangeness and familiarity are basic categories for ordering the world.

Yet, there is much of Simmel which has not been explored in relation to tourism. His examination of the role of number in the determination of the group, for instance, could be usefully applied to the study of the differences between individual and mass tourism. Simmel's study of play as a social form has likewise been virtually disregarded in tourism research, as have also his insights on the metropolitan life, alienation, and vocation. All these are worthwhile avenues of theoretical investigation for those intent on distilling the constants in touristic attitudes and behavior.

Phenomenological theory, beyond Schutz's specific work on the stranger, has similarly been overlooked by tourism researchers. Admittedly, Cohen (1979b) has provided a "phenomenology of touristic experiences" which ranges through five modes from the diversionary to the existential; yet, apart from this attempt, few others explore phenomenological insights in relation to tourism. Apparently, no one has fully investigated Schutz's elaboration of "in-order-to" and "because-of" in touristic motivation. Nor, for that matter, has anyone tried to apply to tourists Schutz's analysis of "projected action," in which time reflexively becomes either the future perfect or past perfect tense. Yet, arguably, this would be a viable paradigm for examining in depth the motivations of potential tourists as well as the discourse of promotional material which plays on these self-same motivations. So too would the phenomenological notion of "I–Thou" relationships prove useful in an appraisal of changing host–guest encounters within the framework of Bergson's "stream of consciousness." The theory is there. It just does not seem to have been utilized.

The same observations go for ethnomethodology. While this "candid camera" approach seeks to investigate the undeclared assumptions of human action and discourse, and in that sense could be appropriate for an examination of touristic stereotypes and clichés, only one example could be found, that of McHugh, Raffel, Foss, and Blum (1974), which has adopted an ethnomethodological perspective for the study of tourism. Yet, such an approach, bereft of its more outrageous claims, could prove worthwhile in analyzing the conduct of both tourists and the tourist industry.

Symbolic interactionism

The last perspective treated here, that of symbolic interactionism, does seem to have caught the imagination of tourism researchers, to a far greater extent than its micro-sociological counterparts.

In its formative years, symbolic interactionism was primarily concerned with the development of the self through its twin components of the "I" and "Me." While the latter comprised the internalized expectations of an individual's attitudes and behavior held by significant and generalized others, the former represented the principle of autonomy, whereby a person accepted or rejected the social definitions of the situation by others. In order to learn the expectations held by others (role), individuals were conceived of as passing through a number of stages, during which they came to appreciate the rules of social interaction. Later, symbolic interactionists examined in more detail the process of role negotiation by individuals, through which definitions of situations were exchanged, accepted, modified or rejected.

The ludic "as if" quality of touristic experience, examined above in a different framework, can also be dealt with from a symbolic interactionist approach. Participants can be viewed as reverting to a childlike stage of existence, in which they oscillate between the quest for instant gratification – pleasure first of the "I" – and the controls imposed by the guardians and custodians of society – safety first of the "Me." It is thus possible to speak of "the tourist as child" (Dann 1989), and to examine the ways in which spontaneous hedonism is checked by the industry through its language of social control (in brochures, travelogs, advertisements, couriers, guides, etc.).

Some students have extended the approach to incorporate a Freudian perspective, in which the "I" and "Me" are translated into the "id," "ego," and "superego." Subsequently, these concepts have been used to define the personality states of "Child," "Parent," and "Adult." Those who have adopted the ensuing transactional analytical approach to the study of tourism (e.g. Dann 1989; Mayo and Jarvis 1981) find that it can be applied to a broad spectrum of relationships. In addition to providing a fuller understanding of native host–foreign guest interaction, transactional analysis can also be applied to encounters involving persons directly or indirectly employed in the tourist industry. Indeed, some airlines and hotels use the transactional approach in the training of their personnel.

Other researchers have explored the dynamics of role negotiation in touristic encounters. This approach has been found particularly useful for the study of asymmetrical one to one relationships. Karch and Dann (1981), for instance, have been able to apply the approach to the study of the serious misunderstandings which arise in tourist–beachboy relationships, predicated as they are on marked differences of class, gender, culture, and race. Achieving a working consensus in such situations can be quite problematic, since both parties bring an entirely different set of expectations to the encounter.

A symbolic interactionist perspective has also been extended to the content of tourist interviews, and in particular to the fantasy component of motivation.

Whereas fantasy is a common enough theme among critics of the industry, symbolic interactionism, perhaps more than any other sociological perspective, permits the researcher to analyze this all important theme in a systematic manner which transcends hitherto gratuitous assertions.

Reference has already been made to the work of Goffman (1959), whose "presentation of self in everyday life" focused on the "I" component of the self in a theatrical context. It was also noted how MacCannell, by employing some insights of Goffman, was able to show how the front–back stage dichotomy can be effectively used to tackle the problem of authenticity in tourist settings. Yet, there is a great deal in Goffman which has been underutilized by tourism researchers. For instance, the concepts of sincere and cynical performances, deference and demeanor, performance blunders and team performances, are all appropriate for analyzing the various actors on the tourism stage.

Finally, the latest developments in symbolic interactionism have turned in the direction of semiotics and semiology. These are strongly influenced by French theoreticians whose works emphasize the exchange, dissemination, and influence of myths and images, rather than the reality they are supposed to represent. Following Barthes (1984), several researchers have applied the semiotic approach to the study of tourism promotional literature. More specifically, some have analyzed the "people content" of brochures and looked at how the industry attempts to control the forthcoming interaction of tourists, hotel staff, and locals, through pictures and verbal descriptions of resorts.

CONCLUSIONS

From this brief overview of sociological theories and their application to tourism research, it is evident not only that are most major theoretical perspectives represented, but that each of these has contributed something towards a sociological understanding of contemporary tourism. Admittedly, they have met with varying degrees of success, and some have turned out to be more viable than others, but it is only by performing such a state of the art analysis that one is in a position to evaluate the relative contribution of various theories to an understanding of tourism. Indeed, and by way of summary, a few general conclusions emerge.

First, it should be clear that there exists no all embracing theory of tourism, since tourism, like any other field of human endeavor, is a target field, comprising many domains and focuses, to which various theoretical approaches can be appropriately applied.

Second, it emerges from the above that no single sociological perspective can reasonably claim a monopoly in providing an understanding of tourism. Rather, the insights contributed by various approaches should be regarded as forming pieces of a jigsaw, which, when assembled, can supply the basis for a pluralistic sociological interpretation of touristic reality. As a matter of fact, some of the best work in tourism has been eclectic, linking elements of one perspective with those of another, rather than opting for an exclusive point of view. At the same time, one may appreciate the point that even the eclectic approach can experience difficulty, as for example in the Vienna Centre's (Bystrzanowski 1989) multitheoretical stance in relation to tourism and social change.

As an extension or corollary to the foregoing, it appears that sociology itself provides only a partial interpretation of the multifaceted phenomenon of tourism. For a more complete picture, it is necessary to combine sociological insights with those from other social science disciplines. Thus, when Theuns (1984) questioned 37 international tourism experts on their attitudes towards research, he found that the need for multidisciplinary basic studies ranked the highest on all agenda items. Interestingly, however, while economists and geographers viewed contributions from sociologists in positive terms, the sentiment was not reciprocated. Although the investigator could not explain why exactly such a situation obtained, one may suggest that the sociologists were probably reacting negatively to what they perceived as the overly quantitative and positivistic approaches of their colleagues in other behavioral disciplines. Yet, surely this calls for greater, rather than less, dialog. If sociological insights on tourism are still fairly rudimentary, such a situation clearly indicates that further collaboration and more mutual openness are required if sociological theorizing on tourism is to progress.

REFERENCES

Barthes, R. 1984 *Mythologies*. London: Paladin Books.
Boorstin, D. 1964 *The Image: A Guide to Pseudo Events in America*. New York: Harper and Row.
Bystrzanowski, J., ed. 1989 *Tourism as a Factor of Change: A Sociocultural Study*. Vienna: European Coordination Centre for Research and Documentation in Social Sciences.
Cohen, E. 1972 Toward a Sociology of International Tourism. *Social Research* 39: 64–82.
—— 1974 Who is a Tourist? A Conceptual Clarification. *Sociological Review* 22: 527–555.
—— 1979a Rethinking the Sociology of Tourism. *Annals of Tourism Research* 6: 18–35.
—— 1979b A Phenomenology of Tourist Experiences. *Sociology* 13: 179–201.
—— 1984 The Sociology of Tourism: Approaches, Issues and Findings. *Annual Review of Sociology* 10: 373–392.
—— 1988a Traditions in the Qualitative Sociology of Tourism. *Annals of Tourism Research* 15: 29–46.
—— 1988b Authenticity and Commoditization in Tourism. *Annals of Tourism Research* 15: 371–386.
Dann, G. 1977 Anomie, Ego-enhancement and Tourism. *Annals of Tourism Research* 4: 184–194.
—— 1981 Tourism Motivation: An Appraisal. *Annals of Tourism Research* 8: 187–219.
—— 1989 The Tourist as Child: Some Reflections. *Cahiers du Tourisme, Série C*, No. 135. Aix-en-Provence: CHET.
Forster, J. 1964 The Sociological Consequences of Tourism. *International Journal of Comparative Sociology* 5: 217–227.
Goffmann, E. 1959 *The Presentation of Self in Everyday Life*. New York: Doubleday.
Gottlieb, A. 1982 Americans' Vacations. *Annals of Tourism Research* 9: 165–187.
Graburn, N. 1989 Tourism: The Sacred Journey. In *Hosts and Guests. The Anthropology of Tourism* (2nd edn), V. Smith, ed. Philadelphia: University of Pennsylvania Press.
Hiller, H. 1976 Escapism, Penetration and Response: Industrial Tourism in the Caribbean. *Caribbean Studies* 16: 92–116.

Karch, C., and G. Dann 1981 Close Encounters of the Third World. *Human Relations* 34(4): 249–268.

Knebel, H. 1960 *Soziologische Strukturwandlungen in Modernen Tourismus*. Stuttgart: Enke.

Lanfant, M. 1980 Tourism in the Process of Internationalization. *International Social Science Journal* 32: 14–42.

Lundberg, D. 1980 *The Tourist Business*. Boston: Cahners Books.

MacCannell, D. 1973 Staged Authenticity: Arrangements of Social Space in Tourist Settings. *American Journal of Sociology* 79: 589–603.

—— 1976 *The Tourist: A New Theory of the Leisure Class*. New York: Schocken.

McHugh, P., S. Raffel, D. Foss, and A. Blum 1974 Travel. In *On the Beginning of Social Inquiry*. London: Routledge.

Machlis, G., and W. Burch 1983 Relations between Strangers: Cycles of Structure and Meaning in Tourist Systems. *Sociological Review* 31: 666–692.

Mayo, E., and L. Jarvis 1981 *The Psychology of Leisure Travel*. Boston: CBI.

Moore, A. 1980 Walt Disney World: Bounded Ritual Space and the Playful Pilgrimage Center. *Anthropological Quarterly* October: 207–217.

Nash, D. 1981 Tourism as an Anthropological Subject. *Current Anthropology* 22: 461–468.

Pearce, P. 1982 *The Social Psychology of Tourist Behaviour*. Oxford: Pergamon.

Schudson, M. 1979 On Tourism and Modern Culture. *American Journal of Sociology* 85: 1249–1258.

Theuns, H. 1984 Tourism Research Priorities. A Survey of Expert Opinions with Special Reference to Developing Countries. *Cahiers du Tourisme, Série C*, No. 96. Aix-en-Provence: CHET.

Turner, L., and J. Ash 1975 *The Golden Hordes: International Tourism and the Leisure Periphery*. London: Constable.

Turner, V. 1973 The Center Out There: The Pilgrim's Goal. *History of Religions* 12: 191–230.

Veblen, T. 1899 *The Theory of the Leisure Class*. New York: Macmillan.

von Wiese, L. 1930 Fremdenverkehr als zwischenmenschliche Beziehung. *Archiv für Fremdenverkehr* 1.

Weber, M. 1968 *The Theory of Social and Economic Organization*. New York: Free Press.

Young, G. 1973 *Tourism: Blessing or Blight?* Harmondsworth: Penguin.

16

MEGA-EVENTS AND MICRO-MODERNIZATION*

On the sociology of the new urban tourism

Maurice Roche

I TOURISM, MEGA-EVENTS AND MODERNIZATION

A Introduction

The international recession of the early 1980s and the renewal of economic growth in the mid and late 1980s in the advanced industrial societies of North America and Europe have had major consequences for many towns and cities in their old established manufacturing regions. In the early 1980s a relatively sudden, unanticipated and apparently uncontrollable collapse of employment occurred in traditional materials and manufacturing industries (e.g. coal, steel, etc.). A process of 'deindustrialization'[1] occurred leading to the graphic popular description of these regions as 'rust belts'.

However, in the mid and late 1980s, 'rust belt' cities have begun to respond to their decline with local economic strategies aimed at boosting employment. These strategies involve efforts to attract new inward investment, to diversify into new service sector industries, and generally to modernize local economic and social infrastructures. One feature of such strategies has often been the at first sight rather implausible looking project of creating, out of unpromising material, a new 'urban tourism' industry.[2] There are usually a variety of components to such strategies including the development of museums, trails, package tours, 'interpretation' services and facilities celebrating local 'industrial heritage'; the development of regionally significant leisure-oriented mass and high-quality retailing complexes and facilities (e.g. Metro Centre Newcastle, Meadowhall Sheffield); the development of waterfronts, canal basins, viewing towers, etc.; and finally the planning and staging of 'mega-events'. In this paper I am concerned particularly with this last component. Mega-events are large-scale cultural or sporting events designed to attract tourists and media attention. Some British examples are Garden Festivals (Liverpool 1984, Stoke 1986, Glasgow 1988, Gateshead 1990), Cultural Festivals (Glasgow 1990) and

* Reproduced with permission from *British Journal of Sociology*, 1992, 43: 563–600.

mega sports events (Sheffield World Student Games 1991, Birmingham and Manchester Olympic Games 1992 and 1996 bids).

The major late twentieth-century phenomenon of mass tourism in general, not to mention urban tourism, has strangely attracted all too little social science interest. So this paper is an initial attempt to do two things. First, I aim to explore one component of the new urban tourism, namely the impacts and functions of mega-events. Second, I will take the opportunity to reflect more widely on the sociological and interdisciplinary (economics etc.) perspectives, concepts and research agendas necessary for an adequate comprehension of mega-events and of modern tourism in general. As far as this latter theme goes, my general line of argument is that dominant approaches to understanding tourism and its effects tend to be narrowly economic, implicitly functionalist or (economic) system-reproducing, and de-contextualizing. Against this I suggest that narrowly economic and functionalist approaches are inadequate to understanding the general processes and direction of socio-economic change which underlies and contextualizes the growth of tourism in late twentieth-century advanced industrial societies. I suggest that contextually sensitive perspectives which recognize structural change and its disorganizing features are potentially of greater importance for the future of tourism and mega-events research. This theme is developed in Section II below.

As far as the former theme goes, concerning the understanding of mega-events and urban tourism, my general line of argument will also be concerned with the problem of contextualization. While it *is* vital to understand mega-events' *economic* impacts and functions, I suggest that this is better done in broad rather than narrow terms and in terms which take account of contexts of structural change and of discontinuity and reorganization in local economies. I elaborate on this theme in Sections III and IV. In Section III I discuss conceptions of mega-events and their impacts in contemporary tourism research, while in Section IV I consider first the urban policy context and then recent attempts in tourism research to redefine and renew the analysis of mega-events and their impacts in ways which might begin to take some account of this urban policy context. But, to set the scene for the exploration of my two main themes, it is first necessary to provide a brief sketch of the nature and significance of modern tourism.

B Tourism in modern society

Mass tourism and travel is a quintessentially modern (late twentieth-century) capitalist industry and socio-cultural phenomenon. In the advanced societies it encompasses the large sub-sectors of business travel and of socially motivated travel (e.g. visiting friends and relatives) as well as all of the various forms of holiday-making and travel-for-pleasure. It is a major consumer of, and therefore a *raison d'être* for, the products of the automobile, aircraft and other vehicle manufacturing and energy supply industries. It requires the construction, maintenance and development of airports and motorways, of tourist attractions and hotels, and of service industries concerned with information and 'interpretation', and with 'people management' and 'hospitality', etc. It is now one of the leading constituent industries in the modern economic order whether at local, national or international levels.[3]

Tourism and travel have grown vigorously and consistently at around 5 per cent per annum throughout the post-war period, whatever the state of the trade cycle. This growth has been fuelled by general long-term economic growth and affluence, the growth of leisure time and discretionary income, and also by technological developments in the transport and tourist-attraction industries (e.g. respectively wide-bodied, long-haul jets and high-speed trains on the one hand and the 'imagineering' of Disney-style theme parks and all-weather resort facilities on the other). This long-term growth of tourism and travel looks set to continue for the conceivable future, at least into the medium term. But this is subject to the proviso that the great and varied demands and *costs* that the industry tends ultimately to impose on both natural and social ecological systems are not seen to outweigh the various economic and other *benefits* it brings. If costs do come to be seen to outweigh benefits then – at least in the advanced industrial societies, where there are popular checks on policy-making, and where ecological conservationist and NIMBY ('not in my back yard') politics are currently popular – it is likely that the brakes will be put on this growth.

For these and other reasons, and whether for good or ill, the development of tourism has long been seen as both a vehicle and a symbol at least of westernization, but also, more importantly, of 'progress' and 'modernization'. This has been particularly the case in Third World countries.[4] But this role as both a symbol and vehicle of economic and socio-cultural change and 'modernization' is potentially just as significant for the advanced industrial societies and begins to suggest some of the research needs and relevant approaches to the study of modern tourism.

C A socio-economic approach to tourism and mega-events

On the one hand, as far as *economic change* and 'modernization' goes the leisure industry in general,[5] but in particular tourism, appears to some commentators to be structurally linked with an emerging *'post-industrial'*/'information society' pattern in the advanced societies;[6] while, on the other hand, as far as *socio-cultural change* and 'modernization' goes tourism has been seen, rightly in my view, as a quintessential expression of *'post-modern'* culture.[7]

Tourism certainly generates distinctive *cultural* mixtures, collages, sign-stews and style-stews in resort areas, not least in dress and in architecture, and it radically accelerates and accentuates the established cosmopolitanism of modern cities. It anticipates in practical consumerist and existential terms the cultural, epistemic and moral relativism and fragmentation reflected on and (misguidedly in my view) positively valued by the currently fashionable movement of 'post-modern' social and cultural theory.[8]

In all of these respects, however, the need to develop sociological and inter-disciplinary studies and analyses of the various and complex *cultural* impacts of tourism on forms of personal and collective self-understanding and self-representation both by tourists and hosts in the modern social order is evidently great.[9] Equally evidently it has for far too long been almost completely overlooked. Also, since culture is necessarily interconnected with polity and economy, a dialogue between research studies and analyses of tourism in post-industrial economic terms and in post-modern cultural terms might ultimately be one of the most profitable directions for social science research to develop along in this area. However, there is a limit to what can be addressed in a paper such as this, so in what follows I will angle my discussion mainly at the *socio-economic* dimension of tourism and mega-events, deferring discussion of their cultural dimensions to the Conclusion.

It is necessary in any case, and particularly in any *sociological* analysis of tourism, in my view, to take as full account as possible of tourism's economic dimensions. This is for the simple reason that probably the main political and social stimuli and motivations for developing a tourism industry at all derive from its assumed potential to generate employment. So, before turning to mega-events more directly it is important to elaborate on both the economic aspects of tourism, particularly tourism employment, and also the general conceptions of *socio-economic* change necessary to adequately understand the nature and significance of tourism's economic aspects in the contemporary period.

II TOURISM AND MACRO-MODERNIZATION: TOURISM'S ECONOMIC DIMENSION

A Socio-economic change and the limitations of functionalism

The development of modern tourism needs to be seen against the background of long-term structural economic change in contemporary capitalism. The advanced industrial societies in the late 1970s and 1980s have embarked upon a process of structural transformation from an industrially based to an information-technology-based *'post-industrial'* (or 'post-Fordist') form of capitalism.[10] This has coincided with the development in this period of *'post-national' capitalism*, in particular major movements of 'globalization'[11] in the international capitalist economy (particularly among and between the advanced, as opposed to the Third World, economies). This is increasingly producing in the late twentieth century a transnational and *multi*-polar, interdependent and highly interactive, capitalist economic *world order*, an order largely unanticipated in conventional (neo-Marxist and other) conceptions of 'imperialism' and 'the world system'.[12]

These claims are evidently large and arguable, but I will assume their validity for the purposes of my discussion here and concentrate on developing their implications in general and for tourism in particular. I will assume, then, that long-term and coincidental processes of post-industrial and post-national transformation are at work in modernity and that they constitute the socio-economic 'deep structures' of our era. The notion of 'modernization' and of a 'modernization problematic' I am using in this paper refers to these deep structures and processes.

The modernization problematic, as I understand it, carries with it important general theoretical and applied/policy implications. Foremost among them are that all

318

'functionalist' or 'system-reproductionist' conceptions of the modern social order are rendered at least essentially limited, at worst invalid. Functionalism or system-reproductionist conceptions are extremely pervasive in modern social science, in politics and in economic affairs.[13] All conventional and 'critical' approaches, whether associated with Right, Centre or Left politics, are equally pervaded by functionalism in so far as they assume the existence and reproduction of industrial 'mass-production' and 'mass-consumption' capitalist economies organized within nation-states. Functionalism in *this* sense pervades Centrist liberal and welfare state-based sociology and social policy. But, just as much, it also pervades both Leftist neo-Marxist sociology and critical social policy on the one hand, and Rightist 'free market/strong state'[14] political economy on the other. In the main forms of economics and economic policy dominant in the twentieth century it pervades equally the Keynesian concept of state-regulated capitalism and the Stalinist concept of state-controlled and command-economy 'socialism'.

All of these traditions of social and economic theory and practice assume the existence or possibility in the modern social order of a high order of functional integration and organization of polities, and cultures based on industrial economies, in relatively independent and self-regulated nation-states. By contrast the modernization problematic challenges these variously complacent or pessimistic functionalist assumptions and focuses on processes of structural change and on the discontinuities and contradictions present in historical periods of major transition and transformation. The surface effects of these processes are first (a) to create situations of perceived crisis and disorganization, situations which appear 'out of control' and so on, and second (b) to thereby create political needs and projects for re-organization, restructuring and the regaining of control by affected polities. Contemporary non-functionalist approaches in socio-economic theory which focus on structural change crisis and disorganization are many and various. They include the multi-crises approaches of critical theory and neo-Marxism, and also the recent development of the ecological critique of industrial society.[15] However, the approaches of 'disorganization' theory,[16] 'post-industrial' theory and recent neo-Marxist interests in 'new times' 'post-Fordist' analysis (see notes 6 and 10) are likely to be of most relevance to the development of our general understanding of the post-industrial, post-national and post-functionalist 'modernization' problematic sketched above.

The rethinking of these general theoretical and policy assumptions is of relevance to the analysis of the economic dimension of modern tourism in a number of ways. First, it challenges both the conventional economics of tourism and also the implicit and explicit functionalism and systems analysis of other main social science approaches to tourism, namely those of social geography[17] and management studies.[18] Second, it emphasizes the value of tourism employment to post-industrializing economies threatened by unemployment (B below). Third, it helps to explain the development of micro-functionalist 'local corporatist' political and economic strategies, particularly with regard to tourism in states affected by post-national forces (also B below). And finally it suggests that there are reasonable grounds for assuming that the quality of tourism employment in post-industrial societies is likely to improve in the long term (c below). We can now take each of these points in turn.

One general implication of the anti-functional, structural change and disorganization emphases of the modernization problematic is that conventional economic accounts of the 'economic impact' of tourism need to be fundamentally reconsidered. These accounts typically use the Keynesian concept of the 'economic multiplier' to analyse the indirect, 'knock-on' or 'ripple' effects of a net increase in national or local income due to tourist spending (or capital spending on the building of tourist attractions).[19] Allowing for various 'leakages' out of the economy under consideration a strong tourism industry is an effective way of achieving analogous economic growth stimulation to that provided by a strong export sector. But multiplier analysis, together with the 'input–output' matrix analysis of local industrial sector exchanges which is usually needed to calculate particular multiplier ratios, tend to assume a relatively static, isolable and self-reproducing national or local economic system. They are thus functionalist in economic and (implicitly) in social terms. They are not designed to explain, describe or measure economic processes at work in economies which are *dynamic*. Dynamic economies are (a) distinctively open, vulnerable and responsive to external forces, or (b) involved internally in both (i) a major reordering of the relationships between old and new industrial sectors and (ii) a technological transformation of production processes within all economic sectors, or (c) both (a) and (b). Multiplier analyses of the economic 'benefits' of tourism (together with cost-benefit analyses which are capable of accounting and assessing social and environmental effects as well as economic effects of tourism) are undoubtedly important tools of analysis and inputs to policy-making. But, since they assume relatively static and functional rather than dynamic socio-economic systems, their relevance, use and validity are seriously limited in the conditions of contemporary structural socio-economic change and disorganization described and implied by the modernization problematic.

B Post-industrialism, post-nationalism and tourism

There are two disorganizing/reorganizing dynamics implicit in the modernization problematic which need to be noted as far as tourism goes, firstly post-industrial economic dynamics and secondly post-national (global–local) political dynamics.

Post-industrial modernization, through the increasing application of information technology and automation to industrial production, tends to reduce the demand for labour relative to output, to increase the vulnerability of economies to structural unemployment, and to make orthodox conceptions of 'full employment' into an unattainable policy ideal.[20] The shift to a post-industrial pattern of particularly economic organization looks set to irrevocably disrupt the long-established twentieth-century conventional wisdom about basic economic relationships which guide modern statecraft and economic policy. Keynesian assumptions about the positive relation between investment and employment are becoming just as vulnerable and untrustworthy as Marxist assumptions about the relation between labour and value.

Just as profoundly, through its threat to employment, post-industrialization also irrevocably disturbs the foundational modern conceptions of citizen and consumer–producer identities and statuses (namely those of the 'self-supporting'/income 'earning', free-spending/free-choosing, socially functional and legitimated 'individual'). Employment-based citizen roles and rights have been vital both to people's material well-being (through income distribution) and to their positive mental health in the modern social order.[21] Post-industrialism threatens to undermine manufacturing employment and familiar patterns of income distribution. But at the same time it is likely to provide good conditions for economic growth. *In principle*, future economic growth should provide the resources at least for the maintenance of generally affluent levels of consumption in spite of changing demands for labour and thus changing income distributions. But *in practice* the major political problems of post-industrialism, namely how to achieve a perceivedly just and demand-effective distribution of available employment and income, will need to be resolved. Assuming they are then such a high-technology society with increased discretionary 'leisure' time is likely to generate a strong tourism market and industry. And as noted earlier, the growth of tourism has indeed been one of the characteristic features of post-war economic change parallelling and complementing the development of post-industrialism.

In the difficult and often contradictory contexts of post-industrialism the need of nation-states to defend and cultivate labour-intensive (or even significantly labour-*using*) industries and sectors is considerable. In these contexts the tourist industry, as a quintessential service industry, apparently has the rare virtues of simultaneously (a) requiring continuous technological development, capital investment and renewal, (b) being commercially attractive both to consumers and investors, and (c) being significantly labour-intensive. As against this optimistic view it is also true that there are currently serious questions about both the quantity and often poor quality of direct employment in the tourist industry[22] and also over how best to research and measure it.[23] We will return to these later (E below).

The second aspect of the modernization problematic which carries implications for tourism research and policy is that of *post-nationalism* and the global–local political dynamics it involves. International tourism markets in, and largely between, the advanced industrial societies have grown spectacularly and apparently inexorably in the post-war period. Next to the spread of western pop and consumer culture this has been one of the major signs of a distinctive new historical period in the development of capitalism and of modern society. In this post-national period the long-established set of relationships holding between global, national and local power centres in the advanced societies since the nineteenth century which privileges the nation-state are being irrevocably disturbed and undermined. There is now a long-term erosion of power under way among the major power holders in contemporary nation-states, including large 'national' capitalist corporations and trades union associations, but particularly central government and the state apparatus. Throughout much of the post-war period these institutions implicitly or explicitly with greater or lesser success tended to coordinate their activities in various forms of national 'corporatism' in support of economic growth and full employment policies. In the late twentieth century such national corporatist arrangements have become increasingly untenable as power erodes away from the national level mainly to transnational levels but also to sub-national levels.

Transnationally national power is eroding in favour firstly of economic and political institutions operating at the global level (multinationals, the UN, the web of non-governmental international organizations, etc.). And secondly (and most palpably in the 1980s and 1990s in the case of Britain and West European states), national power and sovereignty is eroding in favour of multinational groupings (such as the (proto) federation of the European Community). On the other hand, at the periphery of the modern nation-state, away from central government and its capital city/dominant region (i.e. in provincial cities, regions and sub-nationalities), the possibilities for an increased degree of political, cultural and economic autonomy are beginning to be sensed and explored. Post-industrial technology facilitates spatially distributed and decentralized organization in all spheres. And in the European Community in particular there is a belief that localities' power and autonomy will probably be stimulated by the very existence of the EC as a new transnational sphere of activity, and by the economic and other opportunities it offers.

This global/(transnational) local dynamic in contemporary modernization processes can lead to a certain amount of perceived disorganization and loss of control at the centre of national states. It can thus lead also to periodic reactive and probably ultimately unsuccessful attempts to regain and recentralize control by central government.

Britain in the Thatcher decade has experienced something of this kind, an attempt by central government to put the brakes on what it saw as the accumulating powers of local government. The politics of this, together with the underlying dynamics of post-nationalism and of de-industrial/post-industrial restructuring at local level, will be outlined later (Section IV below). They provide the background necessary to properly grasp attempts in the mid and late 1980s to develop urban modernization policies involving mega-events and the development of urban tourism particularly in the 'rust belt' cities of Britain. We will focus on Britain to illustrate the argument, but much of the same kind of analysis could be equally well applied to comparable tourism developments in the USA and Western Europe. While functionalism is breaking down at the macro-level of the national economy (i.e. the decline of Keynesianism, national corporatism, etc.) it is of *increasing* significance at the local level. So in Section IV in particular we will note the emergence of micro-forms of functionalism such as 'local corporatism' in the urban politics behind the new wave of developments in urban tourism.

C Tourism and employment

As noted earlier, the political drive to encourage the tourism industry derives from its potential for generating employment in an era threatened to a new degree by persistent structural unemployment. However, while tourism may be a sector in which relatively high *quantities* of employment can be generated (see note 3), doubts are often expressed about the low *quality* of this employment and about economic strategies which do not appear to be capable of generating 'real' jobs.

In my view low employment quality in tourism (i.e. insecure, part-time, low-paid, etc.) is in large part the product of (a) an immature industry, subject (b) to 'seasonality' (i.e. summer weather-dependent) and (c) to the negative effects of geographic isolation and the mono-culture of 'resorts' and resort areas. It is an admittedly optimistic but nonetheless reasonable speculation that these factors are likely to exert a diminishing influence on the hi-tech and urban tourism industries of the advanced societies in the long term. The seasonality problem is likely to be increasingly tackled (a) by the spread of investment in all-weather tourist attractions and facilities (e.g. new generations of 'domed' multi-purpose leisure centres,[24] indoor arenas, large hotel complexes, leisure-oriented shopping malls, etc.), (b) by the development of the off-season short-breaks market, and (c) by counter-seasonal programming of special and mega-event cycles. Finally the problems typically associated with the isolated monoculture of resort areas and of tourist-only facilities do not apply to essentially cosmopolitan urban areas, where both mega-events and their associated facilities (new stadia, theatres, etc.) are typically designed with the large local resident market (if not always 'the community') as well as the tourism market in mind.

The mass tourism industry is a relatively recent phenomenon. Industrial associations and identity, managerial professionalism, quality training, long-term career structures and trade unionism have all had little time and opportunity to take root. But it is reasonable to expect that they will take root and grow over the coming decades and that the quality and diversity of employment available in the tourism sector, as it matures, should improve and increase. But, along with so many other aspects of contemporary tourism, these points remain speculative and await detailed examination by a new generation of research studies in the sociology and economics of post-industrial tourism, a generation of studies which is overdue academically and badly needed in policy terms.

In this section some of the main economic aspects of contemporary tourism have been indicated together with some of the concepts necessary to contextualize tourism in terms of broader and deeper processes of socio-economic change and 'modernization'. For the rest of this paper I want to focus on one particular component of contemporary tourism, namely mega-events. In Sections III and IV we will look at two aspects of mega-events. Firstly we will consider their *impacts* and issues arising from social science attempts to define, analyse and measure these, particularly economic impacts (Section III). And secondly we will consider their *contexts*, outlining some of the main aspects of contemporary urban politics and

323

policy-making influencing the creation of mega-events and urban tourism strategies (Section IV).

My discussion of mega-event impacts and contexts is intended to be an exploratory one and to be informative about those issues in their own rights. So no attempt will be made here to translate these phenomena systematically and elaborately into the terms and assumptions of the post-industrialist, post-nationalist and modernization perspective outlined in this section. Nevertheless, the 'modernization problematic' indicated clearly has a general background relevance for adequately comprehending mega-event impacts and contexts. It is generally suggestive about directions in which social science research agendas in these areas need to be revised. So we will occasionally refer back to this problematic at points during the following discussion.

III MEGA-EVENTS AND IMPACT ANALYSIS

A Background

Research into mega-events and their economic and other impacts is fairly sparse in the longer-established and otherwise currently burgeoning fields of tourism and leisure studies. It mainly dates from North American studies in the 1970s. In spite of the world-scale of mega-events such as the Olympic Games, the paucity of mega-event research is surprising, and many commentators have noted this.[25]

Early studies in the tourism event field focused on events defined by duration (i.e. 'short-term' events) and by status (i.e. 'prestige', 'hallmark' events). Later work extended to include events that have size or scale as one of their key distinguishing features (i.e. mega-events). It is worth noting that the concept of scale is essentially relative to the size of the community staging the events.[26] What Goeldner and Long refer to as 'mini-events'[27] may nonetheless have an impact as big and as 'mega' in relation to small communities and towns as for instance the Olympic Games usually is in its impact on the city and nation that host it.

Research studies of the economic impact of mega-events tend to reveal three major limitations. That is, little is said in such studies about (a) the urban community context of mega-events, (b) the medium-term temporal–historical context within which mega-events occur and (c) the urban policy processes within which mega-events occur. Thus mega-event research tends to be *de-contextualized* by failing to provide (and worse, being indifferent to the absence of) accounts of these three inter-related sorts of context. In this section we will briefly review some of the inadequacies of the tradition of research on mega-event impacts (i.e. narrowly economistic studies and lack of follow-ups) which tend to *de-contextualize* the latter in the ways suggested above (B below); a contemporary attempt to *re-contextualize* mega-events by developing a concept of 'total impacts', (C below); and some ways to begin to *redefine* mega-events in order to better research and understand them (D below).

B De-contexting mega-events

Economistic research

The early studies in this field took in both relatively small cities (Newport, Rhode Island; Quebec City; Bridgeport, Connecticut; Fort Wayne, Indiana) and large cities (Atlanta, Edinburgh).[28] They tended to use both multiplier analysis and local expenditure surveys to provide measurements of direct and indirect short-term economic effects. Among the weaknesses with the design and methodology of many of these studies[29] were (a) an optimistic tendency to measure only 'gross benefits' without measuring or deducting costs to achieve a 'net benefit' figure, (b) in general an implicit reliance on a relatively static and functionalist conception of the local economic system (and see above Section I), and (c) a generally atheoretical empiricism with little attempt to use detailed case studies to illustrate or test broader conceptualizations and theory in socio-economic analysis. This sort of research tends to de-context mega-events from their urban/community context.

Pre-event impact projections

It might be said to take a good imagination to dream up a unique grand event or a bid to host an established mega-event. Be that as it may it surely takes some degree of courage from local politicians and businesses to actually go ahead and risk their city's name and money on it. They usually proceed on the basis of more or (often) less adequate feasibility studies and cost-benefit studies (often long on benefits and short on cost projections).[30] This context of mega-event policy-making and organization, whether in association with contemporaneous wider economic policy-making or not, can generate pre-event studies and projections which make claims not only about direct and indirect event effects and impacts but also about impacts of concern in the wider and longer-term local economic development policy context.[31]

The documentation surrounding mega-events at the different stages of their development and organization is considerable. If we were to imagine the creation of a 'mega-event library' of such material obviously some of its main components would be (i) initial *bid* documents (whether successful or not) presented to mega-event authorities such as the International Olympic Committee, together with (ii) *feasibility studies* and (iii) *impact projections* for local policy-making purposes. Considering the ever-increasing competition by cities around the world to host the ever-increasing list of sporting, cultural and other such mega-events, the library would soon accumulate a surprisingly large collection of such *pre*-event documents. However, by comparison the collection of *post*-event documents in the mega-event library would be tiny. This is not just because most bids fail, but also because even winning-bid organizers fail to do very much post-event accounting and analysis. Obviously they usually do as much as they are legally required to do, most often in the narrow economic terms of a revenue and expenditure review. But compared with pre-event projections of direct and indirect economic and social impacts immediate post-event accounting is extremely uninformative; while medium- and longer-term follow-up studies are notable by their almost complete absence from the field of mega-event analysis and from our ideal archive. 'Before' and 'after' studies, to test the accuracy of projections and adequacy of the methods they use, are also virtually non-existent.[32]

Pre-event projections evidently form an inadequate basis for the study of mega-events. Not only do they tend to rely over-much on economistic impact analysis, they are also of course totally uninformative about the policy contexts which generated them and the political use to which they were put. There may well be much local journalism to add to the ideal mega-event library, and this would provide much information about day-to-day decision-taking and policy debates in and around most mega-event organizing groups. Nonetheless very few political sociological studies of policy-makers and organizers of mega-events and/or tourism strategies exist.[33] So the image of mega-events that somebody might form after consulting our ideal mega-event library is that of a politically and socially de-contextualized short-term event with little or no medium- or long-term effect or impact.

C Re-contexting mega-events: total impact analysis

Research interests in the field of mega-event impacts have broadened in recent years. Firstly, from the study of mega-events as large-scale but *temporary* tourist attractions, attention is now being given to large-scale *permanent* tourist attractions.[34] This evolution promises to reconnect mega-event study with mainstream tourism studies and particularly to the area of urban tourism. The tourism industry and tourism policy is currently in a very dynamic phase both nationally and internationally. At *local* level this dynamism has manifested itself in major new construction development programmes in the late 1980s and early 1990s involving new museums and art galleries, leisure-oriented mega-shopping complexes, leisure-oriented waterfronts, theme parks, viewing towers, etc.

It is in this field of urban modernization in particular that the future mega-events research agenda needs to be constructed and developed, and I will elaborate on this argument in Section IV below.

Secondly, beyond this potential study of (micro) 'modernization' impacts and contexts of current mega-events, researchers are now also entertaining the concept of 'total-impact' analysis. And this seems to be a more extreme and problematic re-contextualizing response to the de-contextualizing inadequacies of previous mega-events research, as compared with the study of modernization impacts. We need to look a little further at this particular form of re-contextualization before turning back to consider modernization impacts.

The key study laying a framework down for what can be called 'total-impacts' research is Brent Ritchie's analysis of 'hallmark event' impacts in 1984.[35] Besides his work on the Quebec Carnaval, Ritchie and his colleagues devised a research programme for periodic monitoring of the Calgary community's pre-event awareness and attitudes to the Calgary Winter Olympics 1988 in a series of studies undertaken in 1984, 1985 and 1987.[36] While this research did not address total-impact issues its longitudinal perspective and design is unique in the literature and illustrates a key dimension of any prospective research on total and development impacts, that is to say the long time-frame within which studies need to be conducted.

Ritchie (1984) offers a classification of types of hallmark (or mega-)events, types of impact, and types of variables that might be measured as indicators for each type of impact. Hallmark events are defined as

> Major one-time or recurring events of limited duration, developed mainly to enhance the awareness, appeal and profitability of a tourism destination in a short and/or long term. Such events rely for their success on uniqueness, status, or timely significance to create interest and attract attention.[37]

For Ritchie there are seven main *types of hallmark event*: (a) world fairs/expositions (e.g. Expo 67 Montreal, Expo 80 Vancouver, etc.); (b) unique carnivals and festivals (e.g. Quebec Winter Carnaval, Calgary Stampede, etc.); (c) major sports events (e.g. Olympics, Marathons, Grand Prix Racing, etc.); (d) significant cultural and religious events (e.g. Oberammergau Passion Play, Royal Weddings); (e) historical milestones (Anniversaries, Centennials, Bicentennials, etc., e.g. 500th Anniversary of the Discovery of America 1492–1992); (f) classical, commercial and agricultural events (e.g. Royal Winter Fair, Toronto; Florida 82, Amsterdam); and (g) major political personage events (e.g. Presidential inaugurations, Papal visits, Heads of State funerals, etc.). Evidently this catalogue is indicative rather than fully comprehensive. For instance, the mega-event status often achieved by visits, tours and performances by 'mega stars' particularly in the rock and pop world, and in addition the recent very significant mobilizations of popular interest in moral and political issues through rock and pop mega-events (Band Aid, Live Aid, etc.), are not mentioned or catered for. They don't fit easily into categories (b) or (c) so category (g) needs to be expanded beyond the idea of 'political personality' to accommodate them. Failing that, other categories are probably needed to develop the categorical scheme if it is to be adequate to the constantly changing touristic events field of the late twentieth century.

The six main types of event *impact* Ritchie gives are the following: (a) economic, (b) tourism/commercial, (c) physical; (d) socio-cultural, (e) psychological and (f) political, and for each of them both positive and negative possible impacts are noted. The positive impacts (a)–(c) are most often cited and studied while their negative impacts are rarely discussed (as we noted above in Section III). In Ritchie's view 'an honest assessment' of impact should seek to acknowledge costs as well as benefits, and thus, in these cases, should acknowledge (1) consumer price rises during the event and real estate speculation as possible economic costs; (2) a poor reputation because of inadequate facilities or antagonism from local business at the competition as possible tourism/commercial costs; and (3) environmental damage and overcrowding as possible physical costs.

In Ritchie's view impact types (d)–(e) are 'emerging' and 'speculative' areas in which the positive benefits may be 'increase in permanent level of local interest and participation in type of activity associated with event' and 'strengthening of regional traditions and values' (socio-cultural); 'increased local pride and community spirit' and 'increased awareness of non-local perceptions' (psychological); and finally 'enhanced international recognition of region and its values' and 'propagation of political values held by government and/or population' (political). However, the negative impacts which need to be considered include: 'commercialization of activities which may be of a personal or private nature' and 'modification of nature of event/activity to accommodate tourism' (socio-cultural); 'tendency towards defensive attitudes concerning host regions' and 'high possibility of misunderstandings leading to varying degrees of host/visitor hostility' (psychological); and finally, 'economic exploitation of local population to satisfy ambitions of political elite' and 'distortion of true nature of event to reflect values of political system of the day' (political).[38]

To *measure* these various types of impact, Ritchie suggests some indicators and discusses the problem of data collection and interpretation associated with each of them: (a) economic indicators and their problems we have already discussed; (b) tourism/commercial indicators include international and national awareness and knowledge of destination and knowledge of investment opportunities; (c) physical indicators include, for facilities development, additional recreational plant/resources attributable to event and infrastructure development and, for environmental impact, physical degradation/enhancement and overcrowding/crowd management; (d) socio-cultural indicators include measure of compatibility between event and community and of extent to which event contributes to local socio-cultural change and development; (e) psychological indicators include measures of the event's prestige and of hospitality of the host community; and (f) political indicators include the degree of enhancement of the influence of particular political ideologies and of key political actors' roles and careers. Generally none of these indicators including the narrowly economic ones, as we have seen, are seen by Ritchie as being particularly easy to measure. This is particularly so, given their nature, for the political indicators and effects.

A recent attempt to apply a 'total-impact' framework, which appears to be derived from Ritchie's version, is Ahn's pre-event study of the possible impact of the Seoul Olympic Games 1988.[39] It appears to have been a useful schema for ordering the wide range of themes touched on in the available Korean official and journalistic literature on the Seoul Games. Since most of Ritchie's indicators imply post-event and in some cases long-term measurement little can be gleaned from this particular application. Nevertheless, Ahn's clear and wide-ranging conception of the role of tourism, that it 'is regarded as an economic development strategy, a cultural-decay stopper, a peace industry and a money-making venture',[40] implies the need for some such total-impact framework in this case and in general; while the stimulus for the study of such impacts as socio-cultural effects contained in both Ritchie's and Ahn's own version of such a framework leads Ahn to some interesting and dramatic observations on the possible positive and negative effects of the Seoul Olympics on Korean society and culture. On the positive side he comments 'We feel [*sic*] ourselves to be world citizens in the course of the Games ... (and this effect will endure) ... to form an advanced people's consciousness'. On the negative side: 'We can predict that social crimes will increase and the social value system will change. The traditional value system and traditional public morality will be destroyed'.[41]

As we have seen in this section, mega-event studies have evolved from narrow economic impact to total-impact perspectives. More recently they have begun to evolve again towards a focus on what might be called 'development impacts' or 'micro-modernization impacts'. We will consider these latter impacts later (Section IV below). For the moment we can conclude with some brief criticisms of the 'total-impact' approach to re-contextualizing mega-events and researching them. A basic problem with total-impact frameworks is that they present research ideals and it is hard to see how they could be operationalized as a whole in a given study of a particular event. Their demands on research time, interdisciplinary expertise and money would seem to render them prohibitive. Further, while on the one hand they seem to aim at an unrealizable descriptive or empirical exhaustiveness, on the other they do not appear to explicitly test or illustrate any significant modern social theories or explanatory models (e.g. concerning modernization, functional adaptation to social change, urban development or other such issues). We will return to these issues later (Section IV below).

D On redefining mega-events

On the basis of my discussion in this section it is clear that we need to begin to re-define mega-events in sociological terms if adequate research is to be developed in this area of the sociology of tourism. To redefine mega-events for research purposes is (a) to see them as essentially complex and *multi-dimensional events* which require interdisciplinary social and policy study and/or broadly based sociological conceptualization and study and, importantly (in terms of the discussion in this section), (b) to see them as *events-in-context*.

Events-in-context

'Mega-events' refer to urban processes which are extended over time. We can elaborate on these two points as follows.

Mega-events and the policy-making connected with them are in principle a *whole (local/urban) community's business*. To be 'successful' against a range of criteria they need to involve as broad a spectrum of the community as possible.[42] The processes of deciding on mega-events (via local politics), staffing them (as volunteers), supporting them (as paying spectators) and reaping whatever benefits are to be had (through small business opportunities, after-use of facilities, etc.)[43] are 'whole community' matters both in principle and in many respects in practice also. The 'communities' involved with genuine mega-events are usually big cities. Sometimes these are world cities and/or capital cities in which the general impact of mega-events may be reduced to that of a temporary addition to the city's big tourist attractions. But most often the city involved at best has some regional or provincial standing but is usually small enough to be capable of being 'taken over' for a period by the event. Mega-events, then, whatever else they are, are usually *urban phenomena*, and are best conceptualized and studied as such against the background of the broader local social and economic situation in which they occur (see also Section IVA below).

In addition, *the significance of mega-events extends over time* beyond the present event to include the three temporal/historical dimensions of past, present and future, together with the plans and policies, practices and processes that generate and react to the event. Mega-events and their significance depend on all sorts of background processes which usually extend in time well beyond the short period of the event occurrence itself. The background processes can obviously include *pre-event*, the creation and planning of both the event and its physical facilities and associated (transport and accommodation) infrastructures, and *post-event*, the successive waves of direct and indirect economic impact and the possibly longer-term impacts on city image and civic morale. The multi-dimensionality of mega-events includes these temporal, processual and practical dimensions.

Multi-dimensionality

In addition to these temporal multi-dimensions of mega-events, the latter are also essentially *sociologically complex* and multi-dimensional in the following sense. Any mega-event such as an Olympic Games, for example, is simultaneously a *work* experience for the participants, an unusual *leisure* experience for local spectators, a *touristic* experience for visiting spectators, and a *media* phenomenon for media professionals and viewers. In addition, besides these socio-psychological, symbolic and *cultural* dimensions, there are also of course the event's explicit and direct *economic* dimensions. These include the employment and income, ticket sales and revenue, profits and losses, etc., that any mega-event generates.

This range of *practical, socio-cultural* and *economic* aspects of mega-events implies that the proper study of them cannot be said to fall neatly into the sphere of any single discipline or perspective. In this respect the study of mega-events brings into sharp relief the same sort of problems which face the study of tourism in general (as we noted above, Sections I and II).[44] That is, the essential complexity of the mega-event phenomenon calls for the development and application in empirical studies of interdisciplinary as well as broad sociological approaches. We can now turn to some important aspects of the context of contemporary mega-events, namely urban tourism and urban development, or 'micro-modernization'.

IV MEGA-EVENTS AND MICRO-MODERNIZATION: THEMES AND ISSUES

My discussion in this paper suggests that the sociology of contemporary tourism needs to be informed both by the implications of the modernization problematic outlined in Section II, and also by an awareness of the deficiencies of existing approaches to the study of mega-events and their local impacts outlined in Section III. Taken together these two lines of argument suggest the need for a broad conception of the future research agenda in the tourism field. They suggest the need to guide and contextualize the study of mega-events and their impacts in terms of two kinds of contextual studies. First, there are the *policy contexts* of urban tourism policy-making, local economic development policy-making and relevant central–local politics. Second, there are the *structural contexts* (at 'macro'-levels) of the 'micro'-modernization processes and 'in general' relevant aspects of macro-modernization developments.

'Micro-modernization' simply refers to locally variable and specific processes of more general macro-level processes of de-industrial/post-industrial and post-national change, processes of structural disorganization and reorganization (Section II above). These structural context processes, along with the policy context processes, are in principle researchable through specific community city and region case studies. The research agenda for mega-events calls for a new generation of such studies in the 1990s adequate to the intellectual and political challenges of the times, studies which refuse to treat structural and policy changes and their effects in isolation from each other, and which address themselves to the dialectics of tourism (Section V below), not least those between 'contexts' and 'impacts'.

This effectively summarizes the main programmatic points I wish to make in this paper. However, to be properly understood, these schematic points could do with some illustration and elaboration. So in this section, with regard to *contexts*, I will first look at some aspects of the urban policy context connected with micro-modernization. Here, in particular, reference needs to be made to the recent emergence of 'local corporatism' in British urban politics (A below); while, second, as regards *impacts*, we need to consider some of the issues involved in attempting to conceptualize, describe and assess mega-event impacts as (micro-)modernization impacts (B below). In principle the linkages between these two sets of issues would need to be explored and to be made manifest empirically and analytically. However, space prohibits me from doing much more here than outlining aspects of the two sets and merely flagging the need for research to establish the linkages.

A Issues in urban tourism policy analysis: micro-modernization and local corporatism

As we noted earlier (Section I) in recent years local and urban tourism strategies, particularly in 'rust belt' cities, have been encouraged by governments in North America and Europe. They have been undertaken as part of long-term processes of economic reorientation, restructuring and renewal. Tourism is seen, for instance in British government employment and enterprise policy[45] and urban tourism policy,[46] as promising a major stimulus to regional and urban economies. In addition to its intrinsic effects as a growth industry tourism also often requires significant and lasting improvements in urban infrastructures and facilities, particularly labour market and training, communications, transport and 'hospitality' infrastructures. It is assumed that urban tourism can induce a modernization of the local economy and, through its effects on local morale and outside image, that it can positively influence outside and local investment and local labour productivity. But, as we have suggested throughout this paper, this development of urban tourism strategies in the advanced societies needs to be understood in terms of the contexts of local structural change and local politics and also of the national urban policies within which they occur. We will now briefly consider these structural contexts (1 below), with particular reference to the recent British experience of them (2 below).

1 The modernization problematic outlined earlier (Section II above) implies the emergence of structural effects and new political possibilities at the urban or micro-level due to (a) the development and diffusion of post-industrial technologies at macro/nation-state level and (b) the corrosive effects of post-national developments on nation-state power. The former implies that local and peripheral levels and areas have a new and unprecedented capacity in the late twentieth century both to host and/or participate in the powerful 'national' political and economic institutions, organizations and systems hitherto monopolized by capital cities and dominant regions; while the latter, as the corollary of this, implies that that transnational development of power and organization at transnational levels (such as the European Community) will tend to undermine and loosen nationally centralized power, authority and sovereignty. This in turn is likely to give a stimulus to the development of subnational/urban autonomy and empowerment, to fuel centre–local conflicts over the distribution of power and authority with the nation-state, and to engender periods and processes of disorganization and reorganization at and between central and local levels in the state.[47] These dynamics provide the general background for the development in Britain, for instance, of centre–local conflicts in the 1980s.

2 Recently effective forms of local political and economic coordination and 'partnership', identified by some commentators as 'local corporatism',[48] have begun to develop. Micro-modernization in the 1980s and 1990s, particularly in the 'rust belt' cities and regions of North America and Europe – to the extent that it is capable of being subjected to political control – is likely to be a product of varieties of 'local corporatist' arrangement. We can briefly look at British experience in light of these general themes of 'micro-modernization' and 'local corporatism'.

   Urban or micro-modernization in Britain in the mid and late 1980s has been undertaken against an often highly charged and divisive political background[49] and against a background of de-industrialization, with high levels of inner city unemployment and occasional rioting.[50] In the central–local government context the main themes of the 1980s have been those of conflict and central government domination, a decline in the power of the local state and a growth in the power of central government.[51] In addition to the straightforward abolition of a regional tier of government (the Greater London Council, etc.) there has been an unremitting regime of spending controls on local authorities' (LAs) spending, most recently through the disastrous 'poll tax' experiment. Since the 1987 Election in particular a distinctive Thatcherite urban policy has been developed. Privatization and/or responsiveness to market forces and 'consumer choice' have been imposed across the range of local public services and the power of Local Education Authorities (LEAs) has been significantly undermined and reduced. Centrally funded Urban Development Corporations (UDCs) and, recently (1990), Training and Enterprise Councils (TECs) have been imposed on many large cities. The UDCs' roles have been to both cut through and cut back local land-use planning controls and to attract external private sector investment to key inner city areas; while the TECs' role is to take control of local vocational education and training, reducing the influence of LEAs, and attempting to maximize the capacity of local training systems and markets to respond quickly and flexibly to changes in local labour market conditions and local employers' labour skill needs.

Against this conflictual background the dominant theme particularly at local level, and particularly in the late 1980s, has come to be that of *cooperation* between the various major players in the urban policy game. The era of 'corporatism' and 'consensus politics', at its height in the 1960s and 1970s, may well have been long dead at the *national* level in Britain. But in the late 1980s it seems that we are witnessing a resurgence of corporatism and consensus politics at the *local* level (Grant 1985, Judd and Parkinson 1990, see note 48). In the earlier period national corporatism was held, by many pro-market and Right-wing political and economic analysts at least, to be largely responsible for Britain's post-war economic problems, for Britain's traditionalism and its *failure* to *modernize*. But the current era of 'local corporatism' has emerged as a response precisely to the needs *for modernization* and for local economic development.[52]

Contemporary local corporatism, then, involves a range of more or less systematic and institutionalized attempts to promote economic development regeneration and restructuring in Britain's major cities. The major players in the local policy-making process – namely the LAs, the UDCs and TECs (where they are in place), the Chambers of Commerce and often major private and public sector employers – are usually formally linked together in some kind of 'partnership' arrangement. This then provides the local backdrop and incentive for coordinated campaigns:

(a) to attract new inward investment, particularly in 'post-industrial' service sector industries (such as banking and finance, information and communication technologies, leisure-based retailing, hospitality and tourism, etc.);

(b) to upgrade the local communications and transport infrastructure; and

(c) to improve the skills of the local labour market by improving the local education and training infrastructure.

The process of modernization involves the transformation of industrially oriented economies and their physical, technological and social infrastructures in a post-industrial economy direction.[53] Local corporatism is probably the main political and policy-making mechanism through which the profound structural changes associated with late twentieth-century modernization and post-industrialism will be achieved or at least channelled in the 1990s.[54] This is particularly so in Britain where central government is antagonistic to planning long-term change (or at least is antagonistic to being *seen* to be planning).

3 On the basis of this brief account of the urban policy context it is clear that on

any future research agenda for the sociology of mega-events in the 1990s we will need to include at least the following sorts of studies. First, studies are needed of the various forms of local corporatism and local modernization policies; while second, the relation between these on the one hand and local tourism and mega-event policy-making on the other hand needs to be described and better understood than it is currently. One aspect of this second area of research involves developing better descriptions and assessment of mega-event impacts than we have had hitherto. It is this theme of rethinking mega-event impact analysis to which we now briefly turn.

B Issues in mega-event impact analysis: 'micro-modernization impacts'

How can we begin to identify and to research the economic development-relevant (or 'micro-modernization'-relevant), direct and indirect impacts (effects, functions, etc.) of mega-events? The concept of the 'total impact' of mega-events discussed earlier (Section IIIc above) may be of some use here. Ritchie's (1984, see note 35) concept of 'total impacts' is subject to various criticisms of the kind outlined earlier. Nonetheless a selection and combination of *some* of the elements of his total-impact schema helps to begin to map out potential modernization-relevant impacts. These impacts, then, consist in general of the kinds of effects Ritchie classifies as 'economic', 'touristic/commercial' and 'physical' (facilities and infrastructure). Ideally, in terms of the modernization problematic, these effects and functions would need to be described and assessed in terms of (a) their status (*or* lack of it) as goals of a coherent urban policy; (b) their degree of inter-relationship and synergy; and (c) their combined effects over some appropriate period of time beyond the short term.

However, even without these provisos this rough schematization of potential modernization impacts usefully expands upon the conceptualization of the economic dimensions and impacts of mega-events characteristic of tourism research and mega-event studies.[55] It brings on stage factors largely ignored in the early studies, in particular *tourism/commercial* factors like publicity, media coverage and sponsorship, and *physical factors* like facility construction and associated capital expenditure and infrastructural programmes. These two major sets of factors have been recognized in more recent studies of mega-events, although in an *ad hoc* way and without much contextualization in urban policy or modernization-sensitive terms. We can now look at them in a little more detail.

Tourism/commercial factors: media involvement

Early studies of mega-event impacts tended to ignore the role of the media and associated sponsorship and publicity effects. To a certain extent this was because event organizers themselves did not always make such factors a priority or give an account of them. But such a situation would be inconceivable in the 1980s and 1990s both for organizers and for researchers.

Event income

The role of media and sponsorship is most clear in the case of premier mega-events like the Summer and Winter Olympic Games. The sale of television rights has, for many years, contributed a major and (usually) continually inflating proportion of the *gross income* earned by Olympic event organizers, for instance. This, in turn, funds their part of their initial direct expenditure in the host city.[56] Although the Olympics *appear* to guarantee profits to all concerned *in fact* their financial base, because it has become so media dependent, is now vulnerable to an unprecedented degree. This vulnerability was demonstrated in 1988 at the Calgary Winter Olympics where the USA TV network ABC got its figures wrong and its fingers burned. The network paid $309 million for the rights to televise the Games.[57] The network's ability to sell advertising time slots to large companies turned out to be less than it had assumed, and at the end of the day it was left with a loss of £65 million. As a consequence of this a new phenomenon has appeared in the international multinational auction of TV rights to Olympic Games, namely a *reduction* in the sale price rather than the usual increase. The 1992 Winter Olympics at Albertville in France will have to get by on $243 million for its TV rights, considerably down on Calgary.[58] Income from the sale of TV rights is usually the biggest single item of income to the organizers of premier mega-events such as the Olympic Games (bigger than ticket sales, etc.). The Albertville organizers may reasonably have expected and planned for an increase on Calgary income and thus planned a certain scale of spending on facility construction appropriate to this probable increase. Given the reductions in TV rights income it might have been assumed that they would have attempted to constrain their budget. But such is the 'booster' mentality which takes over mega-event organizers that by 1992 their budget had soared to $740 million and the event's finances were in crisis. (See Burrington and Warshaw, 1992, in footnote 58.)

Publicity values and effects

Besides this impact of the sale of TV rights on direct expenditure, media presentation of the event also inevitably provides a certain amount of free and generally positive *publicity* for the host city. The market value of this can, in principle, be estimated, and ideally future studies of event impacts could reasonably be expected to include such estimates in addition to direct media income.

As Ritchie (1984) noted, the *effect* of this media exposure of the host city on potential future tourists' and future investors' awareness, attitudes and behaviour, in principle at least, can also be roughly established and measured using appropriate surveys.[59] But clearly the practical difficulties in studying publicity effects for, let us say, a global Olympics audience over an appropriate time-frame in which the effects might be felt are profound and have not yet been adequately tackled by researchers in this field.

Publicity effects, together with scale of media and sponsorship income, need to be built into our definition of what counts as a mega-event (Section IIID above).

Mega-events ought to be by definition 'newsworthy' so *some* kind of publicity measure, however difficult to achieve, needs to be added to our set of criteria for determining what counts as a mega-event and as a mega-event's impact.

Finally we need to bear in mind that there can even be impacts felt by mega *non*-events. On the publicity benefits of bidding for mega-events and *failing* Travis and Croizé comment dryly (regarding Birmingham's 1986 failed bid for the 1992 Olympics) as follows:

> It has been suggested that advertising time on world media, equivalent to £25 million sterling, was gained by the £5 million bid. Cynics are suggesting that mega-event bids, without the event, give a greater weight of net benefits![60]

Physical factors, infrastructures and longer-term effects

It is vitally important in studying contemporary mega-events to identify something that the early mega-event impacts usually failed to do, namely that mega-events are quite often accompanied by a local building boom of one kind or another. These usually include, for instance, the new sporting and/or cultural facilities the event requires, new transport infrastructures, new hotels and tourist attractions, and improvements in strategically significant telecommunications systems and networks. To provide adequate event facilities may require either a new build-from-scratch programme (e.g. Seoul 1988) or a refurbishment of existing facilities (e.g. Los Angeles 1984). Rather than attempting to define mega-events and their impacts exclusively in terms of visitor numbers or direct spend[61] I believe that, in addition to the 'newsworthiness' criterion noted above, we also need to build in some measure of the scale of associated building and infrastructure spend.

Building-spend has two sets of effect. (1) One set is *prior* to the event and is felt on employment and thus on local income with its multiplier effects, and also on the 'bringing forward' or 'levering' effect it exerts on other capital investments. (2) The other set of effects concerns the *post-event after-use* of the 'legacy' of facilities, and their long-term effects as tourist attractions on tourism income, local employment, business opportunities, profit-making and investment. Providing the physical build due to the event can be accurately identified then it should be possible to calculate *both* pre- and post-event sets of construction, employment and income effects with a greater or lesser degree of accuracy. It is reasonable to expect that future mega-event impact studies will at least include estimates of the economic impacts of both sets of building-spend effects.

The pre-event levering effect of mega-events on other capital investments has been noted in recent research on a number of occasions. In his pioneering review of 30 mega-developments, Armstrong[62] demonstrated that, pre-event, the preparation of a mega-event often encourages both public and private sector to bring forward building plans they may have been going to proceed with anyway. In the case of the 1986/7 American's Cup Race off Perth, an interesting pre-event projection study[63]

336

was able to attribute a significant amount of already planned new building in the region to the fact of the occurrence of the event. The study assumed that if these planned developments were brought forward by at least three years and were in some other way event-related then they could be treated as anticipatory or preparation effects of the event and their costs counted in to the event's direct expenditure effect. In the case of the Sheffield World Student Games 1991, c£25 million is being spent purely on the organization of the cultural and sporting events, while c£150 million was spent on related sports and other facilities. Much of this latter sum, together with other expenditure (e.g. major retail developments brought forward) and their impacts, could be reasonably credited to the World Student Games mega-event as indirect and 'levered' pre-event effects of its preparation. Longer-term effects are obviously more difficult to measure but the need to attempt to measure them is clear. It has been unequivocally recognized and asserted by Travis and Croizé, in their review of contributions to this field by Armstrong and others, that

> if events are the start of new longlife use of major new capital investments then impact assessment should be over the life of the investment.[64]

However, event-related capital investment, particularly in permanent and *visible* tourist attractions, (such as waterfronts, hotels, mega-retailing complexes, etc.), must be associated with a development of local labour force skills and community attitudes, i.e. *invisible* assets. Referring to the local development of hospitality attitudes and skills, and generally local tourism industry skills and interests, Travis and Croizé make the telling point, from a development perspective, that

> Without the accumulation of these invisible assets, which must be fostered long in advance, the mega-event would remain a festival without a future.[65]
> [That is to say, whatever the event's visible, *physical* and architectural legacy.]

In Britain the role of the new Training and Enterprise Councils (TECs) is likely to be a critical one in terms of the development of local labour market skills, while joint public–private sector 'partnership' ventures such as Marketing Bureaux will have important effects on local attitudes as well as 'city images'.

Policy aspects of mega-events seen as modernizing agents

Current studies can be said to flag the 'micro-modernization' or development impacts and the policy-problematic when they address themselves to providing strategies (a) for prolonging the positive effects of mega-events and in general (b) for ensuring that mega-events have a long-term usefulness to their host communities. Some recent proposals in their vein have been made by Ritchie and Yangzhou and Travis and Croizé.[66]

Brent Ritchie and Yangzhou propose among other things:

1 modifying the events' duration (by adding on days);

2 creating lead-in, follow-up events and satellite events involving the local community;

3 post-event utilization of the event facilities for scaled replicas of the event in related fields (e.g. junior/veteran Olympics, etc.);

4 organizing long-term legacy funding for updating the facilities, etc.

Travis and Croizé make similar kinds of proposals to minimize disbenefits and maximize long-term benefits. They also suggest, in developing a programme or cycle of mega-events, the possibility of timetabling mega-events into the off-season to achieve the benefits Ritchie and Beliveau[67] proposed in countering the 'seasonality' factor. The timing of most mega-events tends to ignore this and, occurring at peak season times, they can produce a 'stay-away' effect on normal business travellers and tourists not interested in the event. This effect was recorded at the 1984 Los Angeles Olympics.[68]

According to Travis and Croizé – expressing a view in accord with the emerging 'conventional policy wisdom' of the new local corporatism in organizing mega-events – the public sector should be used to cover infrastructural costs and the private sector to cover as much of the direct costs as possible. They imply that two approaches can be taken in developing mega-events strategies. One aims to maximize local benefits (direct and indirect, short- and long-term) which might nonetheless risk high local costs. The other aims to minimize disbenefits. A benefit-maximizing strategy is one which stages a mega-event to launch a new generation of facilities and infrastructures. A disbenefit-minimizing strategy is one which involves minimal new infrastructure 'but give(s) maximum returns and high multipliers', via mega-events such as motor races.[69]

It is reasonable to infer from this that the propensity to choose one rather than the other mega-event strategy will reflect the relative power and political perspectives of the elements composing the local corporatisms of particular cities and regions. The risky 'local benefit' maximizing strategies are more likely to be adopted in situations where local government has taken more of the initiative, while the 'disbenefit-minimizing (and direct profit-maximizing) strategies' are more likely to be undertaken in situations where the local private sector has more of this initiative. But in the present state of knowledge these are only rough heuristics and hypotheses, to be fleshed out and tested out in case studies and other empirical studies in the political economy and political sociology of urban policy and mega-event organization.

V CONCLUSION

In this conclusion I will briefly summarize the aims and argument of the paper and then offer some initial thoughts about the nature of the sociology of tourism implied and envisaged by the paper. Generally I suggest that the sociology of tourism is best seen as one kind of intellectual response to the existential challenges posed by the touristic culture of modernity and that it needs to be conceptualized in dialectical terms.

A Summary

In spite of the economic and cultural importance of tourism in late twentieth-century society the sociology of tourism in general is undeveloped, and in particular the sociology of the new urban tourism and its characteristic expression, the organization of mega-events, is virtually non-existent. What little sociology of tourism there is tends to be angled towards the cultural and social psychological dimensions of tourism.[70]

In this paper I have aimed to do three things. First I wanted to put together some of the elements of a case for the development of a broadly sociological and interdisciplinary study of contemporary tourism. Second I aimed to indicate the general social theoretical significance of tourism and some of the concepts and methods which seem to me to be both useful and necessary for understanding it. And third I aimed to emphasize the economic (socio-economic/political economic) rather than the cultural aspects of tourism and modern society. I suggested that concepts of post-industrialism and post-nationalism, at both macro (Section II) and micro (Section IV) levels, are useful in conceiving of an economic sociology (and political economy) of contemporary tourism. We need a better understanding of tourism as a major new 'industrial sector' and as a vital factor in economic modernization and structural socio-economic change (Sections II and III). In particular the study of touristic mega-events, together with their urban politico-economic contexts and impacts, seems to me to be particularly important and potentially fruitful. This is not least because they are among the few aspects of tourism to have consistently received a fair amount of attention by social science research in the 1970s and 1980s, albeit of a limited economistic kind. In this context various conceptions and examples of mega-event impact analysis were reviewed and criticized (Section III) and new criteria for more adequate contextualization, and for modernization-sensitive and policy-sensitive impact studies, were suggested (Section IV).

B For a dialectical sociology of tourism

The sociology of tourism I envisage on the basis of my discussion in this paper may appear to be unduly structuralist or political economic. However, I envisage it rather as having an essentially *dialectical* character. This is the understanding implied by the emphasis I have placed on *contextualization* in my discussion – i.e. (a) the contextualization of economistic impact analysis with reference to urban policy and structural change, and also (b), in turn, the contextualization of the latter with reference to macro-structural change and the modernization problematic. So in conclusion I should say a few words about this general dialectical characteristic of the sociology of tourism as I see it.

1 In my view sociology is unavoidably a dialectical discipline. For instance, it requires its practitioners to appreciate *the difference and interdependence* between social facts and social values, between theory and description, and between theory and policy. But further it requires them to appreciate the *unity-in-difference* in social

reality of such complex phenomena as action and structure, continuity and change, consciousness and material conditions, micro- and macro-levels and so on. It is not easy to take full account of these relationships as dialectical ones, that is as requiring *both* dimensions or poles to be addressed. Thus they have often been crudely interpreted as causal or deterministic relationships, (when for instance it is assumed that structures *cause* action, or vice versa, etc.). Or the bi-polarity can be ignored altogether, leading to inadequate one-dimensional accounts of complex social reality in *exclusively* structuralist or actionist terms, etc.

Given that it is not easy to grasp the dialectical character of social reality it can help to think about these relationships using a simple analogy such as that of the figure–ground characteristics of ordinary perceptions. So we can picture dialectical social relationships as being like figure–ground relationships, reversible according to perspective and interest, in this case the perspective and interests of particular researchers and research projects in the sociology of tourism.

Studies which for certain purposes treat e.g. action (or e.g. change or consciousness) as 'figures' need to recognize first that they are essentially *partial*. Second they need to recognize that they *implicitly* treat e.g. structure (or, respectively, continuity or material conditions) as 'grounds' (or as I have noted earlier in Section III as *contexts*); while, third, they need to recognize that alternative studies are always possible which reverse their interests and thus treat e.g. structure, (or continuity and material conditions) as 'figures' and their correlates as 'grounds' or contexts. From this dialectical point of view, a *total* account of any social phenomenon cannot be given unless it is studied in both figure and ground terms.

Furthermore *multi*-dimensional phenomena can also be usefully analysed in these *two*-dimensional dialectical figure–ground terms. For instance, the Weberian notion that societies exhibit *three* dimensions of inequality (class, status, power) and the conventional notion that societies exhibit three structural dimensions (namely, the economic, cultural and political) can be interpreted along these lines, selecting one dimension as the figure and treating the other two as a composite ground or context. We now need to consider the relevance of these theoretical and meta-theoretical observations to the development of a sociology of tourism appropriate to the circumstances of the late twentieth century.

2 Evidently tourism in general, and urban tourism and mega-events in particular, are complex and changing multi-dimensional phenomena with complex and changing impacts and effects. In this paper I have suggested that while they can be studied as narrow *economic* phenomena (e.g. as products sold to tourism consumer markets, as forms of skill sold to tourism labour markets, etc.) they are better studied more broadly as economic phenomena contexted both by *policy-making (political) processes* and *structural socio-economic change* at macro (national/international) but particularly at micro (local/urban) levels. But there is obviously another context beyond the political and the socio-economic contexts, namely the *cultural context*, and I have said very little about this (Introduction, c above).

The cultural dimension of tourism consists of collective (semiological/ideological) and subjective (phenomenological) meanings and values. Of course in

340

post-industrial capitalism meanings are commodified in the economy and they are instrumentalized in the polity. But nonetheless culture retains its relative autonomy both for analytic purposes and in social reality. So, in dialectical terms, the unspoken context or ground of my account of (some aspects of) the *political economy* of the new urban tourism has been my implicit and assumed understanding of *culture* and of tourism as a cultural phenomenon.

It is, then, quite consistent with my emphasis in this paper to envisage research projects which treat culture as the figure rather than as the ground in studies of urban tourism. Clearly much interesting work is possible on urban tourism as a cultural phenomenon and as a 'figure' from perspectives such as post-modern cultural analysis. But it has to be stressed here that *exclusively* cultural studies of tourism are of little use and value. Socio-economic and political economic factors need to be acknowledged as the 'ground' in any study of (urban) tourism and mega-events as cultural phenomena.

3 So, culture, (as the 'ground' of accounts such as mine where urban tourism is a political economic 'figure') can itself evidently be turned into a 'figure' and focus for study. I'd like to conclude by briefly noting why such a study might be intellectually significant and what it involves. At the most general level it seems to me that the *cultural impact* of the modern tourism economy is enormous and that it is threatening to be an all-pervasive impact, making tourists of us all in *every* aspect of our lives, not just when we are 'on holiday'. At both macro- and micro-levels, late twentieth-century modernization processes, particularly in their cultural aspect, seem to be creating a form of civilization favourable not only to the tourism industry but to the touristic world view – in effect a touristic civilization. That is, tourism is not just becoming a strategically important sector in the economy and in economic change, but has for some time been an even more important force for change in the culture of modernity.

A variety of motives for modern people's apparently obsessive interest in travel and touring have been discussed in the sociology of tourism, including frustration with the familiar and the need to 'escape',[71] the desire to find 'roots',[72] the need to revalue 'home' and return to it,[73] and so on. Beyond these questions are further questions about whether modern tourism is essentially a humanizing and educative form of life or an essentially estranged and inauthentic one.[74]

The apparently boundless growth of the tourism industry at all levels is beginning to recreate our domestic environments in order to make them accessible and attractive to strangers. In the late twentieth century people in the advanced societies seem increasingly to be seeking on the one hand moral purpose from respect for nature and from conservation of the plant and animal environments that surround them, and on the other hand a sense of community and unique social identity from place, usually the place where they live, that place usually being in towns or cities. Yet at the same time both natural and urban environments are being increasingly given over to tourism, and the natural and/or unique aspects of place are being marketed and made universally available in tourism consumer markets.

In the light of these general cultural dynamics and problems it seems to me that the need to develop a sociology of contemporary tourism along the lines suggested in this paper is overdue and is pressing. It should have much to tell us about the quality, creativity and value of everyday life in late twentieth-century society.

C The challenge of touristic culture

It may well be that, as the poet says, we need to 'see ourselves as others see us'. The notions of 'others' ('outsiders', 'strangers', 'aliens'), of their views of 'us' and of their disturbing presence in 'our' society are archetypal and perennial.[75] In the late twentieth century western society tourism presents us with all-pervasive, although neutered, versions of these archetypes. For us 'post-moderns', then, the import of Burns' words might be that we ought to 'see ourselves as tourists see us'. But whether we *ought* to or not, indeed whether we *want* to or not, the implacable long-term growth of tourism means that we almost certainly *are* coming to see ourselves this way, and will increasingly do so.

Given this, the critical problems for the touristic civilization which is emerging in the 1990s and early twenty-first century will not be how to heed the poet's advice, but rather how *not* to. That is, *in spite* of touristic culture, how are we to retain (and transmit to future generations) an original and uncompromised grasp of our history, community and identity? We moderns love to travel, but we still take for granted that we can come 'home' and leave the world of tourism behind. Yet touristic civilization will increasingly make 'home' (locality, community, tradition) into a touristic construct.[76] Evidently in the future we will face the classically modern problems of our individual and collective alienation and anomie in disturbing new ways. As tourists we know well enough what 'they' are like and where 'they' belong.

But what are *we* like and where do *we* belong? We had better remember (or discover) before the 'heritage' and 'conservation' industries remember (or invent) it all for us. In the face of touristic civilization we are challenged to remember who we are, where we belong and, most importantly, *to what* we belong. These are, of course, classical sociological themes of identity and community. I assume that sociology aims to develop an understanding of modernity that is classically rooted, critical and contemporary. If so then the study of urban tourism seems to me to be as good a place as any, and a better place than most, in which to begin to understand the challenge of our times.

ACKNOWLEDGEMENTS

This paper in part represents my attempt to make sociological sense out of my own experience both as a tourist and also as a citizen of an old industrial city attempting to use tourism and mega-events to regenerate itself. This paper is not a case study of Sheffield in these respects (on this see Roche, 1991, footnote 64). Nonetheless it does arise out of background research undertaken in connection with my involvement in 1987/9 with the World Student Games (WSG) 1991 Bid Team and also with the Tourism Strategy Joint Officers Group (TJOG) and more generally with the Department of Employment and Economic Development (DEED). My thanks are therefore due to Sheffield City Council and to the members of these groups in particular for the access they gave me to urban tourism policy-making processes.

Gerry Montgomery (WSG Team Leader 1987/8), Jack McBane (WSG), John Taylor (Deputy Director Recreation and TJOG), Keith Cheetham (Director, Publicity and TJOG), David Patmore (Director, Arts and Museums), John Darwin (Deputy Director DEED), and Councillors Peter Price and Don Gow are among many friends and colleagues who, often unwittingly, have given me great help in my research. David Perrow of the University Library gave me considerable help in tracking down economic impact studies. Of course none of these mentioned has any responsibility for the analysis and views expressed in the paper.

NOTES AND REFERENCES

1 F. Blackaby (ed.), *De-industrialisation*, London, Heinemann, 1979.
2 C. M. Law and J. N. Tuppen, *Tourism and Greater Manchester: Final Report of Urban Tourism Research Project*, Salford, University of Salford, 1986; Sheffield City Council, *Going Places: A Tourism Strategy*, Sheffield 1988; English Tourist Board (ETB), *Vision for Cities*, London 1989; ETB, *A Vision for Sheffield*, 1988; M. Parkinson and F. Bianchini, 'Where there's Muck there's Art; Cultural and Urban Renewal', *Times Higher Education Supplement*, 1st December, 1989; J. Roberts, 'Green Mantle: A Critique of Garden Festivals', *Guardian*, 17th June, 1989; R. Hewison, *The Heritage Industry: Britain in a Climate of Decline*, London, Methuen, 1987.
3 House of Commons (Trade and Industry Committee), *Tourism in the UK*, London, HMSO, 1985; Cabinet Office, HMG, *Leisure, Pleasure and Jobs – the Business of Tourism*, London, HMSO, 1985; A. S. Burkart and S. Medlik, *Tourism: Past, Present and Future* (2nd edn), London, Heinemann, 1981, Parts I and II; P. Lavery, *Travel and Tourism*, Cambridge, Elm Publications, 1987, Ch. 1–4; C. Gratton and P. Taylor, *Leisure in Britain*, Hitchin, Herts, Leisure Publications, 1987, Ch. 5 etc.
4 J. M. Bryden, *Tourism and Development*, Cambridge, Cambridge University Press, 1973; E. De Kadt, *Tourism: Passport to Development? Perspectives on the Social and Cultural Effects of Tourism in Developing Countries*, Oxford, World Bank and UNESCO/Oxford University Press, 1979; D. Pearce, *Tourist Development*, Harlow, Longman, 1985; A. Mathieson and G. Wall, *Tourism: Economic, Physical and Social Impacts*, Harlow, Longman, 1987.
5 M. Roche and B. Smart, 'Time, Leisurelineness and Social Futures', in A. Tomlinson (ed.), *Leisure, Polities, Planning and People*, Vol. 4, London, London Leisure Studies Association, 1986; A. J. Veal, *Leisure and the Future*, London, Allen and Unwin, 1987.
6 A. Toffler, *Future Shock*, London, Pan, 1971; P. Murphy, *Tourism: A Community Approach*, London, Methuen, 1985; J-Y. Ahn, 'The Role and Impact of Mega-Events and Attractions on Tourism Development in Asia', in Association Internationals d'Experts Scientifiques du Tourisme (AIEST), *The Role and Impact of Mega-Events and Attractions on Regional and National Tourism Development*, St. Gall, Switzerland, Editions AIEST, 1987; A. Toffler, *The Third Wave*, London, Pan, 1980; A. Toffler, *Preview and Premises*, London, South Green, Pan, 1985; D. Lyon, *The Information Society*, Oxford, Polity Press, 1988.
7 D. MacCannell, *The Tourist: A New Theory of the Leisure Class*, London Macmillan, 1976; see also Boorstin's classic conservative critique of tourism, D. Boorstin, 'The Lost Art of Travel', in his *The Image*, London, Penguin, 1963.
8 J. F. Lyotard, *The Post-Modern Condition*, Manchester, Manchester University Press, 1984; J. Baudrillard, *Selected Writings*, edited by M. Poster, Oxford, Polity Press, 1989; S. Connor,

Post Modernist Culture, Oxford, Blackwell, 1989. For a defence of the kind of humanism post-modernists criticize see M. Roche, *Phenomenology, Language and the Social Sciences*, London, RKP, 1973.

9   For a positive view see J. Krippendorf, *The Holiday Makers: Understanding the Impact of Leisure and Travel*, London, Heinemann, 1989, and for a critique see Mathieson and Wall, 1987, *op. cit.*, ch. 5.

10 On 'post-industrialism' see D. Bell, 'The Social Frameworks of the Information Society', in M. Dertouzos and J. Moses (eds), *The Computer Age: A Twenty Year View*, Boston, MIT Press, 1979; also works by Toffler, 1981, 1985, *op. cit.*, and Lyon, 1988, *op. cit.* On 'post-Fordism' see M. Piore and C. Sabel, *The Second Industrial Divide*, New York, Basic Books, 1986; R. Murray, 'Fordism and Post-Fordism', in S. Hall and M. Jacques (eds), *New Times: The Changing Face of Politics in the 1990s*, London, Lawrence and Wishart, 1989. For a discussion of these general structural changes (in relation to the theory of citizenship) see M. Roche, *Rethinking Citizenship: Welfare, Ideology and Change in Modern Society*, Cambridge, Polity Press, 1992.

11 For a managerialist perspective see T. Levitt, 'The Globalisation of Markets', in his *The Marketing Imagination*, New York, Free Press, 1982, ch. 2. For neo-Marxist perspectives see Hall and Jacques (eds), 1989, *op. cit.*

12 I. Wallerstein and T. Hopkins *et al.*, *World Systems Analysis*, London, Sage, 1982.

13 For a critique of these assumptions in the particular sphere of modern sport and sport policy see M. Roche, 'Sport and Community: Rhetoric and Reality and Contemporary British Sport Policy', in C. Binfield and J. Stephenson (eds), *Sport Culture and Politics*, Sheffield, Sheffield Academic Press, 1993.

14 A. Gamble, *The Free Market and the Strong State: The Politics of Thatcherism*, Basingstoke, Macmillan Educational, 1988.

15 On critical theory see J. Habermas, *Legitimation Crisis*, Boston, Beacon Press, 1973; on neo-Marxist crisis theory see J. O'Connor, *The Meaning of Crisis*, Oxford, Blackwell, 1987; on the ecological critique see J. Porritt, *Seeing Green*, London, Blackwell, 1984.

16 C. Offe, *Disorganised Capitalism*, Oxford, Polity Press, 1985; S. Lash and J. Urry, *The End of Organised Capitalism*, Oxford, Polity Press, 1987.

17 E.g. Murphy, 1985 *op. cit.*, Mathieson and Wall, 1982, *op. cit.*, and Pearce, 1985, *op. cit.*

18 E.g. D. Foster, *Travel and Tourism Management*, London, Macmillan, 1986; also works by Burkart and Medlik, 1981, *op. cit.*, and Lavery, 1985, *op. cit.*

19 M. Roche, 'Mega-Events, Multipliers and Tourism: A Critique of Economic Impact Analysis', 1989 (unpublished), available Policy Studies Centre, Sheffield University.

20 John Keane and John Owens, *After Full Employment*, London, Hutchinson, 1987.

21 M. Roche, 'Citizenship, Social Theory and Social Change', *Theory and Society*, 16: 363–99, 1987; also Roche, 1992, *op. cit.*

22 See for instance, S. Craig-Smith and H. Green, 'Tourism and Job Generation', *Leisure Studies Association Newsletter*, 13: 22–4, 1986; English Tourist Board, *A Vision for England*, London 1987; V. Middleton, *Employment in Tourism*, School of Management, University of Surrey, 1986.

23 D. R. Vaughan, *Estimating the Level of Tourism-related Employment: An Assessment of Two Non-survey Techniques*, London, BTA/ETB, 1986.

24 C. Gratton and C. and P. Taylor, 'Center Parcs: A New Holiday Concept', *Leisure Management*, 7(8), 1987.

25 A. J. Della Bitta, D. Loudon, *et al.*, 'Estimating the Economic Impact of a Short-Term Tourist Event', *Journal of Travel Research*, 16(2): 10, 1977 J. R. Brent Ritchie and J. Yangzhou, 'The Role and Impact of Mega-Events and Attractions on National and Regional Tourism: A Conceptual and Methodological Overview', in AIEST, 1987, *op. cit.*, p. 17; C. Kaspar 'The Role and Impact of Mega-Events and Attractions on Regional and National Tourism Development', Introduction, in AIEST, 1987, *op. cit.*, p. 11; K. Socher

and P. Tschurtschenthaler, 'The Role and Impact of Mega-Events: Economic Perspectives – The Case of the Winter Olympic Games 1964 and 1976 at Innsbruck', in AIEST, 1987, *op. cit.*, p. 103; C. R. Goeldner and P. Long, 'The Role and Impact of Mega-Events and Attractions in North America', in AIEST, 1987, *op. cit.*, p. 121.

26 Goeldner and Long, 1987, *op. cit.*, pp. 119–31; also A. M. O'Reilly, 'The Impact of Cultural Hallmark/Mega-Events on National Tourism Development in Selected West Indian Countries', *Revue de Tourisme*, 4, 1987.

27 Goeldner and Long, *ibid*, p. 121.

28 Respectively, A. J. Della Bitta and D. Loudon, 'Assessing the Economic Impact of Short Duration Tourist Events', *New England Journal of Business and Economics*, pp. 37–45, 1975; Della Bitta, Loudon, *et al.*, 1977, *op. cit.*, B. Ritchie and D. Beliveau, 'Hallmark Events: An Evaluation of a Strategic Response to Seasonality in the Travel Market', *Journal of Travel Research*, 14: 14–20, 1974; W. A. Schaffer and L. Davidson, *The Impact of the P. T. Barnum Festival*, Bridgeport, Barnum Festival Society, 1978; A. Reichert, 'Three Rivers Festival Economic Impact Study', *Indiana Business Review*, 53: 5–9, 1978; W. A. Schaffer and L. Davidson, 'An Economic-Base Multiplier for Atlanta, 1961–1970', *Atlanta Economic Review*, July–August: 52–4, 1973; W. A. Schaffer and L. Davidson, 'The Economic Impact of Professional Football on Atlanta', in S. Ladany (ed.), *Management Science Applications to Leisure-Time*, Amsterdam, North-Holland, pp. 277–96, 1975; D. R. Vaughan, *The Economic Impact of the Edinburgh Festival, 1976*, Edinburgh, Scottish Tourist Board, 1977.

29 Roche, 1989, *op. cit.*

30 T. P. Dungan, 'How Cities Plan Special Events', *The Cornell H. R. A. Quarterly*, pp. 83–9, 1984.

31 Recent examples include pre-event projections for (a) Western Australia's America's Cup Defence 1986, (b) Indianapolis' Pan American Games 1987, (c) Calgary's Winter Olympics 1988, (d) Seoul's Olympic Games 1988, (e) Birmingham's bid for the 1992 Olympics 1985, etc. (Respectively (a) CABR, *America's Cup – Economic Impact: Final Report*, Centre for Applied Business Research (CABR), Nedlands, University of Western Australia, 1986; (b) J. Thiel/SMC Co., 'The Economic Impact of the PanAmerican Games on Indianapolis', (cited in B. Smith, 'Permanent Facilities Here may be PanAm legacy', *The Indianapolis Star*, 11th July, 1986; (c) DPA Group Inc., *Economic Impacts of the XV Olympic Winter Games*, Calgary and Alberta Tourism, 1985; (d) Seoul Olympic Organizing Committee, *Olympic and the Korean Economic*, Seoul; (e) Ove Arup and Partners, *Birmingham Olympics '92: Feasibility Study, Vol. 2. Economic Report*, Birmingham City Council, 1985.

32 A. S. Travis and J. C. Croizé, 'The Role and Impact of Mega-Events and Attractions on Tourism Development in Europe: A Micro-Perspective', in AIEST, 1987, *op. cit.*, p. 59.

33 Colin Michael Hall's various studies of mega-events are probably the most important in this field; see his and others' studies in Geoffrey Syme, *et al.*, (eds), *The Planning and Evaluation of Hallmark Events*, Aldershot, Avebury, 1989, ch. 1, 2, 9, 19, 20, etc.; also his 'The Effects of Hallmark Events on Cities', *Journal of Travel Research*, Fall, pp. 44–5, 1987.

34 See for instance J. L. Armstrong, 'Contemporary Prestige Centres for Art and Culture, Exhibitions, Sports and Conferences: An International Survey', Unpublished PhD, University of Birmingham, UK, 1984; Ritchie and Yangzhou, 1987, *op. cit.*; AIEST, 1987, *op. cit.*

35 Brent Ritchie, 'Assessing the Impact of Hallmark Events', *Journal of Travel Research*, 23(2), 1984: 2–11. Other relevant discussions include Marris' conceptual discussion (T. Marris, 'The Role and Impact of Mega-Events and Attractions on Regional and National Tourism Development: AIEST Resolutions', *Revue de Tourisme*, 4: 3–5, 1987); Ahn's, 1987, *op. cit.*, application of a 'total-impact' framework to a pre-event analysis of the projected impact of the Seoul Games 1988; and Bos, (H. Bos, C. Kamp and J. Zom, 'Events in Holland', *Revue du Tourisme*, 4: 16–19, 1987) on a Dutch national research programme in this field.

36 Brent Ritchie and C. Aitken, 'Assessing the Impacts of the 1988 Olympic Winter Games', *Journal of Travel Research*, 22, (3): 17–25, 1984; Brent Ritchie and C. Aitken, 'Olympulse II Evolving Resident Attitudes towards the 1988 Olympic Winter Games', *Journal of Travel Research*, 23, (3):28–33, 1985; Brent Ritchie and M. Lyons, 'Olympulse II/IV: A Mid-Term Report on Resident Attitudes Concerning the XV Olympic Winter Games', *Journal of Travel Research*, 14: 18–26, 1987.

37 Ritchie, 1984, *op. cit.*, p. 2.

38 Marris suggests a comparable set of eight event impacts on (i) tourism (amount and quality); (ii) economy (multiplier impacts on other industries); (iii) technology (stimulus to modernization and computerization in other industries); (iv) physical environment (conservation, etc.); (v) social life (family values and behaviour patterns); (vi) culture (risk to local traditions); (vii) psychology (local pride, outsider recognition of locality); and (viii) politics (from local to global level). Marris, 1987, *op. cit.*

39 Ahn, 1987, *op. cit.*

40 *Ibid.*, p. 177.

41 *Ibid.*, p. 173.

42 Murphy, 1985, *op. cit.*

43 Travis and Croizé, 1987, *op. cit.*

44 Also Murphy, 1985, *op. cit.*, p. 8.

45 For instance see Cabinet Office, HMG, 1985, *op. cit.*

46 For instance English Tourist Board policy proposals, 1988, *op. cit.*, 1989, *op. cit.*; Dungan, 1984, *op. cit.*; also M. Collinge, 'Tourism: A Catalyst for Urban Regeneration', London, ETB, 1989.

47 See also Tom Nairn, *The Break-up of Britain*, London, Verso, 1977; David Harvey, *Consciousness and the Urban Experience*, Oxford, Blackwell, 1985; Lash and Urry, 1987, *op. cit.*; Hall and Jacques, 1989, *op. cit.*, Section III.

48 K. Young and L. Mills, *Managing the Post-Industrial City*, London, Heinemann, 1983; A. Cawson, 'Corporation and Local Politics', in W. Grant (ed.), *The Political Economy of Corporatism*, London, Macmillan, 1985; R. King, 'Corporatism and the Local Economy', and P. Saunders, 'Corporatism and Urban Service Provision', both in Grant (ed.), *ibid.* See also D. Judd and M. Parkinson (eds), *Leadership and Urban Regeneration*, London, Sage, 1990.

49 R. Hambleton, 'Urban Government under Thatcher and Reagan', SAUS Working Paper No. 76; C. Moore and S. Booth, 'Urban Policy Contradictions', *Policy and Politics*, 14, 1986; M. Stewart, 'Ten Years of Inner City Policy', *Town and Regional Planning Review*, 58, 1987; B. Robson, *Those Inner Cities: Reconciling the Social and Economic Aims of Urban Policy*, Oxford, Clarendon Press, 1988; D. Byrne, *Beyond the Inner City*, Milton Keynes, Open University Press, 1989.

50 Andrew Gamble, *The Decline of Britain* (2nd edn), London, Macmillan, 1985, and Gamble, 1988, *op. cit.*

51 See for instance G. Stoker, *The Politics of Local Government*, London, Macmillan, 1988, also his 'Inner Cities, Economic Development and Social Services', and 'Creating a Local Government for a Post-Fordist Society: the Thatcherite Project?', in J. Stewart and G. Stoker, *The Future of Local Government*, Basingstoke, Macmillan Education, 1989.

52 See studies by Young and Mills, 1983, *op. cit.*, Cawson, 1985, *op. cit.*, King, 1985, *op. cit.*, also by Stoker, 1989, *op. cit.*

53 See studies by Toffler, 1980, *op. cit.*, 1985, *op. cit.*, and Lyon, 1988, *op. cit.*, also Murray, 1989, *op. cit.*

54 Young and Mills, 1983, *op. cit.*, Grant, 1985, *op. cit.*, and Stoker, 1988, *op. cit.*

55 Roche, 1989, *op. cit.*, and *above*, Section III.

56 P. Ueberroth, *Made in America: His Own Story*, New York, William Morrow and Co. Inc., 1985, ch. 4; C. Gratton and P. Taylor, 'The Olympic Games: An Economic Analysis', *Leisure Management*, 8: March: 33–4; 1988; R. Gruneau, 'Commercialization and the

Modern Olympics', in A. Tomlinson and G. Whannel (eds), *Five Ring Circus: Money, Power and Politics in the Olympic Games*, London, Pluto Press, 1984.

57 K. Jones, 'Loser in Numbers Games', *The Independent*, 13th February, p. 30, 1988.

58 J. Samuel, 'CBS win TV War', *The Guardian*, 26th May, 1988. Also see P. Burrington and A. Warshaw, 'Olympic Cash Crisis', *The European*, 3rd January, 1992.

59 Brief reference to these important dimensions of mega-events and their study is provided by Vaughan in his Edinburgh Festival study (1977, *op. cit.*); Socher and Tschurtschenthaler in their report on the two Winter Olympics of 1964 and 1976 held at Innsbruck (1987, *op. cit.*, pp. 110–12); and Gruneau in his review of the commercialization of the Olympics (1984, *op. cit.*).

60 Travis and Croizé, 1987, *op. cit.*, p. 66.

61 *Ibid.*, p. 61.

62 Armstrong, 1984, *op. cit.*

63 CABR, 1986, *op. cit.* (note 31).

64 Travis and Croizé, 1987, *op. cit.*, p. 63. On Sheffield's mega-event see M. Roche, 'Mega-events and Urban Policy: A Study of Sheffield's World Student Games', 1991 (unpublished), available Policy Studies Centre, Sheffield University.

65 *Ibid.*, p. 76.

66 Ritchie and Yangzhou, 1987, *op. cit.*, Travis and Croizé, 1987, *op. cit.*

67 Ritchie and Beliveau, 1974, *op. cit.*

68 Loventhal and Howarth, *The Impact of the 1984 Summer Olympic Games on the Lodging and Restaurant Industry in Southern California – Los Angeles*, Sydney, Australia, 1985.

69 Travis and Croizé, 1987, *op. cit.*, p. 74.

70 MacCannell, 1976, *op. cit.*; Krippendorf, 1989, *op. cit.*; also E. Cohen's various studies in the sociology of tourism, for instance 'A Phenomenology of Tourist Experiences', *Sociology*, 13: 179–202, 1979, 'Rethinking the sociology of tourism', *Annuals of Tourism Research*, 6:18–35, 1979, and 'Traditions in the Qualitative Sociology of Tourism', *Annals of Tourism Research*, 15 (1): 29–46, 1988. On the sociology of mega-events see Hall in Syme *et al.*, 1989, *op. cit.*

71 S. Cohen, and L. Taylor, *Escape Attempts: The Theory and Practice of Resistance of Everyday Life*, London, Penguin, 1978.

72 Cohen, 1979, *op. cit.*

73 Krippendorf, 1989, *op. cit.*

74 Boorstin, 1963, *op. cit.*, and MacCannell, 1976, *op. cit.*

75 A. Schutz, 'The Stranger', in his *Collected Papers*, Vol. II, The Hague, Martinus Nijhoff, 1964; G. Simmel, 'The Stranger', in his *On Individuality and Social Forms*, Chicago, University of Chicago Press, 1971; M. Roche, 'Time and the Critique of Anthropology', *Philosophy of the Social Sciences*, 18(2): 259–62, 1988.

76 Hewison, 1987, *op. cit.*

INDEX